AN ALTERNATIVE APPROACH TO LIE GROUPS AND GEOMETRIC STRUCTURES

An Alternative Approach to Lie Groups and Geometric Structures

ERCÜMENT H. ORTAÇGİL

OXFORD
UNIVERSITY PRESS

OXFORD
UNIVERSITY PRESS

Great Clarendon Street, Oxford, OX2 6DP,
United Kingdom

Oxford University Press is a department of the University of Oxford. It furthers the University's objective of excellence in research, scholarship, and education by publishing worldwide. Oxford is a registered trade mark of Oxford University Press in the UK and in certain other countries

First Edition published in 2018

Published in the United States of America by Oxford University Press
198 Madison Avenue, New York, NY 10016, United States of America

British Library Cataloguing in Publication Data
Data available

Library of Congress Control Number: 2017960678

ISBN 978–0–19–882165–6

Dedicated to the memory of my beloved mother
for her lifetime service, devotion and love

FOREWORD

When I received the first few chapters of this book from the author, I was both surprised and intrigued, asking myself why anyone should write a whole book on such a seemingly naive and trivial subject as *local Lie groups*. The next day, however, I changed my mind entirely.

Since the late 1930s, we have suffered from a rather nasty affliction, the "Syndrome of Globality," under the leadership of Général Nicolas Bourbaki (as played mainly by André Weil, who never raised his sword in Sevastopol at the end of the Crimean War)—although I have always thought that the name of Eudoxus of Cnidus would have been a much better choice for the main actor in this saga. It has seemed as if the only acceptable approach to any mathematical problem or setting is a global one. Of course, this has its charm and beauty, a global Lie group, for instance, being a much more expressive structure than any local counterpart, even possessing a topology and a differentiable structure and no longer being just some finite set of numerical parameters difficult to look for and pinpoint—Humphrey Bogart versus Charlie Chaplin. Then I suddenly and unexpectedly noticed in these chapters an aspect of a most contradictory nature, namely, the definition of a *global* object called a *local* Lie group. With the help of parallelism and a covariant derivative, the author of this book manages to produce a very useful global object that transcends locality. It not only helps by providing a better understanding and greater precision to much of Lie and Cartan's work, but also fits like a glove with many practical applications in physics and technology. It cannot be denied that we live and thrive in a completely local spatial context and hence must adapt our thoughts to this. For many years, the present author has had a great admiration for global objects and for their harmonious interlacing and coexistence, the most spectacular example being Leonardo da Vinci's *Last Supper*, where renaissance art came to coexist for the first time with projective geometry (perspective). I should also mention that other authors have recently focused their attention on locality within a global context, an example being Ronald Brown's approach in his writings on the philosophy of mathematics.

As for the book itself, it is a very agreeable read, containing excellent ideas, with thorough argumentation concerning its purposes, presented mostly through complete proofs. As a final suggestion, I believe that similar consideration should be given to the infinite groups of transformations of Lie and Cartan that nowadays show up in the disguise of Lie groupoids of higher-order Ehresmannian jets or Lie pseudogroups of local transformations in the terminology of Matsushima and Kuranishi.

Antonio Kumpera

ACKNOWLEDGMENTS

I am grateful to Antonio Kumpera, who became my distant teacher with his book years ago and my close friend later, for kindly accepting to write the foreword upon my request and for his many suggestions to improve my exposition.

I am also grateful to Peter Olver for his encouraging taps on my back over the years. Without his support, this book would not be in print.

I express my hearty thanks to Anthony Bloom for reading the whole book and making valuable suggestions. Over the years, he generously shared his ideas and preprints with me, from which I benefited greatly.

While I was stumbling with the first ideas for this book many years ago, some people gave me invaluable support and protection, and I feel a particular debt of gratitude to them. They are my venerable teacher Cahit Arf, Ergün Toğrol (Rector of BU, 1982–92), Yalçın Koç (Dean of Arts and Sciences, 1990–92), and my friends Alp Eden (Head of the Mathematics Department, 1999–2001) and Teoman Turgut (Head of the Feza Gürsey Institute, 2005–2011).

Gregor Weingart pointed out to me the mistake mentioned in Chapter 16 and Jan Draisma helped me with its correction. My friends Ender Abadoğlu, Yılmaz Akyıldız, Yücel and Aydan Eğecioğlu, Alihan Neşeliler, İlhan Özemek and George Simpson made various suggestions and corrections. I thank them all very much.

I also thank Jean-François Pommaret for our intense correspondence in 1994, from which I learned a great deal about jets.

Last, but not least, I express my hearty thanks to Daniel Taber, my editor at OUP, for his friendly and professional assistance during the process of evaluation and preparation of this book.

Ercüment Ortaçgil
ortacgile@gmail.com
Bodrum (Halicarnassus) 2017

CONTENTS

PART I
Fundamental Concepts

Everything should be made as simple as possible but not simpler

Einstein

0

Introduction

The theory of Lie groups is one of the most important mathematical themes of the last century and belongs at the center of modern differential geometry. Starting with their discovery by Sophus Lie in 1872, Lie groups have evolved into a vast subject with links to many branches of mathematics and with many applications in various fields. There are dozens of books (possibly more) on Lie groups and Lie algebras, with varying aims and scopes. When an author writes a book, the most serious question he or she faces is "Why another book?" This question becomes more challenging if the subject is classical and well known like Lie groups. In our case, however, this question has an easy answer: To our knowledge, none of the previous books (including those on transformation groups) have approached the subject from the standpoint of this one. Furthermore, we believe that this new approach is more in accordance with the original work of Sophus Lie and Felix Klein than the present modern theory.

We would like to explain the unique feature of this book briefly in somewhat technical terms. Let us first recall the well-known question: When is a smooth manifold M a Lie group? More precisely, is it possible to define a structure on M in such a way that when the curvature $\mathcal{R}$ of this structure vanishes, M becomes "locally a Lie group?" This question has a well-known answer: For a given M, let G be a Lie group with $\dim G = \dim M$ and with Lie algebra $\mathfrak{g}$. The structure we seek is a $\mathfrak{g}$-valued 1-form ω on M with trivial kernel, called the Maurer–Cartan form, and the curvature $\mathcal{R}$ is defined using the exterior derivative $d\omega$. In this classical approach, the key fact is that the Lie group G is fixed beforehand as a model.

Naturally, the question arises as to whether it is possible to dispense with the model. It is a remarkable fact that the answer is affirmative, and indeed such a model-free approach to geometric structures has been proposed by Blaom [B1, B2] in a very general framework and has also been promoted independently [AO1, AO2]. In the above simplest case, the problem involves starting with some structure on M, examples of which abound in geometry, and then defining its curvature $\mathcal{R}$ in such a way that $\mathcal{R} = 0$ if and only if M "becomes locally *some* Lie group G." We carry out this program in detail in Part I of this book. It turns out that the structure we seek is an absolute parallelism ε on M, which we study in detail in Chapter 1. In Chapter 2, we define the curvature $\mathcal{R}$ of the parallelizable manifold (M, ε), and in Chapter 3, we call (M, ε) a local Lie group (LLG) if $\mathcal{R} = 0$. This is a *global* concept, in contrast to the common modern terminology of "*local* Lie

An Alternative Approach to Lie Groups and Geometric Structures, Ercüment H. Ortaçgil (2018).

group," which is a *local* concept referring to the neighborhood of the identity of a *global* Lie group. An LLG determines a transitive pseudogroup of local diffeomorphisms on M. If M is compact and simply connected, then these local diffeomorphisms extend uniquely to global diffeomorphisms, giving a transformation group $\mathcal{G}$ that acts simply transitively on M. Taking the centralizer $C(\mathcal{G})$ of $\mathcal{G}$ inside $\text{Diff}(M)$, we show in Chapter 4 that $\mathcal{G}$ and $C(\mathcal{G})$ define a Lie group structure on M modulo two choices: (1) a base point that serves as the identity; (2) whether $\mathcal{G}$ (or $C(\mathcal{G})$) should be left (or right) translations. Therefore, a Lie group is a special (globalizable) LLG. In the words of Olver [O4], "This fact reinstates the paradigm of local to global to its historical record," as envisaged by him in [O1]. The crucial fact here is that the Lie group M is not fixed beforehand, that is, this new approach is "model-free." In Chapters 5 and 6, we show that this construction can be carried out equivalently on an infinitesimal level by replacing point transformations by vector fields (called infinitesimal point transformations in the old literature). As a surprising consequence, we show that the well-known tensor calculus, which emerged as an attempt to formulate Riemann's revolutionary ideas, emerges also from Lie theory, however with quite different interpretations of the concepts of torsion and curvature. In Chapter 7, we compare this new approach with the classical one via the Maurer–Cartan form.

The purpose of Part II is to show the usefulness of the theory developed in Part I by establishing some surprising relations, raising some far-reaching questions, and proposing new points of view. For instance, in Chapter 9, we define the analog of the Kodaira–Spencer map well known from the deformation theory of complex structures. In Chapter 11, we establish a concrete but intriguing relation between the cohomology groups of tensorial and trivial representations. In Chapter 12, we recast Chern–Simons theory in a completely different setup. In Chapter 13, we construct a geometric flow that is more elementary than the Ricci flow and, we believe, will simplify the present proof of the Poincaré Conjecture. In Chapter 14, we give a simple proof of the Van Est theorem using the seemingly unrelated idea of horizontal cohomology of a PDE due to Vinogradov [Vn, V]. In Chapter 15, we propose an algebraic prolongation theory of a Klein geometry in analogy with the well-known prolongation theory of a single linear Lie algebra developed by Guillemin, Singer, and Sternberg [KN2, SS]. This new prolongation theory, in which the original idea of taking successive normalizers is due to Pommaret [P1], is considerably more intuitive and geometric than the familiar prolongation theory. Chapters 8–11 and 15 form a logical sequence, while 12–14 are independent of the rest, although all of these chapters require a good understanding of Part I.

Our main purpose in Part III is to generalize Part I in the following way: Define a geometric structure $\mathcal{P}$ and its curvature $\mathcal{R}$ on M in such a way that $\mathcal{R} = 0$ if and only if $\mathcal{P}$ is "locally homogeneous" in the sense that any $p, q \in M$ have neighborhoods U, V such that $\mathcal{P}_{|U}$ can be "identified" with $\mathcal{P}_{|V}$. Naturally, we require this identification to be induced by some local diffeomorphism $f : U \to V$. Therefore, if $\mathcal{R} = 0$, these identifications define a transitive pseudogroup on M as in Part I. Furthermore, if these local diffeomorphisms extend to global diffeomorphisms, then we get a transitive (but not necessarily *simply* transitive as in Part I) transformation group of M, turning M into a homogeneous space $M = G/H$. We fix beforehand only H and $\dim G = \dim \mathcal{P} = \dim H + \dim M$, but not G or its Lie algebra $\mathfrak{g}$; that is, our approach is model-free. Now $\mathcal{P}$ is an absolute parallelism if and only if H is trivial, in which case we recover Part I. At this stage, it is instructive to look at a Riemannian structure. which is locally homogeneous in our sense if and only if the metric has constant curvature. Since we require $\mathcal{R} = 0$ to be equivalent to local homogeneity, just from the outset we discover the surprising (and also annoying) fact that the curvature $\mathcal{R}$ that we seek cannot be the Riemann curvature tensor! The uniformization theorem tells us that a Riemannian structure with $\mathcal{R} = 0$ can be locally homogeneous in only 3 distinct ways. For absolute parallelism, this number is ∞, because we do not impose any condition on the emerging Lie group beforehand, even though such conditions are imposed by the underlying structure of M as in the case of the Poincaré Conjecture in Chapter 13. What is this number for some geometric structure $\mathcal{P}$ yet to be defined? A possible answer, $M = G/H$, gives a representation of $\mathfrak{h}$ on $V = \mathfrak{g}/\mathfrak{h}$ with dimension equal to dim M, where $\mathfrak{h}$ and $\mathfrak{g}$ are the Lie algebras of H and G. This fact gives a hint of the level of difficulty of this question.

As far as we know, there exist essentially three approaches to geometric structures on manifolds:

1. The classical approach via Cartan connections (which generalize the Maurer–Cartan form) on principal bundles (see [CS] and references therein).
2. The more recent approach using groupoids induced by some auxiliary bundles on M [CSa].
3. The intrinsic approach via jet groupoids on M [KS, Ku1, Ly, P1, P2].

In all these approaches, the model is fixed, and in the third, curvature is replaced by the concept of formal integrability. We start our search for a "good" definition following the third approach through the somewhat lengthy and leisurely but very instructive tour of Chapters 16–19. However, our ambitious plan of generalizing Part I to arbitrary geometric structures in a model-free way, which works perfectly well for affine and Riemannian structures, comes to an abrupt halt in Chapter 19 with the unexpected discovery that a projective structure is always flat. The reason is that we need a technical assumption to define curvature, and this assumption forces flatness of a projective structure. We formulate the final definition in Chapter 20, which is inspired by an idea communicated to us by Blaom. This definition depends on the ideas developed in Part I and the concept of a foliation that is compatible with the absolute parallelism. The reader who feels comfortable with Part I can skip directly to this definition in Chapter 20. However, there is an immense amount of structure behind this definition, and it is not really possible to appreciate it without the motivational tour of Chapters 16–19. We finish Chapter 20 by defining seemingly new characteristic classes depending on higher-order jets using Chern–Simons forms, but the question of their nontriviality remains open . . . and this short book ends, unfortunately, at a point where it should actually start.

The prerequisites for this book are quite modest: We assume that the reader is comfortable with the definition of a smooth manifold and some basic structures on it, like vector fields, Lie derivative, differential forms, tensors, etc., and we otherwise start from scratch in Chapter 1. The only technical tool that we use in Part I is the well-known existence and uniqueness theorem for first-order systems of PDEs with initial conditions, which is proved, for instance, in the Appendix of [L]. We use some basic topological notions like simple connectivity and the rudiments of covering spaces in Chapter 3. Therefore, a good undergraduate student should have no difficulty in understanding Part I in depth. The level of abstraction increases slightly in Part II, which requires more active involvement on the part of the reader, but we believe that an undergraduate student with persistence and mathematical maturity can also follow Part II if he or she is willing to take a few facts for granted. However, Part III is the beginning of an adventure in unknown territory and requires a greater degree of sophistication. Our language here becomes more intuitive and descriptive, and some unproved assertions are not entirely trivial, although their detailed proofs would distract us from our main goal, which is to motivate the definition in Chapter 20. We should remark here that the main purpose of this book is to attract the attention of curious and courageous mathematical minds, because what we are doing here is no more than scratching at the topsoil, and much remains to be done to build this new theory on solid ground.

The process of maturation of some ideas in this book was an adventurous, joyful, and even ecstatic journey that spanned a period of more than 30 years, but was at times also deeply solitary, painful, and even frustrating. We would like to remind the reader that the alternative approach adopted in this book is unfortunately, but inescapably, not in much accordance with the mainstream. From our experience over the years, we happened to learn that clinging to our prior knowledge may obstruct the light of the new. So we kindly request the reader to temporarily put aside his or her existing understanding (and hence conditioned views) of some classical concepts (like Lie group, curvature, and representation) and read this book with innocent eyes.

1

Parallelizable Manifolds

In this book, we always assume that M is a smooth manifold; that is, M is a topological manifold (Hausdorff with a countable base) with an atlas whose transition functions are differentiable up to any order. Since we will deal with certain integrability conditions, we also assume that M is connected and $n = \dim M \geq 2$.

We recall that M is called parallelizable if it admits n independent vector fields. In more detail, there exist some vector fields $X_1, \ldots, X_n$ whose values $X_1(p), \ldots, X_n(p)$ form a basis of the tangent space T_pM for all $p \in M$. The purpose of this chapter is to formulate an equivalent definition of parallelizability. As we proceed in this book, it will gradually become clear that this equivalent formulation has some remarkable consequences.

So, let M be a smooth manifold and (a, b) and (c, d) be two elements of $M \times M$. If $b = c$, then we can operate on these two pairs according to the rule

$$(b, d) \circ (a, b) \stackrel{\text{def}}{=} (a, d) \tag{1.1}$$

The set $M \times M$ with the *partial* operation $\circ$ defined by (1.1) is called the pair groupoid on M. Clearly, the pair groupoid can be defined on any set. For reasons to be made clear below, if M is a smooth manifold, we call the pair (a, b) a 0-arrow from a to b. We call a the source and b the target of the 0-arrow (a, b). The formula (1.1) states that two 0-arrows can be composed if (and only if) the target of the first is equal to the source of the second, and the result is a 0-arrow from the source of the first to the target of the second. In this way, (1.1) assumes a very intuitive geometric meaning.

A word of caution with our notation: In many works on groupoids, the pair (b, a) denotes a 0-arrow from a to b, that is, the second argument denotes the source and the first argument denotes the target. With this notation, (1.1) is written as

$$(c, b) \circ (b, a) = (c, a) \tag{1.2}$$

Both notations have their advantages and disadvantages, but we prefer (1.1). Note that (1.1) gives

$$(b, a) \circ (a, b) = (a, a) \tag{1.3}$$

An Alternative Approach to Lie Groups and Geometric Structures, Ercüment H. Ortaçgil (2018).

We define

$$(a,b)^{-1} \stackrel{\text{def}}{=} (b,a) \tag{1.4}$$

so that $(a,b)^{-1} \circ (a,b) = (a,a)$ and $(a,b) \circ (a,b)^{-1} = (b,b)$. Given the 0-arrow (a,b) from a to b, we choose two coordinate systems (U,x) and (V,y) around a and b, respectively. Suppose $a = (a^i)$ and $b = (b^i)$ in these coordinates, with $1 \leq i \leq n = \dim M$. Now (a,b) has the coordinate representation (a^i, b^i). Letting a and b vary inside U and V, we see that any 0-arrow with source in U and target in V has the unique coordinate representation (x^i, y^i). Therefore, choosing coordinates around a, b, d, we can rewrite (1.1) as

$$(y^i, z^i) \circ (x^i, y^i) = (x^i, z^i) \tag{1.5}$$

and (1.4) as

$$(x^i, y^i)^{-1} = (y^i, x^i) \tag{1.6}$$

in coordinates.

We observe that the coordinate expression (a^i, b^i) may have another meaning: It may also represent a 0-arrow whose source and target are in the *same* coordinate system; that is, it may represent (a,b) where $a, b \in (U, x^i)$. In this book, we will often let the target approach the source as a limit in certain coordinate expressions, and this second interpretation will be assumed in such arguments. These remarks apply also to the local expressions for 1-arrows to be defined below.

We let $\mathcal{U}_0$ denote the set of all 0-arrows on M, so $\mathcal{U}_0 = M \times M$. Now $\mathcal{U}_0$ is the first element of a sequence of groupoids $\mathcal{U}_k$, $k \geq 0$ on M, and the definition of $\mathcal{U}_k$ for $k \geq 1$ needs a smooth structure on M. In the first two parts of this book, we will need only $\mathcal{U}_0$ and $\mathcal{U}_1$, except in Chapter 15.

Before we proceed to the definition of $\mathcal{U}_1$, we first define a 1-arrow on M and, as expected, $\mathcal{U}_1$ as a set will be the collection of all 1-arrows on M. Furthermore, $\mathcal{U}_1$ will have the extra structure of a smooth manifold like $\mathcal{U}_0 = M \times M$.

In order to define a 1-arrow from a to b, we fix $(a,b) \in \mathcal{U}_0$ and consider all the *local* diffeomorphisms of M that map a to b. We fix two coordinate systems (U,x) and (V,y) around a and b, respectively, where $a = (a^i)$ and $b = (b^i)$ as above. Let f and g be two such local diffeomorphisms. By choosing U sufficiently small, we may assume that f, g are both defined on U. We declare $f \sim g$ if, in addition to $f(a) = g(a)$, we also have

$$\frac{\partial f^i}{\partial x^j}(a) = \frac{\partial g^i}{\partial x^j}(a) \tag{1.7}$$

Clearly, $\sim$ defined by (1.7) is an equivalence relation on the set of all local diffeomorphisms of M that map a to b. We claim that this equivalence relation is independent of the choice of the coordinate systems (U,x) and (V,y) around a and b. Indeed, let f be a local diffeomorphism that maps (U,x) onto (V,y) and let its derivative have the local representation

$(\partial y^i/\partial x^j)$ with respect to these coordinates. Now we apply a coordinate change $(y) \to (w)$. The chain rule shows that the derivative $(\partial w^i/\partial x^j)$ of f with respect to the coordinates (x) and (w) is given by

$$\frac{\partial w^i}{\partial x^j} = \frac{\partial w^i}{\partial y^a}\frac{\partial y^a}{\partial x^j} \tag{1.8}$$

We have used the Einstein summation convention in (1.8) and we will use it throughout this book. Similarly, a coordinate change $(x) \to (z)$ transforms $(\partial y^i/\partial x^j)$ according to the formula

$$\frac{\partial y^i}{\partial z^j} = \frac{\partial y^i}{\partial x^a}\frac{\partial x^a}{\partial z^j} \tag{1.9}$$

In particular, if the above two coordinate changes are applied simultaneously, we obtain the transformation rule

$$\frac{\partial w^i}{\partial z^j} = \frac{\partial w^i}{\partial y^a}\frac{\partial y^a}{\partial x^b}\frac{\partial x^b}{\partial z^j} \tag{1.10}$$

Now (1.10) shows that the equivalence relation defined by (1.7) is independent of coordinates, because if (1.7) holds in some coordinates, then it holds in all coordinates according to (1.10). A 1-arrow from a to b is simply a geometric name for an equivalence class. So, we make the following important definition:

Definition 1.1 *A 1-arrow from a to b is an equivalence class defined by the equivalence relation* (1.7).

If f is a representative for a 1-arrow from a to b, it is standard to denote this 1-arrow by $j_1(f)_a$ and call it the 1-jet of f with the source at a and the target at $b = f(a)$. To be consistent with our later notation, we will write $j_1(f)^a$ and sometimes also $j_1(f)^{a,b}$, where $b = f(a)$, in order to emphasize the target b, even though this is clear from the notation $j_1(f)^a$. Of course, we may have $j_1(f)^a = j_1(g)^a$ but $f \neq g$ as local diffeomorphisms; that is, an equivalence class has many different representatives. Indeed, f and g may have different derivatives at a of order k for some $k \geq 2$, so $f \neq g$ even though (1.7) may hold, so they will represent the same 1-arrow. In fact, f and g may have the same derivatives of *all* orders at a but we may still have $f \neq g$ as local diffeomorphisms, but this can occur only in the smooth category and not in the analytic category. These preliminary observations provide some intuition regarding the definition of a k-arrow from a to b that we will return to in Chapter 15 and Part III. For instance, it is not difficult at this stage to guess that a 2-arrow from a to b will be defined as an equivalence class on the set of all the local diffeomorphisms that map a to b as follows: We declare $f \sim g$ if, in addition to $f(a) = g(a)$ and $(\partial f^i/\partial x^j)(a) = (\partial g^i/\partial x^j)(a)$, we also have $(\partial^2 f^i/\partial x^j \partial y^k)(a) = (\partial^2 g^i/\partial x^j \partial y^k)(a)$. However, it becomes more and more tedious to show the independence of coordinates directly, and one introduces the jet groups $G_k(n)$ for this and other purposes, as in Part III.

Having defined a 1-arrow from a to b, we let $\mathcal{U}_1^{a,b}$ denote the set of all the 1-arrows from a to b, and we define the set $\mathcal{U}_1 \overset{\text{def}}{=} \cup_{a,b\in M}\mathcal{U}_1^{a,b}$. Then, $\mathcal{U}_1$ is the set of *all* 1-arrows on M. We have the obvious projection map

$$\pi : \mathcal{U}_1 \longrightarrow \mathcal{U}_0 \tag{1.11}$$

which projects a 1-arrow from a to b to the 0-arrow (a, b) from a to b.

To complete the definition of $\mathcal{U}_1$, we need to do two more things: First, we will endow $\mathcal{U}_1$ with the structure of a smooth manifold and, second, we will define a partial composition as well as an inversion on the elements of $\mathcal{U}_1$ in a way compatible with the projection (1.11). In more abstract terms, we will turn $\mathcal{U}_1$ into a differentiable groupoid in such a way that (1.11) will become a homomorphism of groupoids. However, we will not need any prior knowledge of groupoids and our elementary treatment will be self-contained. We refer the interested reader to [MK1] and [MK2] for the general theory of groupoids and also to the excellent expository articles [Wn2] and [Ku2].

So, let $j_1(f)^{a,b}$ be a 1-arrow from a to $b = f(a)$, (U, x) and (V, y) be co-ordinate systems around a and b. We set $\overline{f}^i_j = (\partial y^i/\partial x^j)(a)$, so $\left(\overline{f}^i_j\right)$ is an invertible matrix and therefore an element of $GL(n, \mathbb{R})$. It follows that any 1-arrow from a to b defines an element of $GL(n, \mathbb{R})$ once we fix the coordinates around a, b. Conversely, for any $\left(\overline{f}^i_j\right) \in GL(n, \mathbb{R})$, there exists some local diffeomorphism f satisfying $f(a) = b$ and $\overline{f}^i_j = (\partial y^i/\partial x^j)(a)$: For instance, $y^i = b^i + \overline{f}^i_j(x^j - a^j)$ is such a local diffeomorphism. Now, let W be an open neighborhood of $\overline{f}^i_j = (\partial y^i/\partial x^j)(a)$ in $GL(n, \mathbb{R})$. We declare the set $U \times V \times W$ to be a coordinate system around $j_1(f)^a$ and the triplet $\left(a^i, b^i, \overline{f}^i_j\right)$ to be the coordinates of $j_1(f)^a$ in this system. It follows that any 1-arrow with source in U and target in V is uniquely parametrized by some $(x^i, y^i, f^i_j) \in U \times V \times W$.

Since we will often work with local coordinates in this book, we will identify a 1-arrow with its coordinate representation (x^i, y^i, f^i_j) and will not explicitly refer to the coordinate maps that perform this identification, as we did also for 0-arrows. This identification will also allow us to avoid keeping track of the intersections of some open sets. Keeping this in mind, notice that (1.11) is given by

$$\left(x^i, y^i, f^i_j\right) \longrightarrow \left(x^i, y^i\right) \tag{1.12}$$

in coordinates.

The relation (1.10) together with the local coordinates defined above imply that $\mathcal{U}_1$ is a smooth manifold of dimension $2n + n^2 = 2\dim M + \dim GL(n, \mathbb{R})$. Indeed, the two coordinate changes $(x) \to (z)$ and $(y) \to (w)$ transform the local representation (x^i, y^i, f^i_j) into $(z^i, w^i, g^i_j = (\partial w^i/\partial y^b)f^b_a(\partial x^a/\partial z^j))$ according to (1.10). Consequently, since (z^i) and (w^i) depend smoothly on (x^i) and (y^i), respectively, and since (g^i_j) depends smoothly on (x^i), (y^i), and (f^i_j), we conclude that the transition functions between two such coordinates systems are smooth and therefore that $\mathcal{U}_1$ is a smooth manifold. Furthermore (1.12) shows that (1.11) is a smooth map.

It remains to define the partial composition and inversion. Consider therefore the 1-arrow $j_1(f)^{a,b}$ from a to $b = f(a)$ with the representative f, and the 1-arrow $j_1(g)^{b,c}$ from b to $g(b) = g(f(a)) = c$ with the representative g. We define the composition of these 1-arrows by the formula

$$j_1(g)^{b,c} \circ j_1(f)^{a,b} \stackrel{\text{def}}{=} j_1(g \circ f)^{a,c} \tag{1.13}$$

Note that $\circ$ on the left-hand side of (1.13) refers to the defined composition, whereas on the right-hand side it refers to the ordinary composition of two local diffeomorphisms. Also, (1.13) defines the composition of two equivalence classes in terms of some representatives, and so we must check that (1.13) is well-defined, but this is immediate using (1.10). Therefore, we deduce from (1.13) the composition map

$$\mathcal{U}_1^{b,c} \times \mathcal{U}_1^{a,b} \longrightarrow \mathcal{U}_1^{a,c} \tag{1.14}$$

Choosing coordinates around a, b, c and using the chain rule, (1.13) becomes

$$\left(y^i, z^i, g^i_j\right) \circ \left(x^i, y^i, f_j\right) = \left(x^i, z^i, g^i_a f^a_j\right) \tag{1.15}$$

We note that $\circ$ in (1.13) is a smooth operation. This means that the values on the right-hand side of (1.15) depend smoothly on the values on the left-hand side.

Let I be the identity map of M and consider $j_1(I)^a$, which is $\left(a^i, a^i, \delta^i_j\right)$ in terms of some coordinates around a. Here δ^i_j is the Kronecker delta: $\delta^i_j = 1$ if $i = j$ and $\delta^i_j = 0$ if $i \neq j$. With an abuse of notation, we will simply write $j_1(I)^a = (a^i, a^i, \delta^i_j)$, such notation being used henceforth. We now define

$$\left(j_1(f)^a\right)^{-1} \stackrel{\text{def}}{=} j_1\left(f^{-1}\right)^{f(a)} \tag{1.16}$$

which is easily seen to be well-defined. So, the inverse of a 1-arrow from a to b is a 1-arrow from b to a. Now, (1.16) defines the inversion map

$$\mathcal{U}_1^{a,b} \longrightarrow \mathcal{U}_1^{b,a} \tag{1.17}$$

We have

$$\begin{aligned}\left(j_1(f)^a\right)^{-1} \circ j_1(f)^a &= j_1\left(f^{-1}\right)^{f(a)} \circ j_1(f)^a \\ &= j_1\left(f^{-1} \circ f\right)^a \\ &= j_1(I)^a\end{aligned} \tag{1.18}$$

and similarly

$$j_1(f)^a \circ \left(j_1(f)^a\right)^{-1} = j_1(I)^{f(a)} \tag{1.19}$$

The local formulas for (1.16) and (1.18) now become

$$\left(x^i, y^i, f^i_j\right)^{-1} = \left(y^i, x^i, g^i_j\right), \qquad f^i_a g^a_j = g^i_a f^a_j = \delta^i_j \tag{1.20}$$

and

$$\begin{aligned}\left(x^i, y^i, f^i_j\right)^{-1} \circ \left(x^i, y^i, f^i_j\right) &= \left(y^i, x^i, g^i_j\right) \circ \left(x^i, y^i, f^i_j\right) \\ &= \left(x^i, x^i, \delta^i_j\right)\end{aligned} \tag{1.21}$$

Clearly, the inversion (1.17) is a smooth map by (1.20). Also, (1.5) and (1.15) show that the projection π in (1.11) preserves the composition: Composing and then projecting is the same as projecting first and then composing. Similarly, (1.6) and (1.20) show that π preserves inversion: Taking the inverse and then projecting is the same as projecting first and then taking the inverse.

It is worth paying special attention to the structure of the 1-arrows with the same source and target, namely, $\mathcal{U}_1^{a,a}$ for some $a \in M$. Now, (1.13) and (1.16) show that $\mathcal{U}_1^{a,a}$ is a group. A choice of coordinates around a identifies $\mathcal{U}_1^{a,a}$ with the coordinate expressions (a^i, a^i, f^i_j), which in turn can be identified with the group $GL(n, \mathbb{R})$. Note, however, that there is no such canonical identification.

Having completed the construction of the groupoids $\mathcal{U}_0$ and $\mathcal{U}_1$ as well as the smooth projection homomorphism π in (1.11), we now come to a fundamental point on which this whole book rests.

Suppose there exists a smooth map

$$\varepsilon : \mathcal{U}_0 \longrightarrow \mathcal{U}_1 \tag{1.22}$$

satisfying

$$\pi \circ \varepsilon = I_{\mathcal{U}_0} \tag{1.23}$$

where $I_{\mathcal{U}_0}$ denotes the identity map on $\mathcal{U}_0$. In other words, ε assigns to any 0-arrow from a to b a unique 1-arrow from a to b for any $a, b \in M$ and this assignment is smooth. We further assume that ε preserves the composition and inversion of arrows; that is, we require ε to be a homomorphism of groupoids. We call ε a splitting. If M admits a splitting ε, denoted by (M, ε), then $\varepsilon(\mathcal{U}_0) \subset \mathcal{U}_1$ is a subgroupoid with the obvious meaning. It will turn out in Chapter 2 that $\varepsilon(\mathcal{U}_0)$ is actually a globally defined first-order nonlinear PDE. As preparation for understanding this PDE, we express ε in coordinates. Suppose $\varepsilon(a, b) = j_1(f)^a$ with $f(a) = b$. Choosing coordinates around a, b we can write

$$\varepsilon\left(x^i, y^i\right) = \left(x^i, y^i, \varepsilon^i_j(x, y)\right) \tag{1.24}$$

Note that $\varepsilon^i_j(x,y)$ is a shorthand notation for $\varepsilon^i_j(x^1,\ldots,x^n;y^1,\ldots,y^n)$. The smoothness of ε means that its components ε^i_j are smooth functions of their arguments. Since ε preserves composition, (1.24) and (1.15) imply

$$\begin{aligned}\left(y^i,z^i,\varepsilon^i_j(y,z)\right)\circ\left(x^i,y^i,\varepsilon^i_j(x,y)\right) &= \left(x^i,z^i,\varepsilon^i_a(y,z)\varepsilon^a_j(x,y)\right)\\ &= \left(x^i,z^i,\varepsilon^i_j(x,z)\right)\end{aligned} \tag{1.25}$$

and the second equality holds if and only if we have

$$\varepsilon^i_j(x,z) = \varepsilon^i_a(y,z)\varepsilon^a_j(x,y) \tag{1.26}$$

Clearly, $\varepsilon(a,a) = j_1(I)^a$ for all $a \in M$: We apply ε to the identity $(a,b)\circ(a,a) = (a,b)$ and cancel $\varepsilon(a,b)$. So, we have the local formula

$$\varepsilon^i_j(x,x) = \delta^i_j \tag{1.27}$$

for all $x \in U$. Setting $z = x$ in (1.26) and using (1.27), we also get the local formula

$$\varepsilon^i_a(y,x)\varepsilon^a_j(x,y) = \delta^i_j \tag{1.28}$$

So, (1.27) and (1.28) are consequences of (1.26). Equivalently, if ε preserves composition, then it must also preserve inversion. We remark here that (1.26)–(1.28) are *local* expressions for *global* formulas.

It is a truly remarkable fact that *everything* we do in Parts I and II of this book (*without exception!*) is a direct consequence of (1.26), and it will then become a real challenge to generalize (1.26) in the right way in Part III. The surprise that we will discover in Chapter 20 is that the right way to generalize (1.26) is to add some extra structure to it, this extra structure being vacuous in the case of absolute parallelism.

However, the existence of a splitting puts a very strong condition on M. To understand this condition, let us recall the definition of an absolute parallelism on M. As we have already remarked, suppose M admits n smooth vector fields $\xi_{(1)},\ldots,\xi_{(n)}$ on M, where $n = \dim M$, such that their values $\xi_{(1)}(p),\ldots,\xi_{(n)}(p)$ at p form a basis of the tangent space T_pM for all $p \in M$. In this case, we say that these vector fields define an absolute parallelism on M, or that M is parallelized by these vector fields. Without any reference to the vector fields, we also say that M is parallelizable. Lie groups are the most important examples of parallelizable manifolds. In fact, a Lie group can be parallelized in two ways: by left translations and by right translations. These two parallelizations can be quite "different" if the Lie group is not abelian. We will see in the next two chapters that Lie groups are very special parallelizable manifolds. The key fact here is that we do not actually need the left (right) translations to parallelize a Lie group, but only the 1-arrows of these translations. As another example, any orientable 3-manifold is known to be parallelizable. This fact will be crucial in Chapter 13. We will see in Chapter 20 that the total spaces of some principal bundles relevant for our purposes are also parallelizable. Clearly, M can be parallelized in different ways, in the same way as it can admit different splittings. We now have the following proposition:

Proposition 1.2 *The following are equivalent:*

(i) *M admits a splitting;*

(ii) *M is parallelizable.*

The proof is simple and depends on an equivalent formulation of a 1-arrow from p to q, which is clarified by the following trivial lemma:

Lemma 1.3 *There is a 1–1 correspondence between the following objects:*

(i) *1-arrows from p to q;*

(ii) *linear isomorphisms $T_pM \to T_qM$.*

We will give the proof in some detail in order to fix our notation. Let $j_1(f)^{p,q} = (p^i, q^i, f^i_j)$ be a 1-arrow from p to q and let $\xi_p = \xi^a(\partial/\partial x^a)_{|p}$ be a tangent vector at p that we write in short as $\xi_p = (\xi^i)$. We define the components (η^i) of an object η_q by

$$\eta^i \stackrel{\text{def}}{=} f^i_a \xi^a \tag{1.29}$$

Using the transformation rules for (p^i, q^i, f^i_j) and (ξ^i) under coordinate changes, we easily check that (η^i) are the components of a tangent vector at q, so $\eta_q = \eta^a(\partial/\partial y^a)_{|q}$. Since the map $\xi_p \to \eta_q$ defined by (1.29) is linear, we conclude that $j_1(f)^{p,q}$ defines a linear map (using the same notation) $j_1(f)^{p,q} : T_pM \to T_qM$, which is an isomorphism since the matrix (f^i_j) is invertible. Note that (1.29) is equivalent to $j_1(f)^{p,q}(\partial/\partial x^j)_{|p} = f^a_j(\partial/\partial y^a)_{|q}$, and therefore (f^i_j) is the matrix of $j_1(f)^{p,q}$ with respect to the coordinate bases $(\partial/\partial x^j)_{|p}$ and $(\partial/\partial y^j)_{|q}$ of T_pM and T_qM, respectively. Conversely, given an isomorphism $T_pM \to T_qM$, let (f^i_j) be its matrix with respect to the coordinate basis, and it follows easily that (p^i, q^i, f^i_j) is a 1-arrow from p to q.

Now we can easily prove Proposition 1.2. Assume that ε is a splitting. Fixing some basepoint $e \in M$ and choosing a basis $\xi_{(1)}(e), \ldots, \xi_{(n)}(e)$ of T_eM, we know, by Lemma 1.3, that $\varepsilon(e,x)$ defines a linear isomorphism $\varepsilon(e,x) : T_eM \to T_xM$ for any $x \in M$. We define the n vector fields $\xi_{(1)}, \ldots, \xi_{(n)}$ on M by the formula $\xi_{(i)}(x) \stackrel{\text{def}}{=} \varepsilon(e,x)(\xi_{(i)}(e))$, $1 \leq i \leq n$. These vector fields are smooth and define an absolute parallelism on M.

Conversely, assume that the vector fields $\xi_{(1)}, \ldots, \xi_{(n)}$ parallelize M. For $p, q \in M$, we define a linear isomorphism $\mathcal{E}_{p,q} : T_pM \to T_pM$ by sending the basis $\xi_{(1)}(p), \ldots, \xi_{(n)}(p)$ of T_pM to the bases $\xi_{(1)}(q), \ldots, \xi_{(n)}(q)$ of T_qM, respectively. By Lemma 1.3, $\mathcal{E}_{p,q}$ defines a 1-arrow $\varepsilon(p,q)$ from p to q. Clearly, we have $\mathcal{E}_{q,r} \circ \mathcal{E}_{p,q} = \mathcal{E}_{p,r}$, and therefore ε is a splitting. Note that the composition of these linear maps corresponds to the multiplication of their matrices, thus giving (1.26).

A digression: Our frequent use of local coordinates throughout this book may seem old fashioned and therefore pointless. However, we hope we will be able to convince the reader in this book that sometimes (though not always!) the use of local coordinates may be very rewarding indeed.

There are two objects that emerge naturally from the groupoid $\mathcal{U}_1$ and play an important role in the theory. The idea of the first is already hidden in the proof of Proposition 1.2:

We fix some base point $e \in M$ and define $\mathcal{U}_1^{e,\bullet} \stackrel{\text{def}}{=} \cup_{x \in M} \mathcal{U}_1^{e,x}$; that is, $\mathcal{U}_1^{e,\bullet}$ is the set of all 1-arrows spreading out from e. The projection $\pi : \mathcal{U}_1^{e,\bullet} \to M$ maps the set $\mathcal{U}_1^{e,x}$ onto x. The local representations of 1-arrows shows that $\pi : \mathcal{U}_1^{e,\bullet} \to M$ is a smooth and locally trivial fiber bundle with fibers $\pi^{-1}(x) = \mathcal{U}_1^{e,x}$. Further, the composition $\mathcal{U}_1^{e,x} \times \mathcal{U}_1^{e,e} \to \mathcal{U}_1^{e,x}$ of 1-arrows provides a smooth right action of the group $\mathcal{U}_1^{e,e}$ on the fiber $\mathcal{U}_1^{e,x}$, and $\pi : \mathcal{U}_1^{e,\bullet} \to M$ is a right principal bundle with structure group $\mathcal{U}_1^{e,e}$. We recall that a choice of coordinates in a neighborhood of e identifies the group $\mathcal{U}_1^{e,e}$ with $GL(n, \mathbb{R})$. Fixing such a coordinate system once and for all, we obtain a principal $GL(n, \mathbb{R})$-bundle. Let us call $\pi : \mathcal{U}_1^{e,\bullet} \to M$ the (first-order) frame bundle of M with base point e. In much the same way, we can define the left principal bundle $\pi : \mathcal{U}_1^{\bullet,e} \to M$ with structure group $\mathcal{U}_1^{e,e}$, where $\mathcal{U}_1^{\bullet,e} \stackrel{\text{def}}{=} \cup_{x \in M} \mathcal{U}_1^{x,e}$. We call this bundle the coframe bundle of M with base point e. We note that these principal bundles determine the groupoid $\mathcal{U}_1$, because $\mathcal{U}_1^{p,q} = \mathcal{U}_1^{e,q} \circ \mathcal{U}_1^{p,e}$ for all $p, q \in M$. The splitting ε determines a global section of $\mathcal{U}_1^{e,\bullet} \to M$, namely, $x \to \varepsilon(e, x)$, so this principal bundle is trivial. Therefore, this book is about some trivial principal bundles!!

We next define $\mathcal{A}_1 \stackrel{\text{def}}{=} \cup_{x \in M} \mathcal{U}_1^{x,x}$, and observe that the restriction $\pi : \mathcal{A}_1 \to M$ maps the group $\mathcal{U}_1^{x,x}$ onto the point x. The local representation of a 1-arrow shows again that $\pi : \mathcal{A}_1 \to M$ is a smooth and locally trivial group bundle, and we call a (smooth) section of this bundle a (universal first-order) gauge transformation. The fiberwise group structure allows us to compose two gauge transformations in the obvious way, and therefore the set of gauge transformations forms a (very big) group. Gauge transformations will play a fundamental role in Chapters 8, 9, and 13.

Finally, we should mention, for those readers familiar with the formalism of principal bundles and their associated bundles, that the group bundle $\mathcal{A}_1 \to M$ and the tangent and cotangent bundles $TM \to M$ and $T^*M \to M$ can be constructed as bundles associated with the principal bundles $\mathcal{U}_1^{e,\bullet} \to M$ or $\mathcal{U}_1^{\bullet,e} \to M$. However, we will not make use of this general formalism until Chapter 16.

2

The Nonlinear Curvature

Let M be a smooth manifold with a splitting ε, which we fix once and for all. We denote this data by (M, ε). By the definition of the 1-arrow $\varepsilon(p, q)$ from p to q, there exists a local diffeomorphism f defined near p with $f(p) = q$ such that

$$j_1(f)^{p, f(p)} = \varepsilon(p, f(p)) \tag{2.1}$$

The representative f in (2.1) defines also the 1-arrows $j_1(f)^{x,f(x)}$, for all $x \in \mathrm{Dom}(f)$. However, there is no reason why we should have

$$j_1(f)^{x,f(x)} = \varepsilon(x, f(x)), \qquad x \in \mathrm{Dom}(f) \tag{2.2}$$

even though (2.2) holds at $x = p$ by (2.1). The key fact here is that f in (2.1) depends on (p, q) and the same f may not work for other (p, q)'s. The question arises whether there exists a local diffeomorphism f defined near p satisfying

(i) $f(p) = q$;
(ii) $j_1(f)^{x,f(x)} = \varepsilon(x, f(x))$ for all $x \in \mathrm{Dom}(f)$.

If (i) and (ii) hold, we say that f is a local solution of ε satisfying the initial condition $(p, q) \in \mathcal{U}_0 = M \times M$ and that the 1-arrow $\varepsilon(p, q)$ integrates to the local solution f. Clearly, if f is a local solution of ε, then it satisfies any initial condition $(r, f(r))$ for $r \in \mathrm{Dom}(f)$. Equivalently, a local solution integrates all its 1-arrows. This fundamental concept is worth a definition:

Definition 2.1 *Let $U \subset M$ be an open subset and $f : U \to f(U)$ be a diffeomorphism. Then f is a solution of ε on U if*

$$j_1(f)^{x, f(x)} = \varepsilon(x, f(x)), \qquad x \in U \tag{2.3}$$

Some natural questions arise: Does ε admit solutions? If it does, where are the solutions defined? Are they unique? Our ultimate purpose will be to find conditions that will imply the existence of solutions of ε that are global diffeomorphisms of M.

An Alternative Approach to Lie Groups and Geometric Structures, Ercüment H. Ortaçgil (2018).

To answer the above questions, we express (2.3) in coordinates. Now, the local representation $j_1(f)^{x,f(x)} = (x^i, y^i, (\partial f^i/\partial x^j))$ and (1.24) in Chapter 1 show that (2.3) takes the form $(x^i, f(x)^i, (\partial f^i/\partial x^j)) = (x^i, f(x)^i, \varepsilon^i_j(x, f(x)))$ for all $x \in U$, which is equivalent to

$$\frac{\partial f^i}{\partial x^j} = \varepsilon^i_j(x, f(x)), \qquad x \in U \tag{2.4}$$

Now, (2.4) is a first order nonlinear system of PDEs for f. Assuming that f is a solution of (2.4), we differentiate (2.4) with respect to x^k, which gives

$$\frac{\partial^2 f^i}{\partial x^k \partial x^j} = \frac{\partial \varepsilon^i_j(x, f(x))}{\partial x^k} + \frac{\partial \varepsilon^i_j(x, f(x))}{\partial y^a}\frac{\partial f^a}{\partial x^k} \tag{2.5}$$

Substituting $\partial f^a/\partial x^k$ from (2.4) into (2.5) and alternating k and j, we get

$$\left[\frac{\partial \varepsilon^i_j(x, f(x))}{\partial x^k} + \frac{\partial \varepsilon^i_j(x, f(x))}{\partial y^a}\varepsilon^a_k(x, f(x))\right]_{[kj]} = 0 \tag{2.6}$$

where $[kj]$ means alternation; for instance, $\left[A^i_{kj}\right]_{[kj]} = A^i_{kj} - A^i_{jk}$ (omitting the factor $\frac{1}{2}$). Now (2.6) gives the so-called integrability conditions for (2.4). We define

$$\mathcal{R}^i_{jk}(x, y) \stackrel{\text{def}}{=} \left[\frac{\partial \varepsilon^i_j(x, y)}{\partial x^k} + \frac{\partial \varepsilon^i_j(x, y)}{\partial y^a}\varepsilon^a_k(x, y)\right]_{[kj]} \tag{2.7}$$

where $x \in U$, and $y \in V = f(U)$. Therefore, $\mathcal{R}^i_{kj}(x, f(x)) = 0$, $x \in U$, is a necessary condition for the existence of a solution $f : U \to f(U) = V$. Following the old terminology, we call ε completely integrable if for any $(p, q) \in \mathcal{U}_0$, ε admits a local solution f satisfying $f(p) = q$. Consequently, complete integrability of ε on $(U, x) \times (V, y)$ implies $\mathcal{R}(x, y) = 0$ on $(U, x) \times (V, y)$.

It seems at first that $\mathcal{R}^i_{jk}(p, q)$ is defined on $(U, x) \times (V, y)$, where (U, x) and (V, y) are any two coordinate neighborhoods, and therefore $\mathcal{R}$ has only a local meaning. However, since ε is globally defined on $\mathcal{U}_0 = M \times M$, $\mathcal{R}$ is defined consistently on the overlap $((U, x) \cap (Z, z)) \times \big((V, y) \cap (W, w)\big)$, and therefore $\mathcal{R}(p, q)$ is also defined globally on $\mathcal{U}_0$. What kind of an object is $\mathcal{R}$? Equivalently, how does $\mathcal{R}^i_{kj}(x, y)$ transform if we apply the coordinate changes $(x) \to (z)$ and $(y) \to (w)$?

Lemma 2.2 *We have the transformation rule*

$$\mathcal{R}^i_{kj}(z, w) = \frac{\partial x^c}{\partial z^j}\frac{\partial x^b}{\partial z^k}\mathcal{R}^a_{bc}(x, y)\frac{\partial w^i}{\partial y^a} \tag{2.8}$$

The proof of Lemma 2.2 is a somewhat tedious but straightforward verification that we will leave to the reader. However, it is quite easy to guess (2.8): Since $\mathcal{R}^i_{kj}(x, y) = 0$ *does* have a coordinate-free meaning and only first-order derivatives are involved, $\mathcal{R}^i_{kj}(x,y)$ *must* transform tensorially. Consequently, a coordinate change $(x) \to (z)$ at the source transforms $\mathcal{R}^i_{kj}(x,y)$ as a 2-form in the indices k, j, and a coordinate change $(y) \to (w)$ at the target transforms $\mathcal{R}^i_{kj}(x,y)$ as a tangent vector in the index i, which is the content of (2.8).

The next proposition restates (2.8) in an elegant coordinate-free way:

Proposition 2.3 *$\mathcal{R}(p,q)$ is an element of $\wedge^2\left(T^*_p\right) \otimes T_q$.*

Definition 2.4 *$\mathcal{R}$ is the nonlinear curvature of (M,ε).*

We have already observed that the complete integrability of ε on $\mathcal{U}_0 = M \times M$ implies $\mathcal{R} = 0$ on $\mathcal{U}_0$. Our purpose is to find a local converse to this statement. This is contained in the following proposition:

Proposition 2.5 *Suppose $\mathcal{R}^i_{kj}(x,y) = 0$ on $(U,x) \times (V,y)$. For any $(p,q) \in U \times V$, there exists a neighborhood $\overline{U} \subset U$ of p and a diffeomorphism defined on $\overline{U} \subset U, f(\overline{U}) \subset V$, satisfying (i) $f(p) = q$ and (ii) $j_1(f)^{x,f(x)} = \varepsilon(x,f(x))$ for all $x \in \overline{U}$; that is, f is a solution of ε on $\overline{U}$ with the initial condition $f(p) = q$. Further, f is unique: If g is another such solution on $\overline{W} \subset U$, then $f = g$ on $\overline{U} \cap \overline{W}$.*

Proposition 2.5 is a direct consequence of the well-known local existence and uniqueness theorem for systems of first-order PDEs with initial conditions applied to (2.4) (see, e.g., the Appendix of [L]).

Corollary 2.6 *Suppose $\mathcal{R} = 0$ on $M \times M$. Then, for any $(p,q) \in M \times M$, (2.4) admits a unique local solution f with $f(p) = q$; that is, any 1-arrow $\varepsilon(a,b)$ integrates uniquely to a local solution.*

Corollary 2.7 *Suppose f, g are two solutions of ε on some connected open set $U \subset M$. If $f(p) = g(p)$ for some $p \in U$, then $f \equiv g$ on U.*

Proof We define $A \stackrel{\text{def}}{=} \{x \in U \mid f(x) = g(x)\} \subset U$. Clearly, A is closed in U but also open by Proposition 2.5, so $A = U$. ■

An important property of $\mathcal{R}$ is given by the following proposition:

Proposition 2.8 *$\mathcal{R}(p,p) = 0, p \in M$; that is, $\mathcal{R}$ vanishes on the diagonal $\Delta \subset M \times M$.*

Proof Differentiation of (1.28) with respect to x^k gives

$$\frac{\partial \varepsilon^i_a(x,y)}{\partial x^k}\varepsilon^a_j(y,x) + \varepsilon^i_a(x,y)\frac{\partial \varepsilon^a_j(y,x)}{\partial x^k} = 0 \tag{2.9}$$

Setting $y = x$ and using (1.27), (2.9) becomes

$$\left[\frac{\partial \varepsilon^i_j(x,y)}{\partial x^k}\right]_{y=x} + \left[\frac{\partial \varepsilon^i_j(y,x)}{\partial x^k}\right]_{y=x} = 0 \tag{2.10}$$

Setting $y = x$ in (2.7) and again using (1.27), we get

$$\mathcal{R}^i_{jk}(x,x) = \left[\left[\frac{\partial \varepsilon^i_j(x,y)}{\partial x^k} \right]_{y=x} + \left[\frac{\partial \varepsilon^i_j(x,y)}{\partial y^k} \right]_{y=x} \right]_{[kj]}$$
$$= [0]_{[kj]} = 0 \tag{2.11}$$

in view of (2.10). ■

Proposition 2.8 is not very surprising: The initial conditions (p,p) integrate to the identity diffeomorphism, and therefore the integrability conditions should be vacuous in this case.

For the reader familiar with the theory of connections on principal bundles and their associated vector bundles, this seems to be a good place to comment on some intriguing points: The trivialization ε of the principal bundle $\mathcal{U}^{e,\bullet}_1 \to M$ defines an obvious "flat connection" on the principal bundle $\mathcal{U}^{e,\bullet}_1 \to M$ according to the general theory, whereas ε now has a nonlinear curvature $\mathcal{R}$ that need not vanish! Even more intriguing is the fact that $\mathcal{R}$ determines a "linear curvature" $\mathfrak{R}$ that *need not vanish either*, as we will show in Chapter 6. In fact, we will prove that $\mathcal{R} = 0 \Leftrightarrow \mathfrak{R} = 0$. Clearly, $\mathfrak{R}$ cannot be the curvature of the induced linear connection on the associated vector bundle $TM \to M$, which is flat too. This discrepancy will be clarified in Chapters 5 and 6.

According to Lemma 2.2, if we fix $p \in M$ and some coordinates around p, then $\mathcal{R}^i_{jk}(p,x)$ transforms only in the index i under a coordinate change $(x) \to (y)$, whereas $\mathcal{R}^i_{jk}(x,p)$ transforms only in the indices j,k. This makes $\mathcal{R}$ a little awkward to work with. We will now tame $\mathcal{R}$ in two ways for later use.

We define

$$\overleftarrow{\mathcal{R}}^i_{jk}(x,y) \stackrel{\text{def}}{=} \mathcal{R}^i_{ab}(x,y)\varepsilon^a_j(y,x)\varepsilon^b_k(y,x) \tag{2.12}$$

and

$$\overrightarrow{\mathcal{R}}^i_{jk}(x,y) \stackrel{\text{def}}{=} \varepsilon^i_a(y,x)\,\mathcal{R}^a_{jk}(x,y) \tag{2.13}$$

Clearly,

$$\mathcal{R} = 0 \Longleftrightarrow \overrightarrow{\mathcal{R}} = 0 \Longleftrightarrow \overleftarrow{\mathcal{R}} = 0 \tag{2.14}$$

on $\mathcal{U}_0 = M \times M$. We recall that the chain rule together with the definition of the 1-arrow gives the transformation rules

$$\varepsilon^i_k(x,z) = \varepsilon^i_a(y,z)\frac{\partial y^a}{\partial x^k} \tag{2.15}$$

$$\varepsilon^i_k(z,x) = \frac{\partial x^i}{\partial y^a}\varepsilon^a_k(z,y) \tag{2.16}$$

under a coordinate change $(x) \to (y)$. Note that (2.16) is obtained from (2.15) by inversion, and vice versa. Now, if we fix $p \in M$ and a coordinate system around p once and for all, then $\overleftarrow{\mathcal{R}}^{i}_{jk}(p, x)$ transforms as a tensor in *all indices* i, j, k under a coordinate change $(x) \to (y)$ as follows easily from (2.12), (2.15), and Lemma 2.2. Similarly, $\overrightarrow{\mathcal{R}}^{i}_{jk}(x, p)$ transforms as a tensor in *all indices* upon a coordinate change $(x) \to (y)$. However, the tensors defined in this way obviously depend also on our choice of the point p, a subtle dependence to be clarified in Chapter 6.

3

Local Lie Groups

In [O1], Olver defines a local Lie group and makes the crucial observation that a local Lie group does not necessarily "globalize" to a Lie group. In his words, "The theory of local Lie groups is not a simple consequence of the global theory but has its own set of interesting and delicate geometric structures." The definition of a local Lie group that we will give below is different than the one in [O1]. Nevertheless, it is fair to say that Part I of this book will amply justify the deep insight of [O1]. In fact, we will show in Chapter 4 that a Lie group is a special (globalizable) local Lie group. This fact reinstates the paradigm of local to global to its historical record, as remarked by Olver [O4].

As in Chapter 2, we assume that M is a smooth manifold with a fixed splitting ε and denote this data by (M, ε).

Definition 3.1 *(M, ε) is called a local Lie group (LLG) if $\mathcal{R} = 0$ on $M \times M$.*

A word of caution with regard to the terminology of Definition 3.1: In the theory of Lie groups, a local Lie group refers to a neighborhood of the identity in some Lie group and is therefore a local concept. We observe that a LLG according to Definition 3.1 is a *global* concept.

In this chapter, (M, ε) is a LLG. According to Corollary 2.6, ε admits unique local solutions with arbitrary initial conditions. Let $\mathcal{G}$ denote the set of all local solutions of ε. We will use the notation $(M, \varepsilon, \mathcal{G})$ for the LLG (M, ε).

At this stage, it is useful to recall the definition of a pseudogroup on M. A pseudogroup on M is a collection $\mathcal{S}$ of local diffeomorphisms of M satisfying the following:

1. If $f, g \in \mathcal{S}$ and $\mathrm{Dom}(f) = \mathrm{Ran}(g)$, then $f \circ g \in \mathcal{S}$.
2. If $f \in \mathcal{S}$, then $f^{-1} \in \mathcal{S}$.
3. If $f \in \mathcal{S}$ and $U \subset \mathrm{Dom}(f)$, then $f_{|U} \in \mathcal{S}$.
4. If $U = \cup U_\alpha$ and f is local diffeomorphism defined on U such that $f_{|U_\alpha} \in \mathcal{S}$ for all α, then $f \in \mathcal{S}$.
5. The identity diffeomorphism I belongs to $\mathcal{S}$.

If for any $p, q \in M$ there exists some $f \in \mathcal{S}$ with $f(p) = q$, then $\mathcal{S}$ is called transitive.

An Alternative Approach to Lie Groups and Geometric Structures, Ercüment H. Ortaçgil (2018).

Proposition 3.2 *Let $(M, \varepsilon, \mathcal{G})$ be an LLG. Then the set $\mathcal{G}$ of local solutions of ε is a transitive pseudogroup on M.*

Proof Properties 3–5 and transitivity are immediate from Definition 2.1 and (2.4), so it suffices to check properties 1 and 2 locally in coordinates. For property 1, let $y = y(x)$ and $z = z(y)$ be two solutions of (2.4). We then have

$$\frac{\partial y^i}{\partial x^j} = \varepsilon^i_j\,(x, y)\,, \qquad \frac{\partial z^i}{\partial y^j} = \varepsilon^i_j\,(y, z) \tag{3.1}$$

Therefore,

$$\frac{\partial z^i}{\partial x^j} = \frac{\partial z^i}{\partial y^a}\frac{\partial y^a}{\partial x^j} = \varepsilon^i_a\,(y, z)\,\varepsilon^a_j(x, y) = \varepsilon^i_j\,(x, z) \tag{3.2}$$

in view of (1.26), and therefore $z = z(x)$ belongs to $\mathcal{G}$. Inverting both sides of $\partial y^i/\partial x^j = \varepsilon^i_j(x, y)$ and using (1.28), we get $\partial x^i/\partial y^j = \varepsilon^i_j(y, x)$, and property 2 holds.

Corollary 2.7 shows that $\mathcal{G}$ is a very special pseudogroup: Its local diffeomorphisms are determined on their connected domains by any of their values or 0-arrows. This fact allows us to "analytically continue" elements of $\mathcal{G}$ along paths as follows.

Suppose $(f, U) \in \mathcal{G}$, where $U = \mathrm{Dom}(f)$, $p \in U$, and let $C : x(t)$, $0 \le t \le 1$, be a (continuous) path from $p = x(0)$ to some $q = x(1)$. Let $\{U_\alpha\}$, $1 \le \alpha \le k$, be an open covering of C satisfying

(i) $U_1 = U$

(ii) $(f_\alpha, U_\alpha) \in \mathcal{G}, f_1 = f$

(iii) $f_i = f_j$ on $U_i \cap U_j$ ■

We call (f_α, U_α) a continuation of (f_1, U_1) along C. Let (g_β, V_β), $1 \le \beta \le m$, be a continuation of (g_1, V_1) along C such $f_1 = g_1$ on $U_1 \cap V_1$, and therefore $f_1(p) = g_1(p)$. Suppose $q \in U_r \cap V_s$ for some $1 \le r \le k$ and $1 \le s \le m$. Using Corollary 2.7, we easily show that $f_r = g_s$ on $U_r \cap V_s$ and therefore $f_r(q) = g_s(q)$. In short, two continuations of a local solution around p along a path from p to q define the same values on the path. Even more succinctly, we can state the following proposition:

Proposition 3.3 *A continuation of a local solution around p along a path from p to q is unique if it exists.*

Proposition 3.3 suggests the following important definition:

Definition 3.4 *$(M, \varepsilon, \mathcal{G})$ is complete if the elements of $\mathcal{G}$ can be continued indefinitely along all paths.*

Observe that we define completeness with the assumption $\mathcal{R} = 0$; that is, that (M, ε) is an LLG. It is an interesting question whether completeness can be defined without this assumption.

Some intuitive observations: $(M, \mathcal{E}, \mathcal{G})$ may fail to be complete because while continuing some local solution f along a path from p to q, there may be a point r on the path between p and q such that we may not be able to "continue through r" even though continuation from p up to and *excluding* r is possible. If the continuation up to and *including* r is possible, then we can continue through r. We define the continuation of $f \in \mathcal{G}$ along a path $x(t)$, $0 \leq t \lneqq 1$, as above, except that the covering (U_α) need not be finite. With these intuitions, we easily deduce the following proposition:

Proposition 3.5 *If M is compact, then $(M, \mathcal{E}, \mathcal{G})$ is complete.*

To prove Proposition 3.5, we need the following technical lemma:

Lemma 3.6 *Let $(M, \mathcal{E}, \mathcal{G})$ be a LLG and $(p, q) \in M \times M$. There exists a sufficiently small neighborhood $(p, q) \in U \times V$ of (p, q) with the following property: For any $(x, y) \in U \times V$, the unique local solution $g \in \mathcal{G}$ with the initial condition $g(x) = y$ can be defined on U.*

Lemma 3.6 states that $\mathcal{G}$ is always "locally complete" in a sense to be made clear below. Using this lemma, we can now easily prove Proposition 3.5: Let $(f, W) \in \mathcal{G}, x(t), 0 \leq t \leq 1$, be a path with $x(0) = p \in W$ and $x(1) = q$. We define $t_0 \stackrel{\text{def}}{=} \sup\{t \in [0, 1] \mid f$ has a continuation on $[0, t)\}$. We choose a sequence $t_n \to t_0^-$. Since M is compact, $f(t_n)$ has a subsequence of the form $f(t_{n_k})$ that converges to some $r \in M$. We solve (2.4) for some g defined on V with the initial condition $g(x(t_0)) = r$, where V is as in Lemma 3.6. Since $(t_{n_k}, f(t_{n_k})) \in V \times g(V)$ eventually, we conclude that $g(x(t)) = f(x(t))$ for $0 \leq t \lneqq t_0$, so g gives a continuation of f on $[0, t_0 + \epsilon]$ for some $0 \lneqq \epsilon$. This contradicts the definition of t_0 unless $t_0 = 1$.

To prove Lemma 3.6, we recall that Proposition 2.5 was proved by reducing the PDE (2.4) to an ODE (see the Appendix of [L] for the technical details). We recall that a vector field X on a smooth manifold N is complete if there exists an $\epsilon \gneqq 0$ such that the 1-parameter family of solutions $h(t, x), 0 \leq t, x \in N$ of X are defined on $0 \leq t \leq \epsilon$ *uniformly for all* $x \in N$. However, any $p \in N$ has a neighborhood W such that this condition holds on W. Therefore any vector field is locally complete but not necessarily globally complete. Now the proof of Lemma 3.6 reduces to this local completeness of vector fields once we reduce the PDE (2.4) to an ODE. We will leave the further technical details to the interested reader.

Proposition 3.7 *Let C_1 and C_2 be two paths from p to q and let $f \in \mathcal{G}$ be defined near p. Suppose f can be continued indefinitely along all paths. If C_1 is homotopic to C_2, then the continuations along C_1 and C_2 define the same value at q.*

The proof of Proposition 3.7 is the standard "monodromy" argument well known from complex analysis. We recall here the main idea: We first observe that Proposition 3.7 holds if C_2 is a "small deformation" of C_1 by the standard subdivision argument. For the general case, since C_1 is homotopic to C_2, we can start from C_1 and reach C_2 by a finite number of small deformations. Since each small deformation does not change the value at q, the same holds for the whole deformation.

Now we have the following important proposition:

Proposition 3.8 *Suppose $(M, \varepsilon, \mathcal{G})$ is complete and M is simply connected. Then, for any $(f, U) \in \mathcal{G}$, there exists a unique diffeomorphism $f^{\mathrm{ex}} : M \to M$ such that $f^{\mathrm{ex}}_{|U} = f$. The set $\mathcal{G}^{\mathrm{ex}} \stackrel{\mathrm{def}}{=} \{f^{\mathrm{ex}} \mid f \in \mathcal{G}\}$ is a group of transformations that acts simply transitively on M.*

Proof Let $(f, U) \in \mathcal{G}$ with U connected. We fix some $p \in U$. For any $x \in M$, we choose a path from p to x and define $f^{\mathrm{ex}}(x)$ to be the terminal value of the continuation of f along this path. Such a continuation is possible since (M, ε) is complete. Since M is simply connected, any two paths from p to x are homotopic, and therefore $f^{\mathrm{ex}}(x)$ is independent of the chosen path and also independent of the choice of the point $p \in U$ by Proposition 3.7. Thus, we get a map $f^{\mathrm{ex}} : M \to M$ with $f^{\mathrm{ex}}_{|U} = f$. By Corollary 2.7, the extension f^{ex} is unique. We claim that $f^{\mathrm{ex}} : M \to M$ is a diffeomorphism. It is onto: For any $y \in M$, we choose a path C from $f(p)$ to y. By Proposition 3.2, $(f^{-1}, f(U)) \in \mathcal{G}$. Using completeness, we continue $(f^{-1}, f(U))$ along C and define x to be the terminal value of this continuation at y. Now this continuation defines a continuation of (f, U) along a path from p to x with the terminal value y, so $f^{\mathrm{ex}}(x) = y$ and therefore f^{ex} is onto. A similar argument shows that f^{ex} is also 1–1. Hence, we conclude that $f^{\mathrm{ex}} : M \to M$ is a diffeomorphism. We claim that the set $\{f^{\mathrm{ex}} \mid f \in \mathcal{G}\}$ is closed under composition and inversion. Indeed, f^{ex} is a global solution of (2.4) on M, and therefore $(f^{\mathrm{ex}}, M) \in \mathcal{G}$. The conclusion now follows from Proposition 3.2. Transitivity is clear: Any 0-arrow (p, q) integrates uniquely to a local solution f with $f(p) = q$ that extends uniquely to some f^{ex}. Finally, Corollary 2.7 shows that the group $\mathcal{G}^{\mathrm{ex}}$ acts simply transitively on M. ■

Suppose $(g, U), (f, V) \in \mathcal{G}$, with $g(U) = V$. Note that the formulas $(f \circ g)^{\mathrm{ex}} = f^{\mathrm{ex}} \circ g^{\mathrm{ex}}$ and $(f^{-1})^{\mathrm{ex}} = (f^{\mathrm{ex}})^{-1}$ are implicitly contained in the proof of Proposition 3.8.

Definition 3.9 *Let $(M, \varepsilon, \mathcal{G})$ be an LLG. If $f \in \mathcal{G}$ extends (necessarily uniquely) to a diffeomorphism $f^{\mathrm{ex}} : M \to M$, then f is globalizable. If all $f \in \mathcal{G}$ are globalizable, then $(M, \varepsilon, \mathcal{G})$ is globalizable.*

It follows that a complete and simply connected LLG is globalizable by Proposition 3.8. Obviously, if $(M, \varepsilon, \mathcal{G})$ is globalizable, then it is complete.

For readers familiar with the definition of an abstract Lie group, the standard example of a globalizable LLG arises from a Lie group G as follows: Let L_a be left translation by a and define $L(G) \stackrel{\mathrm{def}}{=} \{L_a \mid a \in G\}$. For any $p, q \in G$, there exists a unique L_a satisfying $L_a(p) = q$, $a = qp^{-1}$. We define $\varepsilon(p, q) \stackrel{\mathrm{def}}{=} j_1(L_a)^p$. Clearly, ε is a splitting that admits L_a as a global solution. We recall that if a transformation group G acts simply transitively on M, then the pair (M, G) is a called a principal homogeneous space (PHS). It follows that $(G, L(G))$ is a PHS. Similarly, $(G, R(G))$ is another PHS. If $(M, \varepsilon, \mathcal{G})$ is globalizable, then $(M, \mathcal{G}^{\mathrm{ex}})$ is yet another PHS. We observe here the crucial fact that a Lie group G gives rise to two PHSs, $(G, L(G))$ and $(G, R(G))$, whereas a globalizable LLG $(M, \varepsilon, \mathcal{G})$ gives rise to only one PHS, $(M, \mathcal{G}^{\mathrm{ex}})$. Indeed, the transformation group $\mathcal{G}^{\mathrm{ex}}$ shuffles the points of M and there is no such thing as shuffling from the right or left, and so this is only a matter of notation, not a fundamental concept. This means that we need to do more work to construct some abstract Lie group G from the PHS $(M, \mathcal{G}^{ex})$, and this will be the subject of the next chapter.

We have seen above that if $(M, \varepsilon, \mathcal{G})$ is complete, then the local solutions of (2.4) can be continued uniquely along paths according to Proposition 3.3, but two continuations along different paths from p to q may give different values at q if these paths are not homotopic. In this case, we can lift $\mathcal{G}$ to the universal covering of M as follows.

Let $\pi : \widetilde{M} \to M$ be a covering space of (M, ε) and take $a, b \in \widetilde{M}$ with $\pi(a) = p, \pi(b) = q$. There exist open neighborhoods U, V of a, b such that $\pi_{|U}, \pi_{|V}$ are diffeomorphisms. Now using the 1-arrow $\varepsilon(p, q)$ from p to q, we define the 1-arrow $\pi^{-1}(\varepsilon)(a, b)$ from a to b by the formula

$$\pi^{-1}(\varepsilon)(a, b) \overset{\text{def}}{=} j_1(\pi_{|V}^{-1})^{q,b} \circ \varepsilon(p, q) \circ j_1(\pi_{|U})^{a,p} \tag{3.3}$$

Using (3.3) and (1.26), we easily check that $\pi^{-1}(\varepsilon)(b, c) \circ \pi^{-1}(\varepsilon)(a, b) = \pi^{-1}(\varepsilon)(a, c)$, and therefore $\pi^{-1}(\varepsilon)$ is a splitting on $\widetilde{M}$. We call $(\widetilde{M}, \pi^{-1}(\varepsilon))$ the lift of (M, ε) to $\widetilde{M}$. If $(M, \varepsilon, \mathcal{G})$ is an LLG and (f, W) is a local solution of ε that integrates the 1-arrow $\varepsilon(p, q)$, then, by restricting W, U, V so that $\pi(U) = W$ and $f(W) = V$, it is clear from (3.3) that the local diffeomorphism

$$\pi^{-1}(f) \overset{\text{def}}{=} \pi_{|V}^{-1} \circ f \circ \pi_{|U} \tag{3.4}$$

is defined on U, maps a to b, and integrates the 1-arrow $\pi^{-1}(\varepsilon)(a, b)$. Therefore, $\mathcal{R}(\varepsilon) = 0$ implies $\mathcal{R}(\pi^{-1}(\varepsilon)) = 0$; that is, if (M, ε) is an LLG, then so is its lift $(\widetilde{M}, \pi^{-1}(\varepsilon))$. In this case, we denote the local solutions of $\pi^{-1}(\varepsilon)$ by $\pi^{-1}(\mathcal{G})$ and say that $(\widetilde{M}, \pi^{-1}(\varepsilon), \pi^{-1}(\mathcal{G}))$ lifts $(M, \varepsilon, \mathcal{G})$.

Proposition 3.10 *If $(M, \varepsilon, \mathcal{G})$ is complete, then so is $(\widetilde{M}, \pi^{-1}(\varepsilon), \pi^{-1}(\mathcal{G}))$.*

Proof Let $a, b \in \widetilde{M}$, $a \in U$, (h, U) be a solution of $\pi^{-1}(\varepsilon)$ on U and let C be a path from a to some $b \in \widetilde{M}$. The path C projects to the path $\pi(C)$ from $\pi(a) = p$ to $\pi(b) = q$. By restricting U if necessary, (h, U) is the lift of *some* local solution (f, W) of ε defined by (3.4), where $\pi(U) = W$. Since (M, ε) is complete, (f, W) has some continuation along $\pi(C)$ from p to q. Since π is a local diffeomorphism, this continuation (or, rather, its restriction to possibly smaller domains) lifts to a continuation of (h, U) from a to b, showing that $(\widetilde{M}, \pi^{-1}(\varepsilon))$ is complete. ■

Propositions 3.10 and 3.8 give the following corollary:

Corollary 3.11 *If $(M, \varepsilon, \mathcal{G})$ is complete and $\pi: \widetilde{M} \to M$ is the universal covering, then $(\widetilde{M}, \pi^{-1}(\varepsilon), \pi^{-1}(\mathcal{G}))$ is globalizable.*

Therefore, we have the PHS $(\widetilde{M}, \pi^{-1}(\mathcal{G})^{ex})$ if $(M, \varepsilon, \mathcal{G})$ is complete.

It is possible to carry the above analysis further. For instance, assuming the completeness of $(M, \varepsilon, \mathcal{G})$, which $f \in \mathcal{G}$ are globalizable? We recall that $f \in \mathcal{G}$ with $f(p) = p$ for some $p \in M$ must be the restriction of I to $\text{Dom}(f)$. Therefore, such $f \in \mathcal{G}$ are surely globalizable. Let $\pi : \widetilde{M} \to M$ be the universal covering space and D be the group of deck transformations on $\widetilde{M}$. We recall that D acts simply transitively on each fiber $\pi^{-1}(p)$ and is isomorphic to the fundamental group $\pi(M, p)$ of M. In particular, the elements of D commute with the projection π. We will leave the proof of the following proposition to the interested reader:

Proposition 3.12 *Suppose $(M, \varepsilon, \mathcal{G})$ is complete. Then the following are equivalent.:*

(i) $f \in \mathcal{G}$ *is globalizable;*

(ii) $\pi^{-1}(f)^{\mathrm{ex}}$ *commutes with* D.

Corollary 3.13 *Suppose $(M, \varepsilon, \mathcal{G})$ is complete. Then $(M, \varepsilon, \mathcal{G})$ is globalizable if and only if $D \subset \pi^{-1}(\mathcal{G})^{\mathrm{ex}}$ is a normal (hence central) subgroup.*

We will conclude this chapter with the following important question:

> **Q:** Given some (M, ε), is there a splitting $\overline{\varepsilon}$ on M with $\mathcal{R}(\overline{\varepsilon}) = 0$? Equivalently, does a parallelizable M admit an LLG structure?

As we will see in Chapter 13, for compact and simply connected 3-manifolds, **Q** is equivalent to the Poincaré Conjecture (PC). Therefore, the answer is affirmative by the recent proof of the PC due to Hamilton and Perelman. This gives an idea of the depth and the level of difficulty of **Q**.

4

The Centralizer

In this chapter, $(M, \varepsilon, \mathcal{G})$ is an LLG as defined in Chapter 3. If $(M, \varepsilon, \mathcal{G})$ is globalizable, then we obtain the PHS $(M, \mathcal{G}^{\text{ex}})$ as we have seen in Chapter 3. For simplicity of notation, we will henceforth denote $(M, \mathcal{G}^{\text{ex}})$ by $(M, \mathcal{G})$. The purpose of this chapter is to define another pseudogroup $C(\mathcal{G})$ and endow M with the structure of an abstract Lie group using $\mathcal{G}$ and $C(\mathcal{G})$, with the assumption that $(M, \varepsilon, \mathcal{G})$ globalizes. The idea is simple: Let X be a finite set and let $P(X)$ be the set of all permutations of X, that is, the set of all 1–1 and onto maps of X. Let $G \subset P(X)$ be a subset that is closed under composition $\circ$ and inversion of permutations and let $C(G) \subset P(X)$ be the set of all permutations that commute with the elements G, that is, the centralizer of G inside $P(X)$. Clearly, $C(G)$ is also closed under composition and inversion. Now, if G acts simply transitively on X, then $C(G)$ also acts simply transitively. To see this, we fix $p, q \in X$ arbitrarily. For any $x \in X$, there exists a unique $f \in G$ with $f(p) = x$. We define $h : X \to X$ by $h(x) \overset{\text{def}}{=} f(q)$. We easily see that h is 1–1 and onto, $h(p) = q$, and h commutes with G. Thus, $C(G)$ acts transitively and we easily check that this action is also simply transitive. Is there any conceptual reason why $C(G)$ also acts simply transitively? To answer this, we fix a base point $e \in X$ and identify the elements of the *abstract group* $(G, \circ)$ with X in such a way that the action of G on X is identified with the, say, left action of G on $G \simeq X$. Now the right action of G on $G \simeq X$ gives another simply transitive action on X, which can be identified with the above action of $C(G)$. Note that there is no canonical choice of left/right.

Now suppose that $\mathcal{R} = 0$ and that $(M, \varepsilon, \mathcal{G})$ globalizes, so we have the PHS $(M, \mathcal{G})$, where $\mathcal{G}$ acts simply transitively on M according to Proposition 3.8. We have the analogy $X \leftrightsquigarrow M$, $G \leftrightsquigarrow \mathcal{G}$, and $P(X) \leftrightsquigarrow \text{Diff}(M)$. We know that the elements of $\mathcal{G}$ are the unique solutions of the nonlinear PDE (2.4). If we define $C(\mathcal{G})$ as the centralizer of $\mathcal{G}$ inside $\text{Diff}(M)$, what is the structure of $C(\mathcal{G})$? For instance, are its elements the solutions of another nonlinear PDE?

Let $f(x, y, z)$ denote the unique local solution $f \in \mathcal{G}$ of (2.4) in the variable $z \in U$ satisfying the initial condition $f(x) = y$. Using Lemma 3.6, we choose U such that $f(x, y, z)$ is defined for all $x, y, z \in U$. If $(M, \varepsilon, \mathcal{G})$ is globalizable, then clearly $f(x, y, z)$ is defined for all $x, y, z \in M$. Now we have

$$\frac{\partial f^i}{\partial x^j}(p, q, x) = \varepsilon^i_j(x, f(p, q, x)), \qquad f \in \mathcal{G}, \quad p, q, x \in U \tag{4.1}$$

An Alternative Approach to Lie Groups and Geometric Structures, Ercüment H. Ortaçgil (2018).

Clearly,

$$f(x,y,x) = y \quad \text{and} \quad f(x,x,y) = y, \qquad x,y \in U \tag{4.2}$$

Now we fix $p, q \in U$ and define the map g on U by

$$g(y) \stackrel{\text{def}}{=} f(p,y,q), \qquad y \in U \tag{4.3}$$

so $g(p) = q$ by (4.2). Now, g is a 1–1 map by Corollary 2.7. Further, it is a diffeomorphism, since the solutions of (2.4) depend smoothly on the initial conditions. Note that $g(U)$ need not be contained in U. We define the 1-arrow $\widetilde{\varepsilon}(p,q)$ from p to q by the formula

$$\widetilde{\varepsilon}(p,q) \stackrel{\text{def}}{=} j_1(g)^{p,q} \tag{4.4}$$

In coordinates, (4.4) is given by

$$(p^i, q^i, \widetilde{\varepsilon}^i_j(p,q)) \stackrel{\text{def}}{=} \left(p^i, q^i, \left[\frac{\partial f^i(p,y,q)}{\partial y^j}\right]_{y=p}\right) \tag{4.5}$$

which is equivalent to

$$\left[\frac{\partial g^i(y)}{\partial y^j}\right]_{y=p} = \widetilde{\varepsilon}^i_j(p,g(p)) \tag{4.6}$$

We claim

$$\frac{\partial g^i(y)}{\partial y^j} = \widetilde{\varepsilon}^i_j(y,g(y)) \qquad y \in U \tag{4.7}$$

that is, (4.6) holds not only at $y = p$ but for all $y \in U$. So, let $\overline{p}, x \in U$, $\overline{q} \stackrel{\text{def}}{=} f(p,\overline{p},q)$, and $y \stackrel{\text{def}}{=} f(\overline{p},x,\overline{q}) = f(p,x,q)$. Therefore,

$$f(p,x,q) = f(\overline{p},x,f(p,\overline{p},q)) \tag{4.8}$$

We differentiate (4.8) at $x = \overline{p}$ using (4.6). This gives

$$\begin{aligned}\left[\frac{\partial g^i(x)}{\partial x^j}\right]_{x=\overline{p}} &= \left[\frac{\partial f^i(p,x,q)}{\partial x^j}\right]_{x=\overline{p}} \\ &= \left[\frac{\partial f^i(\overline{p},x,f(p,\overline{p},q))}{\partial x^j}\right]_{x=\overline{p}} \\ &= \widetilde{\varepsilon}^i_j(\overline{p},f(p,\overline{p},q)) \\ &= \widetilde{\varepsilon}^i_j(\overline{p},g(\overline{p}))\end{aligned} \tag{4.9}$$

which proves the claim, since $\overline{p}$ is arbitrary.

Note that (4.7) is (2.4) with ε replaced by $\widetilde{\varepsilon}$. However, there are two major differences: First, we need the assumption $\mathcal{R} = 0$ to solve (2.4) locally, whereas (4.7) always has the *unique* local solution g defined by (4.3) with this assumption. Second, ε is defined on $M \times M$, whereas $\widetilde{\varepsilon}$ is defined on $U \times U$, but if $(M, \varepsilon, \mathcal{G})$ is globalizable, then $\widetilde{\varepsilon}$ is defined also on $M \times M$.

Proposition 4.1

$$\widetilde{\varepsilon}(p, r) = \widetilde{\varepsilon}(q, r) \circ \widetilde{\varepsilon}(p, q), \qquad p, q, r \in U \tag{4.10}$$

that is, $\widetilde{\varepsilon}$ *is a* local *splitting defined on* $U \times U$.

Proof Let $f \in \mathcal{G}$ be defined on U, where U is as above. Suppose $f(p) = r$ and $f(q) = m$, $p, q, r, m \in U$. In our notation, this means that $f(p, r, x) = f(x) = f(q, m, x)$ for all $x \in U$. Therefore,

$$f(p, r, x) = f(q, f(p, r, q), x) \qquad p, q, r, x, f(p, r, q) \in U \tag{4.11}$$

which generalizes (4.2).

We regard $r = y$ as variable, fix $x = r$ and differentiate (4.11) with respect to y at $y = p$ by the chain rule using (4.6). What we get is

$$\left[\frac{\partial f^i}{\partial y^j}(p, y, r)\right]_{y=p} = \left[\frac{\partial f^i}{\partial z^a}(q, z, r)\right]_{z=q} \left[\frac{\partial f^a}{\partial y^j}(p, y, q)\right]_{y=p} \tag{4.12}$$

which is (4.10), in view of (4.6). ■

To summarize, any $p \in M$ has a neighborhood U such that we have $(U, \widetilde{\varepsilon})$ in the same way as (M, ε). Let $C(\mathcal{G}, U)$ denote the set of local solutions of $\widetilde{\varepsilon}$ on U. As we did for $(M, \varepsilon, \mathcal{G})$ in Chapter 3, we easily show using (4.7) that $C(\mathcal{G}, U)$ is closed under inversion and composition whenever they are defined. Therefore, $C(\mathcal{G}, U)$ is a transitive pseudogroup on U and we have the LLG $(U, \widetilde{\varepsilon}, C(\mathcal{G}, U))$ in the same way as $(M, \varepsilon, \mathcal{G})$. By our choice of U, any $g \in C(\mathcal{G}, U)$ is defined on U, but it is not necessarily the case that $g(U) \subset U$, as remarked above and as indicated by the condition $f(p, r, q) \in U$ in (4.11). If $(M, \varepsilon, \mathcal{G})$ is globalizable, we can choose $U = M$ and denote $C(\mathcal{G}, U)$ by $C(\mathcal{G})$, which is a global transformation group that acts simply transitively on M. From the definition of $C(\mathcal{G})$, it follows that $C(\mathcal{G})$ commutes with $\mathcal{G}$. Indeed, let $g \in C(\mathcal{G})$, $f \in \mathcal{G}$ and $p \in M$. Suppose $g(p) = q$ and $f(p) = r$. Now $g(f(p)) = g(r) = h(p, r, q)$, where $h(p, r, q)$ is the value at q of the unique solution of (2.4) that maps p to r. Therefore, $h(p, r, q) = f(q) = f(g(p))$. Since p is arbitrary, we conclude that $g \circ f = f \circ g$.

To recapitulate, we have the following proposition:

Proposition 4.2 *Let* $(M, \varepsilon, \mathcal{G})$ *be an LLG. Any* $p \in M$ *has a neighborhood* $p \in U$ *such that* $(U, \widetilde{\varepsilon}, C(\mathcal{G}, U))$ *is a transitive LLG. If* $(M, \varepsilon, \mathcal{G})$ *is globalizable, then we have the PHSs* $(M, C(\mathcal{G}))$ *and* $(M, \mathcal{G})$*, where the transformations of* $C(\mathcal{G})$ *and* $\mathcal{G}$ *commute.*

Note that if M is simply connected, then we can extend the local splitting $\widetilde{\varepsilon}$ on $U \times U$ to a global splitting on $M \times M$ as follows: Let $p, q \in M$ and C be any path from p to q. We partition

C into sufficiently small curves by choosing the points $x_1(=p), x_2, \ldots, x_{n-1}, x_n(=q)$ such that $\widetilde{\varepsilon}(x_k, x_{k+1})$ is defined for all k. We define $\widetilde{\varepsilon}(p,q) \stackrel{\text{def}}{=} \widetilde{\varepsilon}(x_{n-1}, q) \circ \cdots \circ \widetilde{\varepsilon}(x_1, x_2)$. Passing to a common refinement shows that the definition of $\widetilde{\varepsilon}(p,q)$ is independent of the partition, and the standard monodromy argument shows that it is also independent of the path.

Henceforth in this chapter, we always assume that $(M, \varepsilon, \mathcal{G})$ is globalizable so that we have the two commuting PHSs $(M, \mathcal{G})$ and $(M, C(\mathcal{G}))$. Fixing some base point $e \in M$, for any $x \in M$, there exists a unique $f \in \mathcal{G}$ $(= \mathcal{G}^{\text{ex}})$ with $f(e) = x$. This gives a 1–1 correspondence between $\mathcal{G}$ and M. The same conclusion holds also for $C(\mathcal{G})$. Note that $f \in \mathcal{G}$ and $g \in C(\mathcal{G})$ satisfying $f(e) = g(e)$ need not be equal on M. There is an obvious bijective map

$$\Theta_e : \mathcal{G} \longrightarrow C(\mathcal{G}) \tag{4.13}$$

that maps $f \in \mathcal{G}$ to the unique g satisfying $g(e) = f(e)$. Clearly, $\Theta_e(g \circ f) = \Theta_e(f) \circ \Theta_e(g)$ according to (4.3) and (4.6).

Now, let $\text{Diff}(M)$ denote the group of global diffeomorphisms of M and let $C_{\text{Diff}(M)}(\mathcal{G})$ denote the centralizer of $\mathcal{G}$ inside $\text{Diff}(M)$. Clearly, we have $C(\mathcal{G}) \subset C_{\text{Diff}(M)}(\mathcal{G})$.

Proposition 4.3 $C(\mathcal{G}) = C_{\text{Diff}(M)}(\mathcal{G}) = \{g \in \text{Diff}(M) \mid g \circ f = f \circ g,\ f \in \mathcal{G}\}$.

Proof We claim that $C_{\text{Diff}(M)}(\mathcal{G}) \subset C(\mathcal{G})$. Let $g \in C_{\text{Diff}(M)}(\mathcal{G})$, $g(p) = q$, and $f \in \mathcal{G}$ with $f(p) = x$. Now

$$\begin{aligned}
& & (g \circ f)(p) &= (f \circ g)(p) \\
&\Longleftrightarrow & g(x) &= f(p, x, q) \\
&\Longrightarrow & \left[\frac{\partial g^i}{\partial x^j}\right]_{x=p} &= \left[\frac{\partial f^i(p,x,q)}{\partial x^j}\right]_{x=p} \\
&\Longleftrightarrow & j_1(g)^p &= \widetilde{\varepsilon}(p, g(p))
\end{aligned} \tag{4.14}$$

using (4.3) and (4.4) in the last step. Since p is arbitrary, it follows that g is a solution of (4.6) on $M \times M$, proving the claim. ∎

If G is a Lie group, we recall that $(G, L(G))$ and $(G, R(G))$ are two PHSs and also globalizable LLGs. Clearly, the elements of $L(G)$ and $R(G)$ commute. Now, given the two PHSs $(M, \mathcal{G})$ and $(M, C(\mathcal{G}))$, our purpose is to turn M into a Lie group by defining some multiplication and inversion on it in such a way that we will have $L(M) = \mathcal{G}$ and $R(M) = C(\mathcal{G})$. Obviously, this Lie group structure cannot be canonical, because it is not possible to distinguish between $\mathcal{G}$ and $C(\mathcal{G})$ since there is a complete symmetry between them. In other words, we can also define this Lie group structure such that $R(M) = \mathcal{G}$ and $L(M) = C(\mathcal{G})$. We warn the reader that this seemingly naive point turns out to be subtle and will have intriguing consequences later, as already mentioned in Chapter 2.

We now resume our construction of the Lie group structure on M. First, we fix a base point $e \in M$. We define an operation

$$* : M \times M \longrightarrow M \tag{4.15}$$

as follows: For $p, q \in M$, there exists a unique $f, g \in \mathcal{G}$ with $f(e) = p, g(e) = q$. We define $p * q \overset{\text{def}}{=} g(p)$. Note that we can also use $C(\mathcal{G})$ instead of $\mathcal{G}$. With our notation in (4.3),

$$p * q \overset{\text{def}}{=} f(e, q, p), \qquad f \in \mathcal{G} \tag{4.16}$$

Since the solution $f(e, q, p)$ depends smoothly on its argument p and the initial condition q, (4.15) is a smooth map. Since function composition is associative, $*$ is associative. Further,

$$e * p = p * e = p, \qquad p \in M \tag{4.17}$$

by (4.2), and e is the identity. Setting $q = f(p, e, e)$ in (4.16) gives

$$p * q = f(e, f(p, e, e), p) = e \tag{4.18}$$

and therefore $p * f(p, e, e) = e$. Similarly, $f(p, e, e) * p = e$. Therefore,

$$p^{-1} = f(p, e, e) \tag{4.19}$$

Since $f(p, e, e)$ depends smoothly on p, inversion is a smooth map.

Thus, we conclude that M is a Lie group with the above composition and inversion. Let $L(M)$ and $R(M)$ be the groups of left and right translations on M. We claim that

$$L(M) = C(\mathcal{G}), \qquad R(M) = \mathcal{G} \tag{4.20}$$

Now, $L_p(x) = p * x = f(e, x, p)$. Therefore,

$$j_1(L_p)^e = j_1(f(e, x, p))_{|x=e} = \widetilde{\varepsilon}(e, p) \tag{4.21}$$

by (4.3). Inverting both sides of (4.21) and replacing p by q gives

$$j_1(L_{q^{-1}})^q = \widetilde{\varepsilon}(q, e) \tag{4.22}$$

Composing (4.21) and (4.22), we deduce (omitting $*$)

$$\begin{aligned} j_1\left(L_{pq^{-1}}\right)^q &= j_1\left(L_p \circ L_{q^{-1}}\right)^q \\ &= j_1\left(L_p\right)^e \circ j_1\left(L_{q^{-1}}\right)^q \\ &= \widetilde{\varepsilon}(e, p) \circ \widetilde{\varepsilon}(q, e) \\ &= \widetilde{\varepsilon}(q, p) \end{aligned} \tag{4.23}$$

According to (4.23), $L_{pq^{-1}}$ is a global solution of (4.7) that maps q to p. However, $g(q, p, x) \in C(\mathcal{G})$ is another such solution, and therefore $L_{pq^{-1}} = g(q, p, x) \in C(\mathcal{G})$ by uniqueness. Therefore, $L(M) = C(\mathcal{G})$. Similarly, we show that $R(M) = \mathcal{G}$.

We could also define $*$ by $p * q \stackrel{\text{def}}{=} f(e, q, p), f \in \mathcal{G}$, or equivalently $p * q \stackrel{\text{def}}{=} g(e, p, q)$, $f \in C(\mathcal{G})$. This choice would give $L(M) = \mathcal{G}$ and $R(M) = C(\mathcal{G})$.

Thus, we obtain the following fundamental proposition:

Proposition 4.4 *Suppose that $(M, \varepsilon, \mathcal{G})$ is a globalizable LLG. Then M has a Lie group structure such that $L(M) = C(\mathcal{G}), R(M) = \mathcal{G}$. This Lie group structure is canonical modulo two choices: (1) the choice of the base point e and (2) The choice of left/right.*

Changing the base point conjugates our transformation groups and produces an "isomorphic" Lie group. We will leave the details to the interested reader. Proposition 4.4 asserts that the modern theory of Lie groups is a special case of the theory of LLGs.

For someone who is familiar with the definitions of an abstract group and a smooth manifold, it takes two lines to define a Lie group. Therefore, one may argue at this point that Proposition 4.4 unnecessarily complicates a simple matter by replacing two lines by four chapters. However, one may also argue that this two-line definition actually expresses a very abstract concept and that it does indeed take four chapters to make it concrete. Needless to say, we favor the second argument in this book and hope that the following chapters will give more evidence supporting our view.

Finally, we will define the adjoint map of an LLG, which will be needed later. Let $e \in U \subset M$ be a basepoint and $x \in U$. We define

$$\mathrm{Ad}(e, x) \stackrel{\text{def}}{=} \varepsilon(x, e) \circ \widetilde{\varepsilon}(e, x) \in \mathcal{U}_1^{e,e}, \qquad x \in U \tag{4.24}$$

where U is sufficiently small that $\widetilde{\varepsilon}$ is defined on $U \times U$. If $(M, \varepsilon, \mathcal{G})$ is globalizable, then $Ad(e, x)$ is defined on the Lie group M and, using the definitions, it is easy to show that

$$\mathrm{Ad}(e, xy) = \mathrm{Ad}(e, x) \circ \mathrm{Ad}(e, y) \tag{4.25}$$

We will end this chapter with a remark about complexification. It seems to us that there is no unified conceptual approach to complexification in differential geometry. We complexify various real objects, like vector spaces, vector bundles, and Lie algebras. A real analytic manifold can be complexified to a complex manifold, but not uniquely. As a special case, a compact Lie group complexifies *uniquely* to a complex Lie group, but this is not true for an arbitrary Lie group (see [Ho]). In all these complexifications, the underlying idea is to substitute a complex variable for a convergent *real* power series. What is the complexification of (M, ε)? A glance at [W] and [Wi] shows that the answer is far from being trivial, and Parts II and III will put more emphasis on this question. Unfortunately, however, this issue will not be pursued further in this book.

5

ε-Invariance

Up to now, we have worked with the "Lie group" and now it is time to deal with the "Lie algebra." In this chapter, (M, ε) denotes, as before, a smooth manifold M with a fixed splitting ε.

We choose some coordinate neighborhood (U, x) and define the functions $\Gamma^i_{jk}(x)$ on U by

$$\Gamma^i_{kj}(x) \overset{\text{def}}{=} \left[\frac{\partial \varepsilon^i_j(x,y)}{\partial y^k}\right]_{y=x} = -\left[\frac{\partial \varepsilon^i_j(x,y)}{\partial x^k}\right]_{y=x} \tag{5.1}$$

where the second equality follows from (2.10). It is extremely important to note that $\Gamma^i_{jk}(x)$ need not be symmetric in j, k, since these indices play quite different roles in (5.1). Note that $\Gamma = (\Gamma^i_{jk})$ is a globally defined geometric object on M, since ε is defined on $\mathcal{U}_0 = M \times M$.

Proposition 5.1 *The components* (5.1) *transform according to*

$$\Gamma^i_{kj}(x) = \frac{\partial^2 x^i}{\partial y^b \partial y^a}\frac{\partial y^b}{\partial x^k}\frac{\partial y^a}{\partial x^j} + \frac{\partial x^i}{\partial y^a}\Gamma^a_{bc}(y)\frac{\partial y^b}{\partial x^k}\frac{\partial y^c}{\partial x^j} \tag{5.2}$$

upon a coordinate change $(x) \to (y)$.

Proof We choose $w \in (U, x)$ and apply the coordinate change $(U, x) \to (V, y)$. Suppose w is now represented by $z \in (V, y)$. The chain rule gives

$$\varepsilon^i_j(w, x) = \frac{\partial x^i}{\partial y^a}\varepsilon^a_b(z, y)\frac{\partial z^b}{\partial w^j} \tag{5.3}$$

where $z, x \in U$ and $w, y \in V$. We fix $z = \overline{x}$, $w = \overline{y}$ and regard x, y as variables in (5.3). Note that $(\partial z^b/\partial w^j)(\overline{y}) = (\partial y^b/\partial x^j)(\overline{y})$ in (5.3) is now fixed too and does not depend on x, y. Now we differentiate

$$\varepsilon^i_j(\overline{x}, x) = \frac{\partial x^i}{\partial y^a}\varepsilon^a_b(\overline{y}, y)\frac{\partial y^b}{\partial x^j}(\overline{y}) \tag{5.4}$$

An Alternative Approach to Lie Groups and Geometric Structures, Ercüment H. Ortaçgil (2018).

with respect to x^k, which gives

$$\frac{\partial \varepsilon^i_j(\overline{x}, x)}{\partial x^k} = \frac{\partial^2 x^i}{\partial y^b \partial y^a} \frac{\partial y^b}{\partial x^k} \varepsilon^a_b(\overline{y}, y) \frac{\partial y^b}{\partial x^j}(\overline{y}) + \frac{\partial x^i}{\partial y^a} \frac{\partial \varepsilon^a_b(\overline{y}, y)}{\partial y^c} \frac{\partial y^c}{\partial x^k} \frac{\partial y^b}{\partial x^j}(\overline{y}) \tag{5.5}$$

We set $x = \overline{x}$ in (5.5) so we also have $y = \overline{y}$. Thus, (5.5) gives (5.2) in view of (5.1) and (1.26). ∎

Let $T^{m,r}_p$ denote the tensor space of type (m, r) at $p \in M$. A tensor field t of type (m, r) on M is a smooth section of the vector bundle $T^{m,r} \overset{\text{def}}{=} \cup_{x \in M} T^{m,r}_x \to M$. With an abuse of notation, we simply write $t \in T^{m,r}$. For instance, some $t \in T^{2,1}$ has components $(t^{ij}_k(x))$ with respect to some coordinates (x), and a coordinate change $(x) \to (y)$ transforms these coordinates according to the well-known rule

$$\frac{\partial y^j}{\partial x^c} \frac{\partial y^i}{\partial x^b} \frac{\partial x^a}{\partial y^k} t^{bc}_a(x) = t^{ij}_k(y) \tag{5.6}$$

Recalling Lemma 1.3, giving a 1-arrow from p to q is the same as giving an isomorphism $T_pM \to T_qM$. Now, the isomorphism $\varepsilon(p, q) : T_pM \to T_qM$ extends to an isomorphism $\varepsilon(p, q) : T^{m,r}_p \to T^{m,r}_q$. For instance, according to (5.6) and the definition of a 1-arrow, $\varepsilon(p, q) : T^{2,1}_p \to T^{2,1}_q$ is given by

$$\varepsilon^j_c(p, q)\varepsilon^i_b(p, q)\varepsilon^a_k(q, p)t^{bc}_a(p) = t^{ij}_k(q) \tag{5.7}$$

where we assume that we have already chosen some coordinates around p and q. As always, (5.7) is the coordinate formulation of a *global* formula on M.

Definition 5.2 *A tensor field t on M is ε-invariant if $\varepsilon(p, q)t(p) = t(q)$ for all $p, q \in M$.*

The following trivial lemma will be very useful:

Lemma 5.3 *The following are equivalent:*

(i) t is ε-invariant;

(ii) for some fixed base point e, we have $\varepsilon(e, x)t(e) = t(x)$ for all $x \in M$.

Proof (i) $\to$ (ii) is trivial, and (ii) $\to$ (i) follows from (1.26). ∎

Using Lemma 5.3, we now fix p in (5.7) arbitrarily, regard $q = x$ as variable, differentiate (5.7) with respect to x^r at $x = p$, substitute from (5.1), and use (1.26). Arranging terms and keeping in mind that p is arbitrary, we get

$$\widetilde{\nabla}_r t^i_{jk} \overset{\text{def}}{=} \frac{\partial t^i_{jk}}{\partial x^r} - \Gamma^i_{ra} t^a_{jk} + \Gamma^a_{rj} t^i_{ak} + \Gamma^a_{rk} t^i_{ja} = 0 \tag{5.8}$$

Therefore (5.7) implies (5.8). Using (5.2), it is easy to show that the left-hand side of (5.8) is a tensor.

Two important remarks are in order here:

1. It is crucial that the differentiation index r always be the *first* index in the terms $\Gamma^*_{r\bullet}$ in (5.8). Since Γ^i_{jk} is not necessarily symmetric, this fact is of the utmost importance.
2. If we negate the last three terms in (5.8), it becomes the covariant derivative formula well known from tensor calculus (see, e.g., [L]). This sign difference arises from the choice we made in (5.1) and is not important.

The formula (5.8) generalizes to all tensor fields in the obvious way. We will write $\widetilde{\nabla} t$ as shorthand for the value of the operator $\widetilde{\nabla}$ on the section t.

Definition 5.4 *Let t be a tensor field on (M, ε). Then t is $\widetilde{\nabla}$-parallel if $\widetilde{\nabla} t = 0$.*

The derivation of (5.8) from (5.7) proves the (i) $\to$ (ii) part of the following important proposition:

Proposition 5.5 *The following are equivalent for a tensor field t:*

(i) t is ε-invariant;

(ii) t is $\widetilde{\nabla}$-parallel.

To prepare for the proof of (ii) $\to$ (i), we first assume that $t = \xi = (\xi^i)$ is a vector field. We rewrite (5.8) in this case as

$$\frac{\partial \xi^i}{\partial x^j} = \Gamma^i_{ja} \xi^a \tag{5.9}$$

We now fix some $p \in M$, an initial condition $\xi(p)$ arbitrarily, and define the global vector field ξ by $\xi(x) \stackrel{\text{def}}{=} \varepsilon(p, x)\xi(p)$ as we did in (5.7). The derivation of (5.8) from (5.7) shows that ξ solves (5.9) on M. Now we differentiate (5.9) with respect to x^k, substitute back from (5.9), and alternate k, j. This gives

$$\left[\frac{\partial \Gamma^i_{ja}}{\partial x^k} + \Gamma^i_{jb} \Gamma^b_{ka} \right]_{[kj]} \xi^a = 0 \tag{5.10}$$

We define

$$\widetilde{\mathfrak{R}}^i_{kj,r} \stackrel{\text{def}}{=} \left[\frac{\partial \Gamma^i_{jr}}{\partial x^k} + \Gamma^i_{ja} \Gamma^a_{kr} \right]_{[kj]} \tag{5.11}$$

so that (5.10) becomes

$$\widetilde{\mathfrak{R}}^i_{kj\,a} \xi^a = 0 \tag{5.12}$$

Since we can choose the initial condition $\xi(p)$ arbitrarily, (5.12) is equivalent to

$$\widetilde{\mathfrak{R}}^i_{kj.a}(x) = 0 \tag{5.13}$$

We have proved the following proposition:

Proposition 5.6 *$\widetilde{\mathfrak{R}} = 0$ on (M, ε), where the tensor $\widetilde{\mathfrak{R}}$ is defined by (5.11).*

We again warn the reader about the order of the differentiation index r in (5.11). Now, we assume $\widetilde{\nabla} t = 0$ and rewrite (5.8) as

$$\frac{\partial t^i_{jk}}{\partial x^r} = \Gamma^i_{ra} t^a_{jk} - \Gamma^a_{rj} t^i_{ak} - \Gamma^a_{rk} t^i_{ja} \tag{5.14}$$

We differentiate (5.14) with respect to x^s, substitute back from (5.14), and alternate s, r. This is a somewhat tedious but straightforward computation. However, the result is well known from tensor calculus: What we get is

$$\widetilde{\mathfrak{R}}^i_{sr,a} t^a_{jk} - \widetilde{\mathfrak{R}}^a_{sr,j} t^i_{ak} - \widetilde{\mathfrak{R}}^a_{sr,k} t^i_{ja} = 0 \tag{5.15}$$

The case of the general tensor field is now clear. Proposition 5.6 now shows that (5.15) is identically satisfied. The local existence and uniqueness theorem for first-order systems of PDEs with initial conditions applied to (5.14) states now that for any given initial condition $(t^i_{jk}(\overline{x}))$ at $\overline{x} \in U$, there exists a local solution $(t^i_{jk}(x))$ of (5.14) defined around $\overline{x}$ satisfying this initial condition at $x = \overline{x}$. Further, the local solution $(t^i_{jk}(x))$ is unique: Two local solutions satisfying the same initial condition at $\overline{x}$ coincide in their common connected domain of definition. However, we know by the (i) $\Rightarrow$ (ii) part of Proposition 5.5 that (5.14) in fact admits a global solution, which is obtained by translating this initial condition to all points of M by ε. By uniqueness, our local solutions must be the restrictions of this ε-invariant solution, and therefore t is ε-invariant. This finishes the proof of (ii) $\rightarrow$ (i) of Proposition 5.5.

Now let $\mathfrak{X}(M)$ denote the Lie algebra of all smooth vector fields on M and let $\widetilde{\mathfrak{X}}_\varepsilon(M) \subset \mathfrak{X}(M)$ denote the subspace of ε-invariant vector fields on M. We will write these in brief as $\mathfrak{X}$ and $\widetilde{\mathfrak{X}}_\varepsilon$. Clearly, $\widetilde{\mathfrak{X}}_\varepsilon$ is a vector space of dimension $\dim M$. The natural question arises whether $\widetilde{\mathfrak{X}}_\varepsilon$ is a Lie algebra. So, let $\xi = (\xi^i)$ and $\eta = (\eta^i)$ be two ε-invariant vector fields. We recall the local formula for the bracket of two vector fields:

$$[\xi, \eta]^i = \xi^a \frac{\partial \eta^i}{\partial x^a} - \eta^a \frac{\partial \xi^i}{\partial x^a} \tag{5.16}$$

By Proposition 5.5, we have $\widetilde{\nabla}\xi = \widetilde{\nabla}\eta = 0$; that is, ξ, η solve (5.9). Thus, (5.16) becomes

$$\begin{aligned}[\xi, \eta]^i &= \xi^a \Gamma^i_{ab} \eta^b - \eta^a \Gamma^i_{ab} \xi^b \\ &= \left(\Gamma^i_{ab} - \Gamma^i_{ba}\right) \eta^b \xi^a \end{aligned} \tag{5.17}$$

We make the following important definition:

Definition 5.7 *The tensor*

$$T^i_{jk} \stackrel{\text{def}}{=} \Gamma^i_{jk} - \Gamma^i_{kj} \tag{5.18}$$

is the torsion of (M, ε).

Therefore, if $\xi, \eta \in \widetilde{\mathfrak{X}}_\varepsilon$, then

$$[\xi, \eta]^i = T^i_{ab}\xi^a\eta^b \tag{5.19}$$

We recall (or check directly using the definitions) that $\widetilde{\nabla}$ is a derivation of the tensor algebra and commutes with contractions. For instance, we have

$$\widetilde{\nabla}_r(t^i_{jk}s^l_m) = (\widetilde{\nabla}_r t^i_{jk})s^l_m + t^i_{jk}(\widetilde{\nabla}_r s^l_m) \tag{5.20}$$

and

$$\widetilde{\nabla}_r(t^i_{ja}s^a_m) = (\widetilde{\nabla}_r t^i_{ja})s^a_m + t^i_{ja}(\widetilde{\nabla}_r s^a_m) \tag{5.21}$$

Now we apply $\widetilde{\nabla}_r$ to (5.19). Since $\widetilde{\nabla}\xi = \widetilde{\nabla}\eta = 0$, (5.20) and (5.21) imply

$$\widetilde{\nabla}_r[\xi, \eta]^i = (\widetilde{\nabla}_r T^i_{ab})\xi^a\eta^b \tag{5.22}$$

Since ξ and η are arbitrary, we obtain the following proposition:

Proposition 5.8 *The following are equivalent:*

(i) $\widetilde{\mathfrak{X}}_\varepsilon$ *is a Lie algebra;*

(ii) $\widetilde{\nabla}T = 0$, *or, equivalently, T is ε-invariant.*

Proposition 5.8 shows that T is of fundamental importance. In fact, it will turn out that T dominates the whole theory!

For $\xi, \eta \in \mathfrak{X}$, we define $T(\xi, \eta) \in \mathfrak{X}$ by (5.19); that is,

$$T(\xi, \eta)^i \stackrel{\text{def}}{=} T^i_{ab}\xi^a\eta^b \tag{5.23}$$

We call $T(\xi, \eta)$ the torsion bracket of $\xi, \eta \in \mathfrak{X}$. We observe that this is an algebraic operation on $\mathfrak{X}$, since it does not involve any derivatives. However, if $\xi, \eta \in \widetilde{\mathfrak{X}}_\varepsilon$, their torsion bracket becomes their true bracket according to (5.19). Clearly, $T(\xi, \eta) = -T(\eta, \xi)$. We define the Jacobi 3-form $\mathfrak{J} = (\mathfrak{J}^i_{mjk})$ by the formula

$$\mathfrak{J}(\xi, \eta, \sigma) \stackrel{\text{def}}{=} T(T(\xi, \eta), \sigma)) + T(T(\eta, \sigma), \xi) + T(T(\sigma, \xi), \eta) \tag{5.24}$$

Note that $J(\xi, \eta, \sigma)$ is alternating in its arguments. Also note that $J(\xi, \eta, \sigma)$ need not vanish even if ξ, η, σ are all ε-invariant. In coordinates, (5.23) and (5.24) give

$$\mathfrak{J}^i_{mjk} = \mathfrak{J}\left(\frac{\partial}{\partial x^m}, \frac{\partial}{\partial x^j}, \frac{\partial}{\partial x^k}\right)^i = T^i_{am}T^a_{jk} + T^i_{aj}T^a_{km} + T^i_{ak}T^a_{mj} \tag{5.25}$$

We end this chapter with two questions and a remark.

Suppose we define the operator ∇ on tensor fields by writing the differentiation index r as the *second* index as $\Gamma^i_{\bullet r}$. For instance,

$$\nabla_r t^i_{jk} \stackrel{\text{def}}{=} \frac{\partial t^i_{jk}}{\partial x^r} - \Gamma^i_{ar} t^a_{jk} + \Gamma^a_{jr} t^i_{ak} + \Gamma^a_{kr} t^i_{ja} \tag{5.26}$$

Using (5.1), we can show that (5.26) is again a tensor. In view of Proposition 5.5, it is natural to expect that t is ∇-parallel if and only if it is $\widetilde{\varepsilon}$-invariant. However, note that $\widetilde{\varepsilon}$ is *not defined* without the assumption $\mathcal{R} = 0$, whereas (5.26) is defined without this assumption! Next, suppose we have $\mathcal{R} = 0$ and define $\widetilde{\Gamma} = (\widetilde{\Gamma}^i_{jk})$ by replacing ε and Γ by $\widetilde{\varepsilon}$ and $\widetilde{\Gamma}$ in (5.1). What is the relation between Γ and $\widetilde{\Gamma}$? Both questions will be clarified in the next chapter.

We recall that $\widetilde{\nabla}$ and ∇ can be defined in a coordinate-free way, and it is standard to interpret them as connections on the vector bundle $TM \to M$. Indeed, let $\xi = (\xi^i), \eta = (\eta^i) \in \mathfrak{X}$. We define

$$(\widetilde{\nabla}_\eta \xi)^i \stackrel{\text{def}}{=} (\widetilde{\nabla}_a \xi^i)\eta^a = \left(\frac{\partial \xi^i}{\partial x^a} - \Gamma^i_{ra}\right)\eta^a \tag{5.27}$$

Now, $\widetilde{\nabla}_\eta$ is a first-order linear differential operator acting on the sections of $TM \to M$. The computation in (5.10) shows that the curvature of $\widetilde{\nabla}$ is $\widetilde{\mathfrak{R}} = 0$. In fact, $\widetilde{\nabla}$ is the "flat linear connection" on the associated vector bundle $TM \to M$ induced by the "flat connection" on the trivial principal bundle $\mathcal{U}^{e,\bullet} \to M$ trivialized by ε as mentioned in Chapter 2. In the same way, we can define the linear connection ∇, whose curvature will be the subject of the next chapter.

6

The Linear Curvature

Definition 6.1 *The tensor*

$$\mathfrak{R}^i_{jk,r} \stackrel{\text{def}}{=} \widetilde{\nabla}_r T^i_{jk} \tag{6.1}$$

is the linear curvature of (M, ε).

We will justify this name below by showing that $\mathfrak{R}$ is the "linearization" of $\mathcal{R}$. As a remarkable fact, we will discover that $\mathfrak{R}$ contains the same information as $\mathcal{R}$.

Let $p \in M$ and $\xi_p = \xi(p) = (\xi^i(p))$, $\eta_p = \eta(p) = (\eta^i(p)) \in T_pM$. We define the point bracket $\{\, , \,\}_p : T_pM \times T_pM \to T_pM$ at p by

$$\{\xi(p), \eta(p)\}_p \stackrel{\text{def}}{=} T^i_{ab}(p)\xi^a(p)\eta^b(p) \tag{6.2}$$

Comparing (6.2) and (5.23), we have

$$T(\xi, \eta)(p) = \{\xi(p), \eta(p)\}_p \qquad \xi, \eta \in \mathfrak{X} \tag{6.3}$$

that is, the torsion bracket is defined using the point brackets on M. Clearly, $\{\, , \,\}_p$ is bilinear and alternating, but does not necessarily satisfy the Jacobi identity. Therefore, $(T_pM, \{\, , \,\}_p)$ need not be a Lie algebra.

Now suppose that the isomorphism $\varepsilon(p, q) : T_pM \to T_qM$ satisfies the condition

$$\varepsilon(p, q)\{\xi_p, \eta_p\}_p = \left\{\varepsilon(p, q)\xi_p, \varepsilon(p, q)\xi_p\right\}_q \qquad \xi_p, \eta_p \in T_pM \tag{6.4}$$

If (6.4) is satisfied for all p, q, we say that ε preserves the point brackets. Using the definition of $\varepsilon(p, q)$ and (6.2), we easily check that ε preserves the point brackets if and only if ε preserves the torsion T, that is, $\varepsilon(p, q)T(p) = T(q)$ or, equivalently, T is ε-invariant. Now, the next proposition is a restatement of Proposition 5.8:

Proposition 6.2 *The following are equivalent:*

(i) *ε preserves the point brackets;*

(ii) *T is ε-invariant;*

(iii) *T is $\widetilde{\nabla}$-parallel;*

An Alternative Approach to Lie Groups and Geometric Structures, Ercüment H. Ortaçgil (2018).

(iv) $\mathfrak{R} = 0$;

(v) $\widetilde{\mathfrak{X}}_\varepsilon$ *is a Lie algebra.*

Suppose $\mathfrak{R} = 0$, so that $\widetilde{\mathfrak{X}}_\varepsilon$ is a Lie algebra. Now $(T_pM, \{\ ,\ \}_p)$ becomes a Lie algebra that is isomorphic to $\widetilde{\mathfrak{X}}_\varepsilon$ for all $p \in M$. Indeed, we define the evaluation map

$$\begin{aligned} e_p : \widetilde{\mathfrak{X}}_\varepsilon &\longrightarrow T_pM \\ : \xi &\longrightarrow \xi(p) \end{aligned} \tag{6.5}$$

We know that $\xi, \eta \in \widetilde{\mathfrak{X}}_\varepsilon$ solve (5.9). Now (5.19) shows that $e_p : (\widetilde{\mathfrak{X}}_\varepsilon, [\ ,\ ]) \to (T_pM, \{\ ,\ \}_p)$ is a homomorphism of Lie algebras. It is surjective, since any initial condition extends to an ε-invariant solution, and injective, since this solution is unique. Now Proposition 6.2 gives the commutative diagram

$$\begin{array}{ccc} \widetilde{\mathfrak{X}}_\varepsilon & = & \widetilde{\mathfrak{X}} \\ {\scriptstyle e_p}\downarrow & & {\scriptstyle e_q}\downarrow \\ (T_pM, \{\ ,\ \}_p) & \xrightarrow{\varepsilon(p,q)} & (T_qM, \{\ ,\ \}_q) \end{array} \tag{6.6}$$

where all maps are isomorphisms of Lie algebras. Let $\xi_{(k)}(p)$, $1 \le k \le n$, be a basis of T_pM. We translate them over M using ε and get the independent vector fields $\xi_{(k)}$ on M. Therefore, $[\xi_{(i)}, \xi_{(j)}](q)$ is a linear combination of $\xi_{(k)}(q)$; that is,

$$[\xi_{(i)}, \xi_{(j)}](q) = \sum_{1 \le a \le n} C^a_{ij}(q)\xi_{(a)}(q), \qquad 1 \le i, j \le n \tag{6.7}$$

The structure *functions* $C^a_{ij}(x)$ are defined on M and depend on the initial basis $\xi_{(k)}(p)$, $1 \le k \le n$. To show this dependence, we write $C^a_{ij}(x; \xi_{(1)}(p), \ldots, \xi_{(n)}(p))$. Now suppose $\mathfrak{R} = 0$. As we did in (5.22), we apply $\widetilde{\nabla}_\eta$ to (6.7), which gives $\sum_{1 \le a \le n} \eta(C^a_{ij}(x))\xi_{(a)}(x) = 0$, where $\eta(f)$ denotes the Lie derivative $\mathcal{L}_\eta f$. Since η is arbitrary, we conclude that the functions $C^m_{ij}(x)$ are constant on M, and we easily deduce

$$C^m_{ij}\left(p; \frac{\partial}{\partial x^1}\bigg|_p, \ldots, \frac{\partial}{\partial x^n}\bigg|_p\right) = T^m_{ij}(p) \tag{6.8}$$

which clarifies the meaning of the tensor T. In particular, ε-invariance of T is equivalent to the structure functions being constant, adding one more equivalent condition to Proposition 6.2.

We recall that the assumption $\mathcal{R} = 0$ gives rise to the LLGs $(M, \varepsilon, \mathcal{G})$ and $(U, \widetilde{\varepsilon}, C(\mathcal{G}, U))$. If the elements of $\mathcal{G}$ globalize, then so do the elements of $C(\mathcal{G})$ and we obtain the PHSs $(M, \mathcal{G})$ and $(M, C(\mathcal{G}))$. We also know that the assumption $\mathfrak{R} = 0$ turns $\widetilde{\mathfrak{X}}_\varepsilon$ into a Lie algebra. However, there is another Lie algebra! To construct it, we need to understand $\mathfrak{R}$ better.

Proposition 6.3

$$\mathfrak{R}^i_{kj,r} = \left[\frac{\partial \Gamma^i_{rj}}{\partial x^k} + \Gamma^i_{aj}\Gamma^a_{rk}\right]_{[kj]} \tag{6.9}$$

We should carefully compare the order of the indices in (6.9) with those in $\widetilde{\mathfrak{R}}^i_{kj,r}$ defined by (5.11). This distinction is extremely important (and very confusing!).

Proof

$$\begin{aligned}
\mathfrak{R}^i_{kj,r} &= \widetilde{\nabla}_r T^i_{kj} \\
&= \frac{\partial}{\partial x^r}\left(\Gamma^i_{kj} - \Gamma^i_{jk}\right) - \Gamma^i_{ra}\left(\Gamma^a_{kj} - \Gamma^a_{jk}\right) + \Gamma^a_{rk}\left(\Gamma^i_{aj} - \Gamma^i_{ja}\right) + \Gamma^a_{rj}\left(\Gamma^i_{ka} - \Gamma^i_{ak}\right) \\
&= \frac{\partial \Gamma^i_{kj}}{\partial x^r} - \frac{\partial \Gamma^i_{jk}}{\partial x^r} + \left[\Gamma^i_{ra}\Gamma^a_{jk}\right]_{[rj]} + \left[\Gamma^i_{ka}\Gamma^a_{rj}\right]_{[kr]} + \left[\Gamma^i_{aj}\Gamma^a_{rk}\right]_{[jk]}
\end{aligned} \tag{6.10}$$

Since $\widetilde{\mathfrak{R}}^i_{rk,j} = 0$ by Proposition 5.6, (5.11) gives

$$\left[\frac{\partial \Gamma^i_{kj}}{\partial x^r}\right]_{[rk]} = \left[\Gamma^i_{ra}\Gamma^a_{kj}\right]_{[rk]} \tag{6.11}$$

Substituting the third and fourth terms on the right-hand side of (6.10) from (6.11), we get

$$\begin{aligned}
\mathfrak{R}^i_{kj,r} &= \frac{\partial \Gamma^i_{kj}}{\partial x^r} - \frac{\partial \Gamma^i_{jk}}{\partial x^r} + \left[\frac{\partial \Gamma^i_{jk}}{\partial x^r}\right]_{[rj]} + \left[\frac{\partial \Gamma^i_{rj}}{\partial x^k}\right]_{[rk]} + \left[\Gamma^i_{aj}\Gamma^a_{rk}\right]_{[jk]} \\
&= \left[\frac{\partial \Gamma^i_{rj}}{\partial x^k}\right]_{[kj]} + \left[\Gamma^i_{aj}\Gamma^a_{rk}\right]_{[jk]}
\end{aligned} \tag{6.12}$$

which is equal to the right-hand side of (6.9). ∎

We recall the definition of ∇ given in Chapter 5. We now rewrite $\nabla_j \xi^i = 0$ in the form

$$\frac{\partial \xi^i}{\partial x^j} = \Gamma^i_{aj}\xi^a \tag{6.13}$$

differentiate this with respect to x^k, substitute back from (6.13), and alternate j, k. The result is $\mathfrak{R}^i_{kj,r} = 0$, and therefore $\mathfrak{R}$ is the curvature of the linear connection ∇ on the associated vector bundle $TM \to M$. In particular, $\mathfrak{R} = 0$ gives the integrability conditions for $\nabla\xi = 0$: If any initial condition $\xi(p)$ at $p \in M$ extends to a local solution of $\nabla\xi = 0$ around p, then $\mathfrak{R}(p) = 0$. Conversely, if $\mathfrak{R} = 0$ on M, then, for any $p \in M$ and initial condition $\xi(p)$, there exists a unique local solution of (6.13) around p satisfying this initial condition.

We now assume $\mathfrak{R} = 0$ on M and want to show that the local solutions of $\nabla\xi = 0$ form a Lie algebra like the Lie algebra $\widetilde{\mathfrak{X}}_\varepsilon$. However, note that the elements of $\widetilde{\mathfrak{X}}_\varepsilon$ are global vector fields on M, whereas the solutions of $\nabla\xi = 0$ will be defined only locally. We denote the set of local solutions of $\nabla\xi = 0$ on U by $\mathfrak{U}(U)$, recalling that for any $p \in U$ all local solutions around p are defined on U if U is sufficiently small.

We proceed as before: Let $\xi = (\xi^i)$, $\eta = (\eta^i) \in \mathfrak{U}(U)$. Using (6.13), we compute

$$\begin{aligned}
[\xi,\eta]^i &= \xi^a \frac{\partial \eta^i}{\partial x^a} - \eta^a \frac{\partial \xi^i}{\partial x^a}'' \\
&= \xi^a \Gamma^i_{ba}\eta^b - \eta^a \Gamma^i_{ba}\xi^b \\
&= \left(\Gamma^i_{ba} - \Gamma^i_{ab}\right)\eta^b\xi^a \\
&= T^i_{ba}\eta^b\xi^a \\
&= -T^i_{ab}\xi^a\eta^b
\end{aligned} \tag{6.14}$$

Applying ∇ to (6.14) shows that $\mathfrak{U}(U)$ is a Lie algebra if and only if $\nabla T = 0$, and we now claim that $\mathfrak{R} = 0$ implies $\nabla T = 0$.

There is an obvious relation between $\widetilde{\nabla}$ and ∇:

$$\begin{aligned}
\nabla_r \xi^i &= \frac{\partial \xi^i}{\partial x^r} - (\Gamma^i_{ar} + \Gamma^i_{ra} - \Gamma^i_{ra})\xi^a \\
&= \widetilde{\nabla}_r \xi^i - T^i_{ar}\xi^a
\end{aligned} \tag{6.15}$$

Similarly, we deduce

$$\begin{aligned}
\nabla_r T^i_{jk} &= \widetilde{\nabla}_r T^i_{jk} - T^i_{ar}T^a_{jk} + T^a_{jr}T^i_{ak} + T^a_{kr}T^i_{ja} \\
&= \widetilde{\nabla}_r T^i_{jk} - T^i_{ar}T^a_{jk} - T^i_{ak}T^a_{rj} - T^i_{aj}T^a_{kr} \\
&= \widetilde{\nabla}_r T^i_{jk} - \mathfrak{J}^i_{rjk}
\end{aligned} \tag{6.16}$$

by (5.25)

Proposition 6.4 (The First Bianchi Identity)

$$\mathfrak{R}(\xi,\eta)(\sigma) + \mathfrak{R}(\eta,\sigma)(\xi) + \mathfrak{R}(\sigma,\xi)(\eta) = \mathfrak{J}(\xi,\eta,\sigma) \tag{6.17}$$

Proof We have

$$\begin{aligned}
\mathfrak{R}^i_{kj,r} &= \widetilde{\nabla}_r T^i_{kj} \\
&= \frac{\partial \Gamma^i_{kj}}{\partial x^r} - \frac{\partial \Gamma^i_{jk}}{\partial x^r} - \Gamma^i_{ra}T^a_{kj} + \Gamma^a_{rk}T^i_{aj} + \Gamma^a_{rj}T^i_{ka}
\end{aligned} \tag{6.18}$$

We substitute (6.18) into

$$\widetilde{\nabla}_r T^i_{kj} + \widetilde{\nabla}_k T^i_{jr} + \widetilde{\nabla}_j T^i_{rk} \tag{6.19}$$

and simplify (6.19) using $\widetilde{\mathfrak{R}} = 0$. A straightforward computation now shows that (6.19) is equal to

$$T^i_{ar}T^a_{kj} + T^i_{ak}T^a_{jr} + T^i_{aj}T^a_{rk} \tag{6.20}$$

which is the desired result. ■

The reader may find it interesting to compare (6.17) with the classical First Bianchi Identity (Theorem 2.5 in [KN1]).

Corollary 6.5 *If $\mathfrak{R} = 0$, then $\nabla T = 0$.*

Indeed, if $\mathfrak{R} = 0$, then the left-hand side of (6.17) vanishes by (6.1) and (6.19). Therefore, $\mathfrak{J} = 0$ by (6.17), and the conclusion follows from (6.16).

Corollary 6.6 *If $\mathfrak{R} = 0$, then the vector space $\mathfrak{U}(U)$ of local solutions of $\nabla\xi = 0$ on U is a Lie algebra.*

Clearly, two local solutions of (6.13) on U and V give a local solution on $U \cup V$ if they agree on $U \cap V$. Therefore, continuations of local solutions along paths can be defined in the obvious way using uniqueness of solutions. We will not do this, because it will eventually turn out that if $(M, \varepsilon, \mathcal{G})$ is globalizable, then all the elements of $\mathfrak{U}(U)$ globalize too.

It is worthwhile looking briefly at the relation between the Lie algebras $\widetilde{\mathfrak{X}}_\varepsilon$ and $\mathfrak{U}(U)$. Let $\xi \in \widetilde{\mathfrak{X}}_\varepsilon$ and $\eta \in \mathfrak{U}(U)$. Recalling that ξ solves (5.9) and η solves (6.13), we easily check that $[\xi_{|U}, \eta] = 0$. In fact, let $C_{\mathfrak{X}(U)}(\widetilde{\mathfrak{X}}_\varepsilon(U)) \stackrel{\text{def}}{=} \{\eta \in \mathfrak{X}(U) \mid [\eta, \xi_{|U}] = 0 \text{ for all } \xi \in \widetilde{\mathfrak{X}}_\varepsilon(U)\}$ be the centralizer of $\widetilde{\mathfrak{X}}_\varepsilon(U)$ inside $\mathfrak{X}(U)$. We can now easily show that the following proposition holds:

Proposition 6.7 $C_{\mathfrak{X}(U)}(\widetilde{\mathfrak{X}}_\varepsilon(U)) = \mathfrak{U}(U)$.

Now we have the infinitesimal analog of (4.13): We fix some base point $e \in U$ and define

$$\Xi : \widetilde{\mathfrak{X}}_\varepsilon \longrightarrow \mathfrak{U}(U) \tag{6.21}$$

by mapping some ξ to the unique local solution of $\nabla\xi = 0$ satisfying the initial condition $\xi(p)$. Clearly, Ξ is a bijection and

$$\Xi[\xi, \sigma] = [\Xi(\xi), \Xi(\sigma)] \tag{6.22}$$

The following is an interesting point: Since $\widetilde{\mathfrak{X}}_\varepsilon \subset \mathfrak{X}$ are Lie algebras with the assumption $\mathfrak{R} = 0$, the centralizer $C_{\mathfrak{X}}(\widetilde{\mathfrak{X}}_\varepsilon)$ is clearly a Lie algebra. However, it is not obvious beforehand that $C_{\mathfrak{X}}(\widetilde{\mathfrak{X}}_\varepsilon)$ is finite-dimensional with $\dim C_{\mathfrak{X}}(\widetilde{\mathfrak{X}}_\varepsilon) \leq \dim \widetilde{\mathfrak{X}}_\varepsilon$, as follows from Proposition 6.7. Suppose we define $N(\widetilde{\mathfrak{X}}_\varepsilon) \stackrel{\text{def}}{=} \{\eta \in \mathfrak{X} \mid [\eta, \xi] \in \widetilde{\mathfrak{X}}_\varepsilon \text{ for all } \xi \in \widetilde{\mathfrak{X}}_\varepsilon\}$ = the normalizer of $\widetilde{\mathfrak{X}}_\varepsilon$ inside $\mathfrak{X}$. Is $N(\widetilde{\mathfrak{X}}_\varepsilon)$ finite-dimensional? Note that we can ask this question also for $(M, \mathcal{G})$, assuming $\mathcal{R} = 0$. We will make some progress with these questions in Chapter 15.

Now we come to a fundamental point: What is the justification for calling $\mathfrak{R}$ the "linear curvature"? That is, what is the relation between the nonlinear curvature $\mathcal{R}$ and $\mathfrak{R}$? The above arguments show that the "duality" between $(M, \varepsilon, \mathcal{G})$ and $(U, \widetilde{\varepsilon}, C(\mathcal{G}, U))$ exists infinitesimally also between $\widetilde{\mathfrak{X}}_\varepsilon$ and $\mathfrak{U}(U)$, with the assumptions $\mathcal{R} = 0$ and $\mathfrak{R} = 0$. We have not defined the Lie algebras of the LLGs $(M, \varepsilon, \mathcal{G})$ and $(U, \widetilde{\varepsilon}, C(\mathcal{G}, U))$ yet, but however we define them, it is natural to expect that these two Lie algebras will be in some sense "isomorphic" to $\widetilde{\mathfrak{X}}_\varepsilon(U) \simeq \mathfrak{U}(U)$.

The picture so far gives enough hint for suspecting that

$$\mathcal{R} = 0 \Longleftrightarrow \mathfrak{R} = 0 \tag{6.23}$$

for (M, ε). Note that $\Longrightarrow$ is the analog of the piece of modern folklore: "A Lie group has a Lie algebra." However, $\Longleftarrow$ is more subtle and is the analog of another piece of folklore, namely, the modern Cartan formulation of Lie's Third Theorem, which states that "a Lie algebra determines the Lie group locally." However, note that (6.23) is a global assertion.

Before we proceed to the proof of (6.23), we first recall $\overleftarrow{\mathcal{R}}(p, q)$ defined by (2.12). As shown in Chapter 2, if we fix $p \in M$ and a coordinate system around p once and for all, then $\overleftarrow{\mathcal{R}}^i_{jk}(p, x)$ transforms as a tensor in all indices under a coordinate change $(x) \to (y)$. Therefore, $\widetilde{\nabla}_r \overleftarrow{\mathcal{R}}^i_{jk}(p, x)$ is defined as a tensor for fixed p.

Proposition 6.8

$$\widetilde{\nabla}_r \overleftarrow{\mathcal{R}}^i_{jk}(p, x) = \mathfrak{R}^i_{jk,r}(x) \tag{6.24}$$

In particular, Proposition 6.8 asserts that the left-hand side of (6.24) is independent of p.

The power of (6.24) is evident from the following corollary:

Corollary 6.9 $\mathcal{R} = 0$ *on* $M \times M \Longleftrightarrow \mathfrak{R} = 0$ *on* M.

The $\Longrightarrow$ part of this corollary is trivial in view of (6.24). For $\Longleftarrow$, suppose $\mathfrak{R} = 0$, so $\overleftarrow{\mathcal{R}}^i_{jk}(p, x)$ is $\widetilde{\nabla}$-parallel by (6.24). By Proposition 5.5, $\overleftarrow{\mathcal{R}}^i_{jk}(p, x)$ is ε-invariant. Since $\overleftarrow{\mathcal{R}}^i_{jk}(p, p) = 0$ by Proposition 2.8, we conclude that $\overleftarrow{\mathcal{R}}^i_{jk}(p, x) = 0$ for all x. Since p is arbitrary, $\mathcal{R} = 0$ on $M \times M$ by (2.14).

We now turn to the proof of Proposition 6.8 itself:

Proof We will prove (6.24) in two steps:

Step 1: $\widetilde{\nabla}_r \overleftarrow{\mathcal{R}}^i_{jk}(p, x)$ is independent of p.

Step 2: $\left[\widetilde{\nabla}_r \overleftarrow{\mathcal{R}}^i_{jk}(p, x)\right]_{x=p} = \mathfrak{R}(p)$.

Clearly, Steps 1 and 2 imply (6.24).

Step 1: A straightforward computation using (1.26), (2.7), and (2.12) gives the important formula

$$\overleftarrow{\mathcal{R}}^i_{jk}(x, y) = \overleftarrow{\mathcal{R}}^i_{jk}(z, y) + \varepsilon^i_a(z, y)\, \overleftarrow{\mathcal{R}}^a_{bc}(x, z)\varepsilon^b_j(y, z)\varepsilon^c_k(y, z) \tag{6.25}$$

which clarifies the subtle dependence mentioned at the end of Chapter 2. In (6.25), we fix $x = x_0, z = z_0$ and define the tensor

$$A^i_{jk}(y) \overset{\text{def}}{=} \varepsilon^i_a(z_0, y)\overleftarrow{\mathcal{R}}^a_{bc}(x_0, z_0)\varepsilon^b_j(y, z_0)\varepsilon^c_k(y, z_0) \tag{6.26}$$

on (U, y). Using (1.26) we easily check that

$$A^i_{jk}(y) = \varepsilon^i_a(w, y)A^a_{bc}(w)\varepsilon^b_j(y, w)\varepsilon^c_k(y, w) \tag{6.27}$$

Therefore A is ε-invariant by (6.27) and so $\widetilde{\nabla} A = 0$. Now applying $\widetilde{\nabla}$ to (6.25), we deduce

$$\widetilde{\nabla}_r \overleftarrow{\mathcal{R}}^i_{jk}(x_0, y) = \widetilde{\nabla}_r \overleftarrow{\mathcal{R}}^i_{jk}(z_0, y) \tag{6.28}$$

which proves Step 1.

Step 2: Our proof will be a direct computation using the definitions. In Chapter 9, we will outline an elegant conceptual proof using the equivalence of the local exactness of the nonlinear and linear Spencer sequences. First, we derive some formulas that will be useful later. We differentiate (1.26) with respect to z^k and get

$$\frac{\partial \varepsilon^i_j(x, z)}{\partial z^k} = \frac{\partial \varepsilon^i_a(y, z)}{\partial z^k}\varepsilon^a_j(x, y) \tag{6.29}$$

We rewrite (1.26) as

$$\varepsilon^j_r(z, x) = \varepsilon^j_b(y, x)\varepsilon^b_r(z, y) \tag{6.30}$$

Multiplying (6.29) and (6.30), summing over j, and using (1.28), we obtain

$$\frac{\partial \varepsilon^i_a(x, z)}{\partial z^k}\varepsilon^a_r(z, x) = \frac{\partial \varepsilon^i_a(y, z)}{\partial z^k}\varepsilon^a_r(z, y) \tag{6.31}$$

It follows that the left-hand side of (6.31) does not depend on x. Setting $x = z$ and using (5.1), we deduce

$$\frac{\partial \varepsilon^i_a(x, z)}{\partial z^k}\varepsilon^a_r(z, x) = \Gamma^i_{kr}(z) \tag{6.32}$$

Note that we could also define Γ by (6.32). Clearly, we also have

$$\varepsilon^i_a(x, z)\frac{\partial \varepsilon^a_j(z, x)}{\partial z^k} = -\Gamma^i_{kj}(z) \tag{6.33}$$

We now differentiate (1.26) with respect to z^k and x^r. This gives

$$\frac{\partial^2 \varepsilon^i_j(x,z)}{\partial x^r \partial z^k} = \frac{\partial \varepsilon^i_a(y,z)}{\partial z^k} \frac{\partial \varepsilon^a_j(x,y)}{\partial x^r} \tag{6.34}$$

Evaluating (6.34) at $x = z = p$ and using (5.1), we get

$$\left[\frac{\partial^2 \varepsilon^i_j(x,z)}{\partial x^r \partial z^k} \right]_{x=z=p} = -\Gamma^i_{ka}(p)\Gamma^a_{rj}(p) \tag{6.35}$$

In view of (2.7) and (2.12), $\overleftarrow{\mathcal{R}}^i_{jk}(p,y)$ is given by

$$\overleftarrow{\mathcal{R}}^i_{jk}(p,y) = \left[\frac{\partial \varepsilon^i_b(p,y)}{\partial x^a} \varepsilon^a_k(y,p)\varepsilon^b_j(y,p) + \Gamma^i_{kj}(y) \right]_{[kj]} \tag{6.36}$$

We are now ready for Step 2: Since $\overleftarrow{\mathcal{R}}^i_{jk}(p,p) = 0$, we have

$$\left[\widetilde{\nabla}_r \overleftarrow{\mathcal{R}}^i_{jk}(p,y) \right]_{y=p} = \left[\frac{\partial}{\partial y^r} \overleftarrow{\mathcal{R}}^i_{jk}(p,y) \right]_{y=p} \tag{6.37}$$

In view of (6.36), the right-hand side of (6.37) is equal to

$$\frac{\partial}{\partial y^r} \left[\frac{\partial \varepsilon^i_b(p,y)}{\partial x^a} \varepsilon^a_k(y,p)\varepsilon^b_j(y,p) + \Gamma^i_{kj}(y) \right]_{[kj],y=p} \tag{6.38}$$

We compute the four terms arising from (6.38) separately.

Clearly,

$$\left[\frac{\partial \Gamma^i_{kj}(y)}{\partial y^r} \right]_{[kj],y=p} = \frac{\partial T_{kj}}{\partial y^r}(p) \tag{6.39}$$

Using (6.35), we compute

$$\begin{aligned} \left[\frac{\partial^2 \varepsilon^i_b(p,y)}{\partial y^r \partial x^a} \varepsilon^a_k(y,p)\varepsilon^b_j(y,p) \right]_{[kj],y=p} &= \left[\frac{\partial^2 \varepsilon^i_j(p,y)}{\partial y^r \partial x^k} \right]_{[kj]y=p} \\ &= -\Gamma^i_{ra}(p)T^a_{kj}(p) \end{aligned} \tag{6.40}$$

Now,

$$\left[\frac{\partial \varepsilon^i_b(p,y)}{\partial x^a} \frac{\partial \varepsilon^a_k(y,p)}{\partial y^r} \varepsilon^b_j(y,p) \right]_{[kj],y=p} = \Gamma^i_{aj}(p)\Gamma^a_{rk}(p) - \Gamma^i_{ak}(p)\Gamma^a_{rj}(p) \tag{6.41}$$

Similarly,

$$\left[\frac{\partial \varepsilon^i_b(p,y)}{\partial x^a}\varepsilon^a_k(y,p)\frac{\partial \varepsilon^b_j(y,p)}{\partial y^r}\right]_{[kj],y=p} = \Gamma^i_{kb}(p)\Gamma^b_{rj}(p) - \Gamma^i_{jb}(p)\Gamma^b_{rk}(p) \tag{6.42}$$

Adding the four equations (6.39)–(6.42), we get

$$\begin{aligned}\left[\widetilde{\nabla}_r \overleftarrow{\mathcal{R}}^i_{jk}(p,y)\right]_{y=p} &= \frac{\partial T_{kj}}{\partial y^r}(p) - \Gamma^i_{ra}(p)T^a_{kj}(p) + \Gamma^a_{rk}(p)T^i_{aj}(p) + \Gamma^a_{rj}(p)T^i_{ka}(p)\\ &= \left[\widetilde{\nabla}_r T^i_{jk}(y)\right]_{y=p}\\ &= \mathfrak{R}^i_{jk,r}(p)\end{aligned} \tag{6.43}$$

as required. ■

Having Corollary 6.9 at our disposal, we can now fill in the missing pieces of the puzzle. We start by answering the second question at the end of Chapter 5: Suppose $\mathfrak{R} = 0$ and therefore $\mathcal{R} = 0$. Thus, we have the local splitting $\widetilde{\varepsilon}$ and define the object $\widetilde{\Gamma} = (\widetilde{\Gamma}^i_{jk})$ using $\widetilde{\varepsilon}$ exactly as we defined Γ using ε by (5.1):

$$\widetilde{\Gamma}^i_{jk}(x) \stackrel{\text{def}}{=} \left[\frac{\partial \widetilde{\varepsilon}^i_k(x,y)}{\partial y^k}\right]_{y=x} \tag{6.44}$$

Proposition 6.10

$$\widetilde{\Gamma}^i_{jk}(x) = \Gamma^i_{kj}(x) \tag{6.45}$$

Proof

$$\begin{aligned}\widetilde{\Gamma}^i_{jk}(p) &= \left[\frac{\partial \widetilde{\varepsilon}^i_k(p,y)}{\partial y^j}\right]_{y=p}\\ &= \frac{\partial}{\partial y^j}\left[\left[\frac{\partial f^i(p,x,y)}{\partial x^k}\right]_{x=p}\right]_{y=p}, \qquad f \in \mathcal{G}\\ &= \frac{\partial}{\partial x^k}\left[\left[\frac{\partial f^i(p,x,y)}{\partial y^j}\right]_{y=p}\right]_{x=p}\\ &= \left[\frac{\partial}{\partial x^k}\varepsilon^i_j(p,f(p,x,p))\right]_{x=p}\\ &= \left[\frac{\partial\left(\varepsilon^i_j(p,f(p,x,p)\right)}{\partial y^a}\frac{\partial f^a(p,x,p)}{\partial x^k}\right]_{x=p}\end{aligned}$$

$$= \Gamma^i_{aj}(p) \frac{\partial x^a}{\partial x^k}$$
$$= \Gamma^i_{kj}(p)$$ ■

Now we define ∇ using $\widetilde{\Gamma}$ exactly as we defined $\widetilde{\nabla}$ using Γ:

$$\nabla_r \xi^i \stackrel{\text{def}}{=} \frac{\partial \xi^i}{\partial x^r} - \widetilde{\Gamma}^i_{ra} \xi^a$$
$$= \frac{\partial \xi^i}{\partial x^r} - \Gamma^i_{ar} \xi^a \tag{6.46}$$

and we get the previous definition of ∇, as expected. Consequently, if $\mathfrak{R} = 0$, then a (local) tensor field t is ∇-parallel if and only if it is $\widetilde{\varepsilon}$-invariant. The proof is identical to the proof of Proposition 5.5. We emphasize again the remarkable fact that ∇ is defined without the assumption $\mathfrak{R} = 0$.

To be consistent with our notation $\widetilde{\mathfrak{X}}_\varepsilon$, we now make the definition

$$\mathfrak{U}(U) \stackrel{\text{def}}{=} \mathfrak{X}_{\widetilde{\varepsilon}}(U) \tag{6.47}$$

It is tempting to define $\widetilde{\mathfrak{X}}_\varepsilon$ as the Lie algebra of $(M, \varepsilon, \mathcal{G})$. However, this is not a very good idea: Let ξ be a vector field on M that integrates to the 1-parameter group $\Xi(t, x)$ of local diffeomorphisms, with $x \in U, t \in (-\epsilon, \epsilon)$. Hence, $\Xi(0, x) = x$ and

$$\left[\frac{d\Xi^i(t,x)}{dt}\right]_{t=0} = \xi^i(x) \tag{6.48}$$

Suppose $\Xi(t, x) \in \mathcal{G}$ for all t; that is,

$$\frac{\partial \Xi^i(t,x)}{\partial x^k} = \varepsilon^i_k(x, \Xi(t,x)) \tag{6.49}$$

We differentiate (6.49) at $t = 0$:

$$\frac{d}{dt}\left[\frac{\partial \Xi^i(t,x)}{\partial x^k}\right]_{t=0} = \left[\frac{\partial \varepsilon^i_k(x,y)}{\partial y^a}\right]_{y=x} \left[\frac{d\Xi^i(t,x)}{dt}\right]_{t=0}$$

$$\Longleftrightarrow \qquad \frac{\partial \xi^i(x)}{\partial x^k} = \Gamma^i_{ak}(x)\xi^a(x) \tag{6.50}$$

Comparing (6.50) with (6.13), it follows that $\xi \in \mathfrak{X}_{\widetilde{\varepsilon}}(U)$ and not $\xi \in \widetilde{\mathfrak{X}}_\varepsilon(U)$. In short, the vector fields that integrate to $\mathcal{G}$ belong to $\mathfrak{X}_{\widetilde{\varepsilon}}(U)$. The converse is also true by the uniqueness of solutions: All vector fields in the Lie algebra $\mathfrak{X}_{\widetilde{\varepsilon}}(U)$ integrate to $\mathcal{G}$; that is, solutions of (6.50) integrate to the solutions of (6.49). Similar statements apply to $\widetilde{\mathfrak{X}}_\varepsilon$ and $C(\mathcal{G})$. Therefore, we deduce the following proposition:

Proposition 6.11 *Suppose $\mathcal{R} = 0$ and $(M, \varepsilon, \mathcal{G})$ is globalizable. Then $\mathfrak{X}_{\widetilde{\varepsilon}}$ is the Lie algebra of the infinitesimal generators of the PHS $(M, \mathcal{G})$, and $\widetilde{\mathfrak{X}}_{\varepsilon}$ is the Lie algebra of the infinitesimal generators of the PHS $(M, C(\mathcal{G}))$.*

If $(M, \varepsilon, \mathcal{G})$ is not globalizable, then we define the complete presheaf $U \to \mathfrak{X}_{\widetilde{\varepsilon}}(U)$ as the Lie algebra of $(M, \varepsilon, \mathcal{G})$.

We will end this chapter with two remarks and a question.

(1) Consider the two ODEs

$$\frac{dx^i}{dt} = \varepsilon^i_a(e, x(t))\frac{dx^a}{dt}(0) \tag{6.51}$$

and

$$\frac{dx^i}{dt} = \widetilde{\varepsilon}^i_a(e, x(t))\frac{dx^a}{dt}(0) \tag{6.52}$$

for the curve $x(t)$ starting from e at $t = 0$. Note that (6.51) is defined on (M, ε), whereas (6.52) requires $\mathfrak{R} = 0$. Differentiating (6.51) at $t = 0$ gives

$$\frac{d^2x^i}{dt^2}(0) - \Gamma^i_{ab}(e)\frac{dx^a}{dt}(0)\frac{dx^b}{dt}(0) = 0 \tag{6.53}$$

Differentiating (6.52) at $t = 0$ also gives(6.53), in view of Proposition 6.10. So, the "geodesic equation" (6.53) is the derivative of the "1-parameter subgroup" equations (6.51) and (6.52). We will leave it to the interested reader to harvest some consequences of this fact in relation to the exponential map.

(2) A straightforward computation shows that

$$\begin{aligned}\widetilde{\nabla}_r\widetilde{\nabla}_k\xi^i - \widetilde{\nabla}_k\widetilde{\nabla}_r\xi^i &= -\widetilde{\mathfrak{R}}^i_{rk,a}\xi^a + T^a_{rk}\widetilde{\nabla}_a\xi^i \\ &= T^a_{rk}\widetilde{\nabla}_a\xi^i\end{aligned} \tag{6.54}$$

Thus, the left-hand side of (6.54) depends on the derivatives of ξ. Another straightforward computation gives

$$\nabla_r\nabla_k\xi^i - \nabla_k\nabla_r\xi^i = -\mathfrak{R}^i_{rk,a}\xi^a - T^a_{rk}\nabla_a\xi^i \tag{6.55}$$

Now, the operator $\widetilde{\nabla}$ on vector fields extends in a natural way to an operator on all tensor fields in a coordinate-free way. The modern generalization of (6.54) is

$$\widetilde{\nabla}_\sigma\widetilde{\nabla}_\eta - \widetilde{\nabla}_\eta\widetilde{\nabla}_\sigma - \widetilde{\nabla}_{[\eta,\sigma]} = -\widetilde{\mathfrak{R}}(\sigma, \eta) = 0 \tag{6.56}$$

Similarly, (6.55) is generalized as

$$\nabla_\sigma\nabla_\eta - \nabla_\eta\nabla_\sigma - \nabla_{[\eta,\sigma]} = -\mathfrak{R}(\sigma, \eta) \tag{6.57}$$

However, we observe that substitution of $\sigma = (\partial/\partial x^r)$ and $\eta = (\partial/\partial x^k)$ into (6.56) *does not give* (6.54). A similar statement applies to (6.57) and (6.55).

Some meditation on the above seemingly contradictory fact reveals a subtle point: The classical tensor calculus was introduced in order to formalize Riemann's revolutionary ideas, and the modern theory of connections is generalized from Riemannian geometry. Lie theory is amalgamated with this generalization in the form of abstract Lie groups and their representations using the essentially topological formalism of principal bundles and their associated vector bundles. We hope to have made it clear by now that tensor calculus could have had an alternative origin in Lie theory, and if this had happened, then the concepts of torsion and curvature would have quite different meanings today.

(3) We will end with a question of quite a different nature, but one that is still very much puzzling to us: According to Definition 6.1, the linear curvature is the covariant derivative of the torsion, which we write symbolically as $\widetilde{\nabla}^{(1)} T = \mathfrak{R}$. We have shown that $\widetilde{\nabla}^{(1)} T = \mathfrak{R} = 0$ gives rise to the theory of LLGs, which is a generalization of the theory of Lie groups and Lie algebras. It is therefore natural to ask when does $\widetilde{\nabla}^{(2)} T = 0$ imply $\widetilde{\nabla}^{(1)} T = 0$ and does (M, ε) satisfying $\widetilde{\nabla}^{(2)} T = \widetilde{\nabla}^{(1)} \mathfrak{R} = 0$ have an interesting theory? The reader will notice here the creeping of "symmetric spaces" and their higher-order analogs into the picture through the condition $\widetilde{\nabla}^{(1)} \mathfrak{R} = 0$.

7

The Structure Object

The purpose of this chapter is to clarify the relation between our alternative approach to Lie groups culminating in Proposition 4.4 and the classical approach via the Maurer–Cartan form.

Let (M, ε) be given. We fix some basepoint $e \in M$ and a coordinate system around e once and for all. Let (U, x) be an arbitrary coordinate patch, $U \subset M$. We define the components $(w^i_j(x))$ of a geometric object w on U by

$$w^i_j(x) \stackrel{\text{def}}{=} \varepsilon^i_j(x, e) \tag{7.1}$$

Since, under a coordinate change $(x) \to (y)$ at the source, $\varepsilon^i_j(x, e)$ transforms as $\varepsilon^i_a(x, e)(\partial x^a / \partial y^j) = \varepsilon^i_j(y, e)$, the components $(w^i_j(x))$ transform according to

$$w^i_a(x)\frac{\partial x^a}{\partial y^j} = w^i_j(y) \tag{7.2}$$

Note that $w^i_j(x)$ transforms only in the index j and not in i. Therefore, even though $w^i_j(x)$ looks a like section $\zeta = (\zeta^i_j(x))$ of $T^* \otimes T \to M$ that transforms in all indices, it is *not* a linear map on the tangent space. Also note that the matrix $(w^i_j(x))$ is invertible. The inverse $\overline{w}$ of w is another geometric object with components $(\overline{w}^i_j(x))$ that transform as

$$\frac{\partial y^i}{\partial x^a}\overline{w}^a_j(x) = \overline{w}^i_j(y) \tag{7.3}$$

Thus, a splitting ε defines the geometric object w that is unique modulo the choice of the basepoint. Since $\varepsilon(x, y) = \varepsilon(e, y) \circ \varepsilon(x, e)$, we deduce

$$\overline{w}^i_a(y)w^a_j(x) = \varepsilon^i_j(x, y) \tag{7.4}$$

Therefore, if we are given the geometric object w whose components transform according to (7.2), then we define $\varepsilon(x, y)$ by (7.4) and we can easily check that $\varepsilon(y, z) \circ \varepsilon(x, y) = \varepsilon(x, z)$;

An Alternative Approach to Lie Groups and Geometric Structures, Ercüment H. Ortaçgil (2018).

that is, ε is a splitting. Now, let (M, w) denote a smooth manifold M endowed with such a geometric object w. Hence, we deduce the following proposition:

Proposition 7.1 *(M, w) determines (M, ε) by (7.4), and (M, ε) determines (M, w) by (7.1).*

We now rewrite (7.4) in the form

$$w^i_j(x) = \varepsilon^a_j(x, y) w^i_a(y) \tag{7.5}$$

The simple formula (7.5) states a very important fact: The subgroupoid $\varepsilon(\mathcal{U}_0) \subset \mathcal{U}_1$ consists of those 1-arrows in $\mathcal{U}_1$ that preserve the geometric object w. This situation is analogous to the well-known fact that the orthogonal frames of a Riemannian manifold (M, g) are those frames that preserve the metric g.

Definition 7.2 *w is the structure object of (M, ε). It is unique modulo the choice of the base point.*

We can easily remove the arbitrariness in Definition 7.2 by identifying e with the origin o of $\mathbb{R}^n$ and considering 1-arrows with the source at o and target in M. Equivalently, we can remove the arbitrariness of the base point in the principal bundle $\mathcal{U}^{e,\bullet} \to M$ by passing from the groupoid to the principal bundle viewpoint.

Now (7.5) states that w is ε-invariant, and therefore

$$\widetilde{\nabla}_r w^i_j = \frac{\partial w^i_j}{\partial x^r} + \Gamma^a_{rj} w^i_a = 0 \tag{7.6}$$

which is obtained by differentiating (7.5) at $y = x$. Note that $\widetilde{\nabla}_r w^i_j$ is computed like the covariant derivative of a 1-form in the index j, since w^i_j does not transform in the index i.

For a vector field $X = (X^i)$, the Lie derivative $\mathcal{L}_X w$ is given by the formula

$$(\mathcal{L}_X w)^i_j = \left(\frac{\partial w^i_j}{\partial x^a} X^a + w^i_b \frac{\partial X^b}{\partial x^j} \right) \tag{7.7}$$

Like (7.6), $\mathcal{L}_X w$ is computed like the Lie derivative of a 1-form. We can easily check the formula

$$\mathcal{L}_X \mathcal{L}_Y w - \mathcal{L}_Y \mathcal{L}_X w - \mathcal{L}_{[X,Y]} w = 0 \tag{7.8}$$

In fact, the Lie derivative $\mathcal{L}_X \theta$ is defined for any geometric object θ (not necessarily one of first order like w), and (7.8) holds if we replace w by θ. Now, (7.8) shows that the set of vector fields satisfying $\mathcal{L}_X \theta = 0$ is always a Lie algebra. The book [Y], written in 1957, is dedicated to the study of this subject. This Lie algebra $\mathcal{L}[\theta]$ need not be locally transitive. In our case, it will be locally transitive if and only if $\mathfrak{R} = 0$. Indeed, if $\mathfrak{R} = 0$ and $X \in \mathfrak{X}_{\widetilde{\varepsilon}}(U)$, then $\partial X^b / \partial x^j = \Gamma^b_{aj} X^a$, by (6.13). Substituting into (7.7) gives

$$
\begin{aligned}
(\mathcal{L}_X w)^i_j &= \left(\frac{\partial w^i_j}{\partial x^a} + w^i_b \Gamma^b_{aj}\right) X^a \\
&= \left(\widetilde{\nabla}_a w^i_j\right) X^a \\
&= 0
\end{aligned} \tag{7.9}
$$

by (7.6). Therefore, $\mathcal{L}_X w = 0$ for all $X \in \mathfrak{X}_{\widetilde{\varepsilon}}(U)$, and $\mathcal{L}[w]$ is locally transitive. The converse also holds and is easy to show. We recall here that the zero on the right-hand side of (7.8) is interpreted in modern differential geometry as a lack of curvature, like (6.56).

Let $\xi(p) = (\xi^i) \in T_pM$. We define the tangent vector $\eta = (\eta^i)$ at T_eM by the formula

$$
\eta^i \stackrel{\text{def}}{=} w^i_a(p)\xi^a \tag{7.10}
$$

Thus, we can interpret w as a 1-form on M with values in T_eM. Now we recall the point bracket $\{\,,\,\}_e$ defined on T_eM by (6.2). We recall that $(T_eM, \{\,,\,\}_e)$ is not a Lie algebra unless $\mathfrak{R} = 0$. Let dw denote the exterior derivative of w. Then,

$$
(dw)^i_{jk}(x) = \frac{\partial w^i_j(x)}{\partial x^k} - \frac{\partial w^i_k(x)}{\partial x^j} \tag{7.11}
$$

which is a 2-form with values in T_eM (omitting the factor $\frac{1}{2}$). Since, by (6.33),

$$
\frac{\partial \varepsilon^i_j(x,e)}{\partial x^k} = -\varepsilon^i_c(x,e)\Gamma^a_{kj}(x) \tag{7.12}
$$

(7.1), (7.11), and (7.12) give

$$
\begin{aligned}
(dw)^i_{jk}(x) &= -\varepsilon^i_a(x,e)\Gamma^a_{kj}(x) + \varepsilon^i_a(x,e)\Gamma^a_{jk}(x) \\
&= w^i_a(x) T^a_{jk}(x)
\end{aligned} \tag{7.13}
$$

Therefore, we have the following proposition:

Proposition 7.3 (The Maurer–Cartan Equation)

$$
(dw)^i_{jk}(x) - w^i_a(x) T^a_{jk}(x) = 0 \tag{7.14}
$$

Using the torsion bracket (5.22) we can rewrite (7.14) in a more invariant form as

$$
dw(X,Y) - w(T(X,Y)) = 0 \tag{7.15}
$$

We observe the intriguing fact that we have the Maurer–Cartan equation (7.15), but there is no group in sight, since we have not yet assumed $\mathfrak{R} = 0$ and the zero on the right-hand side of (7.15) has nothing to do with the curvature ($= \mathfrak{R}$), at least from the present standpoint. We now assume $\mathfrak{R} = 0$ so that we can replace $T(X,Y)$ in (7.15) with the true bracket $[X,Y]$ for $X, Y \in \widetilde{\mathfrak{X}}_\varepsilon$. Note that we should change the "−" in (7.15) to a "+" if we use $\mathfrak{X}_{\widetilde{\varepsilon}}$ instead. Now $(T_eM, \{\,,\,\}_e)$ becomes a Lie algebra, which is isomorphic to

$\widetilde{\mathfrak{X}}_\varepsilon$ according to (6.2) and (6.14). Therefore, w becomes a 1-form with values in the Lie algebra $(T_eM, \{\ ,\ \}_e)$, with the assumption $\mathfrak{R} = 0$. Since $\mathfrak{R} = 0$ implies $\mathcal{R} = 0$ by Corollary 6.9, we get the pseudogroup $\mathcal{G}$. Assuming that $\mathcal{G}$ globalizes, M becomes a Lie group by Proposition 4.4, and we recover the well- known setup with w as the Lie-algebra-valued 1-form on the Lie group M.

What condition does $\mathfrak{R} = 0$ impose on w? First, since $\mathcal{R} = 0$, (7.5) gives

$$w^i_j(x) = \frac{\partial f^a}{\partial x^j} w^i_a(f(x)), \qquad f \in \mathcal{G} \tag{7.16}$$

So, the pseudogroup $\mathcal{G}$ preserves w. Next, we recall that for $X, Y \in \widetilde{\mathfrak{X}}_\varepsilon$, the bracket $[X, Y] = T(X, Y)$ localizes at any point in M. Since T is ε-invariant, the tangent vector $\varepsilon^i_a(p, e)\{X(p), Y(p)\}^a_p$ at e is independent of the point $p \in M$. To express this constancy condition in coordinates, we substitute the left-hand side of (7.4) into (2.7)and express the condition $\mathcal{R}(x, y) = 0$ in terms of w. Rearranging terms, the result is

$$\overline{w}^a_j(y)\overline{w}^b_k(y)\left(\frac{\partial w^i_b(y)}{\partial y^a} - \frac{\partial w^i_a(y)}{\partial y^b}\right) = \overline{w}^a_j(x)\overline{w}^b_k(x)\left(\frac{\partial w^i_b(x)}{\partial x^a} - \frac{\partial w^i_a(x)}{\partial x^b}\right) \tag{7.17}$$

Thus, the expression on the right-hand side of (7.17) does not transform under a coordinate change $(x) \to (y)$ and therefore must be a function $c^i_{jk}(x)$. Therefore,

$$(dw)^i_{jk}(x) = \left(\frac{\partial w^i_j(x)}{\partial x^k} - \frac{\partial w^i_k(x)}{\partial x^j}\right) = c^i_{ab}(x) w^a_j(x) w^b_k(x) \tag{7.18}$$

and, as in Chapter 6, we can easily show that the functions $c^i_{jk}(x)$ are constant by applying $\widetilde{\nabla}$ to (7.18).

If θ is a geometric object of any order, then the local transitivity condition of the Lie algebra $\mathcal{L}[\theta]$ imposes certain integrability conditions on θ, like (7.18) (transitivity in our sense is stronger than the point transitivity on the base in the general case, see Chapter 20). For instance, if θ is a metric, then we get the constant-curvature condition. These integrability conditions, which were studied in detail in [Y], were interpreted later in [P1] and [P2] from the standpoint of jets and formal integrability.

It is amusing to contemplate the reasons why these integrability conditions are not taken as the basis of the concept of curvature in modern differential geometry. Surely, the most important reason is that Riemann lived before Lie, an example of how mathematics is like a living organism evolving in time. Nevertheless, we should recall here that the only mathematical formula in Riemann's famous *Habilitationsschrift* is the constant-curvature condition, which is equivalent to local transitivity. Another reason is that Cartan was a much more dominant figure in his time than his contemporary Vessiot, showing that mathematics is also a human endeavor with many psychological and sociological aspects, rather than a collection of absolute facts.

PART II
Some Consequences

Sell your cleverness and buy bewilderment

Rumi

8

The Nonlinear Spencer Sequence

Let Diff(M) denote the group of global diffeomorphisms of M and $\mathrm{Diff}_{\mathrm{loc}}(M)$ the set of local diffeomorphisms of M. Checking the axioms in Chapter 3, we see that $\mathrm{Diff}_{\mathrm{loc}}(M)$ is a pseudogroup. We call $\mathrm{Diff}_{\mathrm{loc}}(M)$ the universal pseudogroup on M and denote it simply by $\mathrm{Diff}_{\mathrm{l}}$.

Now suppose $(M, \varepsilon, \mathcal{G})$ is an LLG. We have the obvious inclusion map

$$\mathcal{G} \longrightarrow \mathrm{Diff}_{\mathrm{l}} \tag{8.1}$$

which maps any local solution of (2.4) to itself. In analogy with the notation used in homological algebra for exact sequences, we can now define an object $\mathbf{1}$ on M and write (8.1) in the form

$$\mathbf{1} \rightsquigarrow \mathcal{G} \longrightarrow \mathrm{Diff}_{\mathrm{l}} \tag{8.2}$$

where $\rightsquigarrow$ is not really a map, but this point is not important.

Now, (2.4) can be written in two (in fact more) equivalent forms. The first is

$$\frac{\partial f^i}{\partial x^j} - \varepsilon^i_j(x, f(x)) = 0 \tag{8.3}$$

and the second is

$$\frac{\partial f^i}{\partial x^a}\varepsilon^a_j(f(x), x) - \delta^i_j = 0 \tag{8.4}$$

Now, $(x^i, f^i(x), \partial f^i/\partial x^j)$ is a 1-arrow from x to $f(x)$. Similarly, $(x, f(x), \varepsilon^i_j(x, f(x))$ is another such 1-arrow. However, the sum and the difference of two 1-arrows from p to q have no meaning independent of coordinates unless $p = q$. Therefore, (8.3) has a meaning only if

An Alternative Approach to Lie Groups and Geometric Structures, Ercüment H. Ortaçgil (2018).

$f(x) = x$, in which case it becomes trivial. However, (8.4) does have a coordinate-free meaning: Indeed,

$$\begin{aligned}&\left(x^i, f^i(x), \frac{\partial f^i}{\partial x^j}\right) \circ (x, f(x), \varepsilon^i_j(x, f(x))^{-1}\\ =&\left(x^i, f^i(x), \frac{\partial f^i}{\partial x^j}\right) \circ (f^i(x), x^i, \varepsilon^i_j(f(x), x))\\ =&\left(f^i(x), f^i(x), \frac{\partial f^i}{\partial x^a}\varepsilon^a_j(f(x), x)\right)\end{aligned} \tag{8.5}$$

is a 1-arrow with source and target at $y = f(x)$. Clearly, $(f^i(x), f^i(x), \delta^i_j)$ is another such 1-arrow. Now, a 1-arrow $(h^i_j(y))$ with source and target at y (omitting the source and target from our notation once they are understood) transforms as

$$h^i_j(x) = \frac{\partial x^i}{\partial y^b} h^b_a(y) \frac{\partial y^a}{\partial x^j}$$

and is therefore an *invertible* section of the vector bundle $T \otimes T^* \rightarrow M$; that is, $(h^i_j(y))$ is an invertible linear map at the tangent space at y. It follows that the left-hand side of (8.4) is a section of $T \otimes T^* \rightarrow M$. However, it need not be invertible; that is, it need not be another 1-arrow. This analysis of the expression on the left-hand side of (8.4) is valid for any local diffeomorphism $y = f(x)$. However, the equality in (8.4) holds if and only if $y = f(x)$ solves (2.4), that is, it belongs to $\mathcal{G}$.

Thus, we have proved the following proposition:

Proposition 8.1 *We have the sequence*

$$\mathcal{G} \xrightarrow{i} \text{Diff}_1 \xrightarrow{\mathcal{D}_1} T^* \otimes T \tag{8.6}$$

If $f : U \longrightarrow f(U) = V$, then $\mathcal{D}_1 f$ is a section of $T^* \otimes T \rightarrow M$ over V and is defined locally by the formula

$$\mathcal{D}_1 : f^i \longrightarrow \frac{\partial f^i}{\partial x^a}\varepsilon^a_j(f(x), x) - \delta^i_j \tag{8.7}$$

Further, (8.6) is exact at Diff_1 : $\mathcal{D}_1 \circ i = 0$, and if $\mathcal{D}_1(f) = 0$, then $f \in \mathcal{G}$.

As we remarked above, $\mathcal{D}_1$ is defined without assuming $\mathcal{R} = 0$, but in the absence of this assumption, its kernel need not be $\mathcal{G}$ and in fact it may consist only of the identity map.

Now our purpose is to extend (8.6) one step to the right by checking the integrability conditions for $\mathcal{D}_1$. So, we assume $\mathcal{D}_1(f) = \xi$, or, equivalently,

$$\frac{\partial f^i}{\partial x^a}\varepsilon^a_j(f(x), x) - \delta^i_j = \xi^i_j \tag{8.8}$$

Thus, ξ is a section of $T^* \otimes T \rightarrow M$ defined over $f(U) = V$. Note that (8.8) implies $\det(\xi + I) \neq 0$ on V, that is, that $\xi + I$ is an invertible section of $T^* \otimes T \rightarrow M$ over V. Now, (8.8) is equivalent to

$$\frac{\partial f^i}{\partial x^j} = \xi^i_a \varepsilon^a_j(x, f(x)) + \varepsilon^i_j(x, f(x)) \tag{8.9}$$

In order to find the integrability conditions for (8.9), we differentiate it with respect to x^k and substitute back from (8.9). This gives

$$\begin{aligned} \frac{\partial^2 f^i}{\partial x^k \partial x^j} &= \frac{\partial \xi^i_a}{\partial y^b}(\xi^b_c \varepsilon^c_k(x,y) + \varepsilon^b_k(x,y))\varepsilon^a_j(x,y) \\ &\quad + \xi^i_a \left(\frac{\partial \varepsilon^a_j(x,y)}{\partial x^k} + \frac{\partial \varepsilon^a_j(x,y)}{\partial y^b} \left(\xi^b_c \varepsilon^c_k(x,y) + \varepsilon^b_k(x,y) \right) \right) \\ &\quad + \left(\frac{\partial \varepsilon^i_j(x,y)}{\partial x^k} + \frac{\partial \varepsilon^i_j(x,y)}{\partial y^b} \left(\xi^b_c \varepsilon^c_k(x,y) + \varepsilon^b_k(x,y) \right) \right) \end{aligned} \tag{8.10}$$

We alternate k, j in (8.10) and substitute (2.7). This gives

$$\begin{aligned} -\xi^i_a \mathcal{R}^a_{kj}(x,y) - \mathcal{R}^i_{kj}(x,y) = \Bigg[&\xi^b_c \frac{\partial \xi^i_a}{\partial y^b} \varepsilon^c_k(x,y) \varepsilon^a_j(x,y) + \frac{\partial \xi^i_a}{\partial y^b} \varepsilon^b_k(x,y) \varepsilon^a_j(x,y) \\ &+ \xi^i_c \frac{\partial \varepsilon^a_j(x,y)}{\partial y^b} \xi^b_c \varepsilon^c_k(x,y) + \frac{\partial \varepsilon^i_j(x,y)}{\partial y^b} \xi^b_c \varepsilon^c_k(x,y) \Bigg]_{[kj]} \end{aligned} \tag{8.11}$$

We multiply both sides of (8.11) by $\varepsilon^k_m(y,x)\varepsilon^j_s(y,x)$, sum over k, j, and substitute from (2.12) and (6.32). Now (8.11) becomes

$$\begin{aligned} -\xi^i_a \overleftarrow{\mathcal{R}}^a_{ms}(x,y) - \overleftarrow{\mathcal{R}}^i_{ms}(x,y) &= \left[\xi^k_m \frac{\partial \xi^i_s}{\partial y^k} + \frac{\partial \xi^i_s}{\partial y^m} + \xi^i_a \Gamma^a_{bs} \xi^b_m + \Gamma^i_{bs} \xi^b_m \right]_{[ms]} \\ &= \left[\frac{\partial \xi^i_s}{\partial y^k} \left(\xi^k_m + \delta^k_m \right) + \xi^b_m \Gamma^a_{bs} \left(\xi^i_a + \delta^i_a \right) \right]_{[ms]} \end{aligned} \tag{8.12}$$

Note that all the variables on the right-hand side of (8.12) are evaluated at y. In particular the left-hand side of (8.12) does not depend on x.

To summarize, if (8.12) holds identically, then we can solve (8.8) with arbitrary initial conditions, the converse being obvious with $y = f(x)$ from the derivation of (8.12) from (8.8). Note that this derivation does not need the assumption $\mathcal{R} = 0$ either. The problem is that (8.12) does not depend only on ξ unless $\mathcal{R} = 0$. However, if $\mathcal{R} = 0$, then (8.12) is identically satisfied if and only if the right-hand side of (8.12) vanishes, and this depends only on ξ. So we define an operator $\mathcal{D}_2$ by the right-hand side of (8.12):

$$\mathcal{D}_2 : (\xi^i_j) \longrightarrow \left[\frac{\partial \xi^i_j}{\partial y^a} \left(\xi^a_k + \delta^a_k \right) + \xi^b_k \Gamma^a_{bj} \left(\xi^i_a + \delta^i_a \right) \right]_{[kj]} \tag{8.13}$$

Since $\mathcal{D}_2(\xi) = 0$ has an invariant meaning, $\mathcal{D}_2(\xi)$ must be a tensor and therefore a section of $(\wedge^2 T^*) \otimes T \to M$ over V. This fact can be checked directly using (8.13).

Thus, we have proved the following proposition:

Proposition 8.2 *Let $(M, \varepsilon, \mathcal{G})$ be an LLG. We have the sequence*

$$\mathcal{G} \xrightarrow{i} \mathrm{Diff}_1 \xrightarrow{\mathcal{D}_1} T^* \otimes T \xrightarrow{\mathcal{D}_2} (\wedge^2 T^*) \otimes T \tag{8.14}$$

which is locally exact also at $T^* \otimes T$ in the following sense: We have $\mathcal{D}_2 \circ \mathcal{D}_1 = 0$. Further, if ξ is a section of $T^* \otimes T \to M$ over V satisfying $\det(\xi + I) \neq 0$ and $\mathcal{D}_2(\xi) = 0$, then, for any $p \in V$ and $q \in M$, there exists a unique local diffeomorphism f satisfying $f(q) = p$ and $\mathcal{D}_1(f) = \xi$. In particular, if $\mathcal{D}_2(\xi) = 0$ on M, then $p \in M$ is arbitrary.

Definition 8.3 *The sequence (8.14) is the nonlinear Spencer sequence of the LLG $(M, \varepsilon, \mathcal{G})$.*

We recall once more that both operators $\mathcal{D}_1$ and $\mathcal{D}_2$ are defined without the assumption $\mathcal{R} = 0$. From (8.8) and (8.12), we deduce

$$(\mathcal{D}_2 \circ \mathcal{D}_1)(f(x)) = -\frac{\partial f^i}{\partial x^a} \varepsilon^a_j(f(x), x)\, \overleftarrow{\mathcal{R}}(x, f(x)) \tag{8.15}$$

Therefore, $\mathcal{R}$ is an obstruction to the local exactness of (8.14) also at the third term. If $\mathcal{R} = 0$, then we get $\mathcal{G}$, and (8.14) becomes locally exact also at the second term, as explained in Proposition 8.2. The important fact here is that we can use (8.15) to *define* $\mathcal{R}$.

We can construct (8.14) also for $(M, \varepsilon, C(\mathcal{G}))$, replacing ε by $\widetilde{\varepsilon}$. These two sequences can be identified by an isomorphism induced by the isomorphism (4.13). In particular, (8.14) is defined for any Lie group G using $(G, L(G))$ or $(G, R(G))$.

Two questions arise naturally:

Q1 What are the geometric meanings of the conditions $\det(\xi + I) \neq 0$ and $\mathcal{D}_2(\xi) = 0$?

Q2 What is the geometric meaning of the local/global exactness of (14) at $T^* \otimes T$?

Note that it is not possible to define the cohomology group of (8.14) at $T^* \otimes T$, because the sections satisfying $\det(\xi + I) \neq 0$ do not have an additive structure.

To answer the first part of **Q1**, we recall the definition of a gauge transformation given in Chapter 1. A gauge transformation $\nu \in \mathcal{A}_1$ (the same abuse of notation) assigns to any $p \in M$ a 1-arrow with source and target at p that is the same as an invertible linear map at the tangent space at p. So, $\nu = (\nu^i_j(x)$ locally on (U, x), where $\nu^i_j(x)$ is an invertible matrix that depends smoothly on x. Some $\nu \in \mathcal{A}_1$ defines a diffeomorphism of $\mathcal{U}_1$ as follows. For $\alpha(p, q) \in \mathcal{U}_1^{p,q}$, we define $(\nu\alpha)(p, q) \in \mathcal{U}_1^{p,q}$ by the formula

$$(\nu\alpha)(p, q) \stackrel{\text{def}}{=} \nu(q) \circ \alpha(p, q) \circ \nu^{-1}(p) \tag{8.16}$$

Therefore, $\nu \in \mathrm{Diff}(\mathcal{U}_1)$. However, ν is a very special diffeomorphism: It maps the fiber $\mathcal{U}_1^{p,q}$ over M onto itself. Conversely, it is easy to show that any diffeomorphism of $\mathcal{U}_1$ with this property is a gauge transformation. Therefore, we call a gauge transformation also a "vertical diffeomorphism" of $\mathcal{U}_1$. The "horizontal diffeomorphisms" of $\mathcal{U}_1$ are "prolongations" of the elements of Diff(M) in a sense to be explained in Chapter 17.

Now let ε be a splitting on M and $\nu \in \mathcal{A}_1$. We define another splitting $\nu(\varepsilon) = \varepsilon'$ whose 1-arrows are given by

$$(\nu(\varepsilon))(p,q) = \varepsilon'(p,q) \stackrel{\text{def}}{=} \nu(q) \circ \varepsilon(p,q) \circ \nu^{-1}(p) \tag{8.17}$$

We can easily check that $\nu(\varepsilon)$ is another splitting. Therefore, some $\nu \in \mathcal{A}_1 \subset \text{Diff}(\mathcal{U}_1)$ maps the submanifold $\varepsilon(\mathcal{U}_0) \subset \mathcal{U}_1$, which is also a subgroupoid, to the submanifold $\nu(\varepsilon)(\mathcal{U}_0) \subset \mathcal{U}_1$, which is another subgroupoid. Let $\mathcal{E}$ denote the set of all splittings on M. The following lemma shows that we have a lot of flexibility with the vertical diffeomorphisms:

Lemma 8.4 *$\mathcal{A}_1$ acts transitively on $\mathcal{E}$.*

Indeed, let $\varepsilon, \varepsilon' \in \mathcal{E}$. We fix a base point e, define $\nu \in \mathcal{A}_1$ by $\nu(p) \stackrel{\text{def}}{=} \varepsilon'(e,p) \circ \varepsilon(p,e)$, and easily check that $\nu(\varepsilon') = \varepsilon$.

The above proof shows that $\mathcal{A}_1$ in fact acts *simply* transitively if we replace the groupoid $\varepsilon(\mathcal{U}_0) \subset \mathcal{U}_1$ with its principal bundle $\varepsilon(\mathcal{U}_0)^{e,\bullet} \subset \mathcal{U}_1^{e,\bullet}$ by fixing a base point e.

Now suppose $\nu \in \mathcal{A}_1$. Then $\nu - I \stackrel{\text{def}}{=} \xi$ is a section of $T^* \otimes T \to M$ satisfying $\det(\xi + I) \neq 0$. Conversely, any section of $T^* \otimes T \to M$ satisfying $\det(\xi + I) \neq 0$ defines some $\nu \in \mathcal{A}_1$ by $\nu \stackrel{\text{def}}{=} \xi + I$. So the formula

$$\nu = \xi + I \tag{8.18}$$

gives a 1–1 correspondence between $\mathcal{A}_1$ and sections of $T^* \otimes T \to M$ satisfying det $(\xi + I) \neq 0$. This answers the first part of **Q1**.

The second part of **Q1** now becomes: What does $\mathcal{D}_2(\xi) = 0$ mean in terms of ν defined by (8.18)?

Let (M,ε) and $\nu \in \mathcal{A}_1$ be given. We define (M,ε') by (8.16), where $\varepsilon' = \nu(\varepsilon)$. We denote (M,ε') also by $\nu(M,\varepsilon)$. Note that everything done in Part I for (M,ε) applies to (M,ε'). For any $p \in M$, we have now two point brackets at p defined by (6.2). We denote them by $\{\,,\,\}_p^{\varepsilon}$ and $\{\,,\,\}_p^{\varepsilon'}$ to show the dependence on the splitting. The question now is whether $\nu(p)$ preserves these point brackets, that is, whether we have

$$\nu(p)\{X,Y\}_p^{\varepsilon} = \{\nu(p)X, \nu(p)Y\}_p^{\varepsilon} \qquad \text{for all} \quad X, Y \in T_pM \tag{8.19}$$

Proposition 8.5 *Suppose $\nu \in \mathcal{A}_1$ and $\xi \in T^* \otimes T$ are related by (8.18). Let (M,ε) be given and (M,ε') be defined by $\varepsilon' = \nu(\varepsilon)$. Then the following are equivalent:*

(i) $\nu(p)$ preserves the point brackets at p;

(ii) $\mathcal{D}_2(\xi)(p) = 0$.

Proof We first write (8.17) in coordinates as

$$\nu_a^i(y)\varepsilon_b^a(x,y)\eta_j^b(x) = (\varepsilon')_j^i(x,y) \tag{8.20}$$

where $\eta = \nu^{-1}$, that is, $\eta_a^i(y)\nu_j^a(y) = \nu_a^i(y)\eta_j^a(y) = \delta_j^i$. We differentiate (8.20) with respect to y^r at $y = x$ using (5.1) and get

$$\frac{\partial \nu_b^i(x)}{\partial x^r}\eta_j^b(x) + \nu_a^i(x)\Gamma_{rb}^a(x)\eta_j^b(x) = (\Gamma')_{rj}^i(x) \tag{8.21}$$

Omitting x, we multiply both sides of (8.21) by $\nu_m^r \nu_k^j$ and sum over r, j. This gives

$$\frac{\partial \eta_k^i}{\partial x^a}\nu_m^a + \nu_a^i\Gamma_{ck}^a\nu_m^c = (\Gamma')_{ab}^i\nu_m^a\nu_k^b \tag{8.22}$$

We set $\nu = I + \xi$ in (8.22), which becomes

$$\frac{\partial \xi_k^i}{\partial x^a}\left(\xi_m^a + \delta_m^a\right) + \left(\xi_a^i + \delta_a^i\right)\Gamma_{ck}^a\left(\xi_m^c + \delta_m^c\right) = (\Gamma')_{ab}^i\nu_m^a\nu_k^b \tag{8.23}$$

or, equivalently,

$$\frac{\partial \xi_k^i}{\partial x^a}\left(\xi_m^a + \delta_m^a\right) + \left(\xi_a^i + \delta_a^i\right)\Gamma_{ck}^a\xi_m^c = (\Gamma')_{ab}^i\nu_m^a\nu_k^b - \nu_a^i\Gamma_{mk}^a \tag{8.24}$$

We alternate k, m in (8.24) and get

$$\mathcal{D}_2(\xi)_{mk}^i = (T')_{ab}^i\nu_m^a\nu_k^b - \nu_a^i T_{mk}^a \tag{8.25}$$

But (i) is equivalent to (8.19), which is equivalent to the vanishing of the right-hand side of (8.25). This proves the claim. ■

Proposition 8.5 answers the second part of **Q1**. Note that its proof does not require $\mathcal{R} = 0$.

The rest of this chapter is devoted to providing an answer to **Q2**. We start with the following proposition:

Proposition 8.6 *Let (M, ε) and $\nu \in \mathcal{A}_1$ be given and let (M, ε') be defined by $\varepsilon' = \nu(\varepsilon)$. Assume that ν preserves point brackets on M. Then $\mathcal{R}(\varepsilon) = 0 \Longleftrightarrow \mathcal{R}(\varepsilon') = 0$.*

Proof We rewrite (8.17) as

$$\varepsilon'(p,q) \circ \nu(p) = \nu(q) \circ \varepsilon(p,q) \tag{8.26}$$

According to (8.26), the action of ν commutes with the 1-arrows of ε and ε'. Now, $\mathcal{R}(\varepsilon) = 0 \Longleftrightarrow T^{\varepsilon}$ is ε-invariant $\Longleftrightarrow$ 1-arrows of ε preserve $T^{\varepsilon} \Longleftrightarrow$ 1-arrows of ε' preserve $T^{\varepsilon'}$, since ν commutes with the 1-arrows $\Longleftrightarrow$ $T^{\varepsilon'}$ is ε'-invariant $\Longleftrightarrow$ $\mathcal{R}(\varepsilon') = 0$. ■

Henceforth, we will always assume that ν preserves the point brackets, so $\mathcal{D}_2(\xi) = 0$ by Proposition 8.5, and $\mathcal{R}(\varepsilon) = 0$, so $\mathcal{R}(\varepsilon') = 0$ by Proposition 8.6. We will denote this data in brief by

$$\nu(M, \varepsilon, \mathcal{G}) = (M, \varepsilon', \mathcal{G}') \tag{8.27}$$

We now pose the following question: How should we define the concepts of "local equivalence" and "global equivalence" of two LLGs on the same base M?

We start with the more intuitive global equivalence. To aid our intuition, we assume that $(M, \varepsilon, \mathcal{G})$ is globalizable and deal with the PHS $(M, \mathcal{G})$. So as not to go astray with abstractions, we reduce the problem to the simplest possible case: Suppose X is a finite set, $P(X)$ the group of permutations of X, and $G, H \subset P(X)$ two subgroups of the same order that act transitively on X. How should we define the equivalence of the "Klein pairs" (X, G) and (X, H)? The requirement of an abstract isomorphism between the groups G, H is not the right answer, because we should take into account also the actions of G and H on X. There seems to be only one possible definition: (X, G) and (X, H) are equivalent if there exists a bijection $\phi : X \to X$, that is, $\phi \in P(X)$, such that $\phi \circ h \circ \phi^{-1} \in G$ for all $h \in H$. Equivalently, $G, H \subset P(X)$ are conjugate inside $P(X)$. With this assumption, the map $H \to G$ that maps h to $\phi \circ h \circ \phi^{-1}$ is an abstract isomorphism. Therefore, $(X, G) \sim (X, H)$ implies $G \simeq H$ as abstract groups. The converse is false unless G, H act simply transitively, as in our case. Indeed, suppose G, H act simply transitively and $\alpha : H \to G$ is an abstract isomorphism. We fix some $e \in X$ and define $\phi : X \to X$ as follows: For $x \in X$, there exists a unique $f \in H$ with $f(e) = x$. If $\alpha(f)(e) = y$, we define $\phi(x) = y$. We easily check that ϕ is a bijection and gives an equivalence $(G, X) \sim (H, X)$.

Keeping the above motivation in mind, we rewrite the PDE (8.8) for the local diffeomorphism $y = f(x)$ in coordinate-free form as

$$\nu^{-1}(q) \circ j_1(f)^{p,q} = \varepsilon(p, q) \tag{8.28}$$

where $f : U \to f(U) = V, p \in U$ and $q = f(p)$. Recall that the integrability conditions for (8.28) are given by $\mathcal{D}_2(\xi) = 0$, which we always assume to hold. Since $\varepsilon(p, q)T(p) = T(q)$ and $\nu(p)T(p) = T'(p)$, (8.28) shows that $j_1(f)^{p,q}T(p) = T'(q)$, or, equivalently, $j_1(f)^{p,q}$ defines the isomorphism of Lie algebras

$$j_1(f)^{p,q} : (T_p(M), \{\, , \}_p^{\varepsilon}) \longrightarrow (T_q(M), \{\, , \}_q^{\varepsilon'}) \tag{8.29}$$

We recall that $(T_p(M), \{\, , \}_p^{\varepsilon}) \simeq \widetilde{\mathfrak{X}}_{\varepsilon}(p) \simeq \mathfrak{X}_{\widetilde{\varepsilon}}(p) \simeq \widetilde{\mathfrak{X}}_{\varepsilon}(M) \simeq \mathfrak{X}_{\widetilde{\varepsilon}}(M)$ by (6.6) and (6.14). Similar isomorphisms hold for $(T_q(M), \{\, , \}_q^{\varepsilon'})$. Since $\widetilde{\mathfrak{X}}_{\varepsilon}(M)$ integrates to $C(\mathcal{G})$ and $\mathfrak{X}_{\widetilde{\varepsilon}}(M)$ integrates to $\mathcal{G}$ by Proposition 6.11, the local solution f of (8.28) maps $C(\mathcal{G})_{|U}$ onto $C(\mathcal{G}')_{|f(U)}$ and $\mathcal{G}_{|U}$ onto $\mathcal{G}'_{|f(U)}$. Combining this picture with the above motivation from permutations, we make the following important definition:

Definition 8.7 *A local diffeomorphism f satisfying $\mathcal{D}_1(f) = \xi$ in (8.14) is a local equivalence between the LLGs in (8.27).*

For $\nu = \text{Id}$ and $\xi = 0$, a local equivalence f of $(M, \varepsilon, \mathcal{G})$ is clearly a solution of (2.4). It follows from Proposition 8.2 that we can choose p and $q = f(p)$ arbitrarily for a local equivalence f. Furthermore, f is uniquely determined on its domain by the initial condition $f(p) = q$ and therefore can be "analytically continued" as in Chapter 3. Repeating the arguments from Chapter 3, we deduce the following proposition:

Proposition 8.8 *If M is compact and simply connected, then any local equivalence extends uniquely to a global diffeomorphism $f^{\text{ex}} : M \to M$ that is a solution of* (8.28). *We call f^{ex} a global equivalence between the two LLGs in* (8.27).

Now a very intriguing point arises: Suppose $(M, \varepsilon, \mathcal{G})$ and $\nu(M, \varepsilon, \mathcal{G}) = (M, \varepsilon', \mathcal{G}')$ are both globalizable and therefore define the PHSs $(M, \mathcal{G})$ and $(M, \mathcal{G}')$. According to Proposition 4.4, $(M, \mathcal{G})$ and $(M, \mathcal{G}')$ define two Lie group structures M_1 and M_2 on M with isomorphic Lie algebras given by (8.29). According to the general theory of abstract Lie groups, a Lie group is determined by its Lie algebra and its fundamental group. Obviously, M_1 and M_2 have the same fundamental group, since their underlying space is M. This fact shows that the Lie groups M_1 and M_2 are globally isomorphic. One is now tempted to conclude that any local equivalence between M_1 and M_2 extends to a global equivalence, and therefore the "cohomology" of (8.14) is always trivial. This argument is incorrect, for a subtle reason: Let $F : M_1 \to M_2$ be the abstract Lie group isomorphism inducing the isomorphism (8.29) on the Lie algebras. Replacing f on the left-hand side of (8.28) by F, it is easy to show that $\nu^{-1}(q) \circ j_1(F)^{p,q}$ maps $T(p)$ to $T(q)$; that is, $\nu^{-1} \circ j_1(F)$ preserves the torsion T on M. However, there are many such 1-arrows on M other than those defined by ε, and therefore we need not have $\nu^{-1} \circ j_1(F) = \varepsilon$ as in (8.28). Indeed, suppose $f^{p,q}T(p) = T(q)$ and $g^{p,q}T(p) = T(q)$. Now $(g^{p,q})^{-1} \circ f^{p,q}T(p) = T(p)$ and consequently $(g^{p,q})^{-1} \circ f^{p,q} \in \text{Aut}(\mathfrak{X}_\varepsilon(p)) \subset \mathcal{U}^{p,p}$. It follows that the set of all 1-arrows from p to q that map $T(p)$ to $T(q)$ is in 1–1 correspondence with the automorphism group $\text{Aut}(\mathfrak{X}_\varepsilon(p))$ of the Lie algebra $\mathfrak{X}_\varepsilon(p)$. We will show in Chapter 15 that the groupoid of all such 1-arrows integrates to a pseudogroup $S(\mathcal{G}) \supset \mathcal{G}$ whose local diffeomorphisms are solutions of a *second-order* nonlinear PDE. To summarize, the abstract isomorphism $F : M_1 \to M_2$ does not necessarily restrict even to a local equivalence as given by Definition 8.7, since it does not take ν and ε into account.

The above argument puts more emphasis on the meaning of the "cohomology" of (8.14) at $T^* \otimes T$. So, we fix the LLG $(M, \varepsilon, \mathcal{G})$ and consider the set $O(M, \mathcal{G})$ consisting of all LLGs on M that are locally equivalent to $(M, \varepsilon, \mathcal{G})$. We identify two such LLGs if they are both globally equivalent to $(M, \varepsilon, \mathcal{G})$ (and hence to each other). We denote this identification space by $O(M, \mathcal{G})/\sim$. Now, the set $O(M, \mathcal{G})/\sim$ seems to be a reasonable definition (or substitute) for the "cohomology" of (8.14) at $T^* \otimes T$. Note that $O(M, \mathcal{G})/\sim$ reduces to a point for compact and simply connected M by Proposition 8.8. The set $O(M, \mathcal{G})/\sim$ has a canonical basepoint, namely, the equivalence class $[(M, \varepsilon, \mathcal{G})]$ of the LLG $(M, \varepsilon, \mathcal{G})$ that we start with.

At this stage, it is instructive to take a glimpse at the analogous construction of $O(M, \mathcal{G})/\sim$ for complex manifolds. Let N be a complex manifold and let $\mathfrak{G}$ be the pseudogroup of holomorphic local diffeomorphisms of N. Since any two complex structures on $\mathbb{C}^n$ are equivalent, it follows that any two complex structures $(N, \mathfrak{G})$ and $(N, \mathfrak{G}')$ on N

are locally equivalent, with the obvious definition of local equivalence. Therefore, the set $O(N, \mathfrak{G})$ is the set of *all* complex structures on N. Now, if we define the set $O(N, \mathfrak{G})/\sim$ as above, it follows that $O(N, \mathfrak{G})/\sim$ is the moduli space of complex structures on N, a familiar but very subtle object. In some favorable cases, it is known that $O(N, \mathfrak{G})/\sim$ is another complex manifold. This fact suggests that the set $O(M, \mathcal{G})/\sim$ has a very rich structure and may even be another LLG! To make progress with this nontrivial problem, we should first define a "tangent vector" to the set $O(M, \mathcal{G})/\sim$ at the point $[(M, \varepsilon, \mathcal{G})]$. This will be done in the next chapter.

A final word about the history: The nonlinear Spencer sequence was first constructed in [KS] for a very broad class of pseudogroups. This sequence was studied extensively in [Po2] and other works of this author. However, much of the work in the literature is concerned with the issue of stabilization of the order of jets in the nonlinear/linear Spencer sequences and not with their cohomology.

9

Deformations

Given some LLG $(M, \varepsilon, \mathcal{G})$, we recall from Chapter 8 that $O(M, \mathcal{G})/\sim$ denotes the set of all LLGs that are locally equivalent to $(M, \varepsilon, \mathcal{G})$ modulo global equivalence. The main purpose of this chapter is to define the tangent space $T_{[(M,\varepsilon,\mathcal{G})]}(O(M, \mathcal{G})/\sim)$ of the set $O(M, \mathcal{G})/\sim$ at the point $[(M, \varepsilon, \mathcal{G})]$ and identify this vector space with some "known" cohomology group in the picture.

Suppose we have a gauge transformation (or a vertical diffeomorphism) $\nu(t) \in \mathcal{A}_1$ for $t \in (-\epsilon, \epsilon)$. In coordinates, $(\nu(t)(x))^i_j = (\nu^i_j(t, x))$ depends smoothly on both t and x. So, for $t \in (-\epsilon, \epsilon)$, $\nu(t, p)$ is a 1-arrow with source and target at $p \in M$, or, equivalently, an invertible linear map at the tangent space T_pM. We write $\nu(t, x)$ for $\nu(t)(x)$ and assume $\nu(0, p) = p$ for all $p \in M$. In coordinates, $\nu^i_j(0, x) = \delta^i_j$ for all $x \in U$. So, for fixed p, $\nu(t, p)$ is a curve in the group $\mathcal{U}_1^{p,p} \simeq GL(n, \mathbb{R})$. To recapitulate, we state the following definition:

Definition 9.1 *$\nu(t)$ is a (small, first-order) deformation if*

(i) $\nu(t) \in \mathcal{A}_1$ *for* $t \in (-\epsilon, \epsilon)$ *and* $\nu(t, x)$ *is smooth;*

(ii) $\nu(0) = I$.

In short, a deformation is a smooth 1-parameter family of gauge transformations starting from the identity for $t = 0$. We observe that a deformation is defined on any smooth manifold M. Since we can act on any first-order geometric object (for instance, a metric g, a symplectic form ω, or an almost-complex structure J) by some deformation, we can deform any first-order geometric object on M using a deformation.

Let (M, ε) be given and $\nu(t)$ be a deformation. Using the notation of Chapter 8, we define $\nu(t)(M, \varepsilon) = (M, \varepsilon(t))$, where $\nu(t)(\varepsilon) = \varepsilon(t)$, $\varepsilon = \varepsilon(0)$ according to (8.16). Henceforth in this chapter we will always assume the following:

1. $\nu(0)(M, \varepsilon) = (M, \varepsilon(0)) = (M, \varepsilon)$ is the LLG $(M, \varepsilon, \mathcal{G})$; that is, $\mathcal{R}(0) = \mathcal{R} = 0$.
2. $\mathcal{D}_2(0)(\xi(t)) = \mathcal{D}_2(\xi(t)) = 0$ for all small t where $\nu(t) = I + \xi(t)$ as in (8.18).

By Propositions 8.5 and 8.6, (1) and (2) imply the existence of the 1-parameter family of LLGs $\nu(t)(M, \varepsilon, \mathcal{G}) = (M, \varepsilon(t), \mathcal{G}(t))$ starting from $(M, \varepsilon, \mathcal{G})$ for $t = 0$. Furthermore,

An Alternative Approach to Lie Groups and Geometric Structures, Ercüment H. Ortaçgil (2018).

$\nu(t)(M, \varepsilon, \mathcal{G})$ is locally equivalent to $(M, \varepsilon, \mathcal{G})$ for all t by some local diffeomorphism $f(t) \in \mathrm{Diff}(M)$ satisfying $\mathcal{D}_1(0)(f(t)) = \mathcal{D}_1(f(t)) = \xi(t)$. Therefore, we may think of $\nu(t)(M, \varepsilon, \mathcal{G})$ as a "curve" in the set $O(M, \mathcal{G})$ starting from the point $(M, \varepsilon, \mathcal{G}) \in O(M, \mathcal{G})$. We observe that everything done in Part I for $(M, \varepsilon, \mathcal{G})$ is valid for the LLG $\nu(t)(M, \varepsilon, \mathcal{G})$ for all t. For simplicity of notation, we will denote $\mathcal{D}_1(0)$, $\mathcal{D}_2(0)$, $\mathcal{R}(0)$, $\widetilde{\nabla}(0)$, etc. by $\mathcal{D}_1$, $\mathcal{D}_2$, $\mathcal{R}$, $\widetilde{\nabla}$, etc. as we have already done above.

For a point $a \in O(M, \mathcal{G})$, let $[a]$ denote its equivalence class in $O(M, \mathcal{G})/\sim$. Therefore, the curve $\nu(t)(M, \varepsilon, \mathcal{G})$ in $O(M, \mathcal{G})$ defines the curve $[\nu(t)(M, \varepsilon, \mathcal{G})]$ in $O(M, \mathcal{G})/\sim$.

Definition 9.2 *$\nu(t)(M, \varepsilon, \mathcal{G})$ is an le-path in $O(M, \mathcal{G})$ (passing through $(M, \varepsilon, \mathcal{G}) \in O(M, \mathcal{G})$ at $t = 0$). The le-path $\nu(t)(M, \varepsilon, \mathcal{G})$ is a ge-path if $\nu(t)(M, \varepsilon, \mathcal{G})$ is globally equivalent to $(M, \varepsilon, \mathcal{G})$ for all t, so that $[\nu(t)(M, \varepsilon, \mathcal{G})]$ is the constant path at $[(M, \varepsilon, \mathcal{G})]$.*

Now let $\nu(t)(M, \varepsilon, \mathcal{G})$ be an le-path. Since $\mathcal{D}_2(\xi(t)) = 0$, we have some local diffeomorphism $f(t, x)$ (called a local equivalence in Chapter 8) satisfying $\mathcal{D}_1 f(t, x) = \xi(t, x)$. We can choose $f(t, x)$ smoothly by choosing the initial conditions smoothly as follows: We fix some $p \in M$, choose a (smooth) path $C(t)$ starting from p at $t = 0$ and choose the unique f that maps p to $C(t)$. Therefore, $f(t, x)$ is a 1-parameter family of local diffeomorphisms defined near p satisfying $f(t, p) = C(t)$ and $f(0, x) = x$, $x \in M$. If $\nu(t)(M, \varepsilon, \mathcal{G})$ is a ge-path, then $f(t, x)$ is a *global* diffeomorphism (called a global equivalence in Chapter 8) for all small t.

We let $\mathrm{LE}(M, \varepsilon, \mathcal{G})$ denote the set of all le-paths through $(M, \varepsilon, \mathcal{G})$ and $\mathrm{GE}(M, \varepsilon, \mathcal{G})$ the set of all ge-paths through $(M, \varepsilon, \mathcal{G})$. Clearly, $\mathrm{GE}(M, \varepsilon, \mathcal{G}) \subset \mathrm{LE}(M, \varepsilon, \mathcal{G})$.

Now let $\nu(t)$ be an arbitrary deformation. We define a section ζ of $T^* \otimes T \to M$ whose value at p is given by

$$\zeta(p) \overset{\text{def}}{=} \left[\frac{\partial \nu(t, p)}{\partial t}\right]_{t=0} \tag{9.1}$$

So $\nu(t, p)$ is a curve in $\mathcal{U}_1^{p,p}$ whose tangent at $t = 0$ is $\zeta(p)$ for all $p \in M$. We do not assume $\nu(t, p) = \exp(t\zeta(p))$.

Definition 9.3 *An infinitesimal deformation is a section of the vector bundle $T^* \otimes T \to M$.*

Thus, a deformation $\nu(t)$ defines an infinitesimal deformation ξ according to (9.1). Conversely, any infinitesimal deformation ξ arises from a deformation ν: Given ζ, we define $\nu(t, x) \overset{\text{def}}{=} \exp(\zeta(x)t)$, which is defined in fact for $t \in (-\infty, \infty)$, and clearly (9.1) holds. We recall that deformations and their infinitesimal deformations are defined on any smooth M.

Definition 9.4 *If $\nu(t)(M, \varepsilon, \mathcal{G})$ is an le-path, we will call the deformation ζ defined by (9.1) an infinitesimal le-deformation (defined by $\nu(t)$).*

We define an infinitesimal ge-deformation similarly. We may think an infinitesimal le-deformation as the "tangent vector" of the "curve" $\nu(t)(M, \varepsilon, \mathcal{G})$ at the point $(M, \varepsilon, \mathcal{G}) \in O(M, \mathcal{G})$.

Now let $\nu(t)(M, \varepsilon, \mathcal{G})$ be an le-path, so that $\mathcal{D}_2(\xi(t)) = 0$ for all t, where $\nu(t) = I + \xi(t)$. We have

$$\left[\frac{\partial \xi(t,x)}{\partial t}\right]_{t=0} = \left[\frac{\partial \nu(t,x)}{\partial t}\right]_{t=0} = \zeta(x) \tag{9.2}$$

Now, (8.13) gives

$$(\mathcal{D}_2\xi(t,x))^i_{mk} = \left[\frac{\partial \xi^i_k(t,x)}{\partial x^a}\nu^a_m(t,x) + \nu^i_a(t,x)\Gamma^a_{ck}(x)\xi^c_m(t,x)\right]_{[mk]} = 0 \tag{9.3}$$

Differentiating (9.3) at $t = 0$, substituting from (9.2), and recalling that $\xi(0,x) = 0$, we get

$$\left[\frac{\partial \zeta^i_k(x)}{\partial x^m} + \Gamma^i_{ak}(x)\xi^a_n(x)\right]_{[mk]} = 0 \tag{9.4}$$

Using (9.4) we define the operator

$$\begin{aligned} D_2 &: T^* \otimes T \longrightarrow (\wedge^2 T^*) \otimes T \\ &: (\xi^i_j) \longrightarrow \left[\frac{\partial \xi^i_k}{\partial x^m} - \Gamma^i_{am}\xi^a_k\right]_{[mk]} \end{aligned} \tag{9.5}$$

We have proved the following proposition:

Proposition 9.5 *Let $\nu(t)(M, \varepsilon, \mathcal{G})$ be an le-path. Then the infinitesimal le-deformation ξ defined by (9.1) satisfies $D_2(\xi) = 0$.*

By the above derivation, the linear operator D_2 is the "linearization" of the nonlinear operator $\mathcal{D}_2$.

Now let $\nu(t)(M, \varepsilon, \mathcal{G})$ be a ge-path; that is, $[\nu(t)(M, \varepsilon, \mathcal{G})] = [(M, \varepsilon, \mathcal{G})]$ for all t. As we observed above, there exists a 1-parameter family of *global* diffeomorphisms $f(t,x)$, $f(0,x) = x$, satisfying $\mathcal{D}_1 f(t,x) = \xi(t)$, or

$$\frac{\partial f^i(t,x)}{\partial x^a}\varepsilon^a_j(f(t,x),x) - \delta^i_j = \xi^i_j(t,x) \tag{9.6}$$

Differentiation of (9.6) at $t = 0$ gives

$$\frac{\partial}{\partial x^j}\left[\frac{df^i(t,x)}{dt}\right]_{t=0} - \Gamma^i_{aj}(x)\left[\frac{df^a(t,x)}{dt}\right]_{t=0} = \zeta^i_j(x) \tag{9.7}$$

We define the *global* vector field $X = (X^i)$ on M by

$$X^i(x) \stackrel{\text{def}}{=} \left[\frac{\partial f^i(t,x)}{\partial t}\right]_{t=0} \tag{9.8}$$

and (9.7) becomes

$$\nabla_j X^i = \xi^i_j \tag{9.9}$$

We define

$$\begin{aligned} D_1 : T &\longrightarrow T^* \otimes T \\ : (X^i) &\longrightarrow \nabla_j X^i \end{aligned} \tag{9.10}$$

Note that $D_1 = \nabla$ as defined by (6.13).

We have proved the following proposition:

Proposition 9.6 *Let $\nu(t)(M, \varepsilon, \mathcal{G})$ be a ge-path with the infinitesimal ge-deformation ξ. Then $D_1 X = \xi$ for some vector field $X \in \mathfrak{X}(M)$.*

Now we have the sequence

$$T \xrightarrow{D_1} T^* \otimes T \xrightarrow{D_2} (\wedge^2 T^*) \otimes T \tag{9.11}$$

Note that T denotes the space of sections of $T \to M$ in (9.11) (with our abuse of notation), which is equal to $\mathfrak{X}(M)$. A straightforward computation using (9.10) and (9.5) now shows that

$$[\,(D_2 \circ D_1)\,(X)]^i_{jk} = -\mathfrak{R}^i_{jk,a} X^a \tag{9.12}$$

Therefore $D_2 \circ D_1 = 0$, since $\mathfrak{R} = 0$. The sequence (9.11) is the initial part of the linear Spencer sequence and will be studied in detail in Chapter 11. We recall that the kernel of D_1 is the sheaf $U \to \mathfrak{X}_{\widetilde{\varepsilon}}(U)$. Therefore, if $(M, \varepsilon, \mathcal{G})$ is globalizable, (9.11) becomes

$$0 \longrightarrow \mathfrak{X}_{\widetilde{\varepsilon}} \longrightarrow T \xrightarrow{D_1} T^* \otimes T \xrightarrow{D_2} (\wedge^2 T^*) \otimes T \tag{9.13}$$

We observe that (9.11), (9.12), and (8.15) give a conceptually elegant interpretation of Step 2 of Proposition 6.8 as follows: Let $X = (X^i)$ be an ε-invariant vector field with 1-parameter group of transformations $f(t,x)$. Substituting $f(t,x)$ into the defining formula (8.7) of $\mathcal{D}_1$ and differentiating at $t = 0$, we get D_1. Therefore, D_1 is the linearization of $\mathcal{D}_1$ in the same way as D_2 is the linearization of $\mathcal{D}_2$. Finally, we linearize (8.15) and deduce (6.24) using (9.12). In particular, we deduce the important fact that (8.14) is locally exact at $T^* \otimes T$ if and only if (9.11) is locally exact at $T^* \otimes T$. We will leave the further details to the interested reader.

Let $H^1(D, T)$ denote the cohomology group defined by (9.11) at $T^* \otimes T$. Note that T in $H^1(D, T)$ stands for the "representation bundle" $T \to M$ and not for the torsion T. Definition 9.4 and Propositions 9.5-9.6 imply the following important corollary:

Corollary 9.7 *There exists a map*

$$\mathrm{KS} : \mathrm{LE}(M, \varepsilon, \mathcal{G}) \longrightarrow H^1(D, T) \tag{9.14}$$

that sends an le-path to its infinitesimal deformation defined by (9.1). KS *vanishes on* $\mathrm{GE}(M, \varepsilon, \mathcal{G}) \subset \mathrm{LE}(M, \varepsilon, \mathcal{G})$.

Definition 9.8 (9.14) *is the Kodaira–Spencer map of the LLG* $(M, \varepsilon, \mathcal{G})$.

The KS map is constructed first to study the moduli space of complex structures on a compact complex manifold. The definition of a deformation used in [Kd] is more subtle, and we refer the reader to this classic book, where Kodaira blends the mathematics with a passionate and extremely interesting history of the deformation theory of complex manifolds and the developments around it.

Now the idea is clear: Let $T_{(M,\varepsilon,\mathcal{G})}O(M, \mathcal{G})$ be the set of the "tangent vectors" of all possible le-paths at $(M, \varepsilon, \mathcal{G}) \in O(M, \mathcal{G})$. For $a, b \in T_{(M,\varepsilon,\mathcal{G})}O(M, \mathcal{G})$, we define $a \asymp b$ if $a - b$ is the "tangent vector" of some ge-path. Using the equivalence relation $\asymp$, the final step is to define $T_{[(M,\varepsilon,\mathcal{G})]}(O(M, \mathcal{G})/\sim) \stackrel{\text{def}}{=} T_{(M,\varepsilon,\mathcal{G})}O(M, \mathcal{G})/\asymp$. This construction will allow us to define

$$\mathrm{KS} : T_{[(M,\varepsilon,\mathcal{G})]}(O(M, \mathcal{G})/\sim) \longrightarrow H^1(D, T) \tag{9.15}$$

and then study the injectivity and surjectivity of KS in order to identify the left-hand side of (9.15) with the cohomology group on its right-hand side.

However there are three serious problems:

1. We will compute $H^1(D, T)$ in Chapter 11. It turns out that $H^1(D, T)$ is isomorphic to $n = \dim M$ copies of the de Rham group $H^1_{\mathrm{dR}}(M, \mathbb{R})$, which is known to depend only on the homotopy type of M. Therefore, $H^1(D, T)$ does not seem to be related to some "deformation cohomology" as we would expect!
2. If $\nu_1(t)(M, \varepsilon, \mathcal{G})$ and $\nu_2(t)(M, \varepsilon, \mathcal{G})$ are two le-paths with infinitesimal deformations ξ_1 and ξ_2, there is no obvious le-path starting from $(M, \varepsilon, \mathcal{G})$ with the infinitesimal deformation $\xi_1 + \xi_2$. Indeed, let $\nu_1(t) = (\nu_1)^i_j$ and $\nu_2(t) = (\nu_2)^i_j$ be two *arbitrary* deformations with infinitesimal deformations ζ_1 and ζ_2. It follows easily from (9.1) that $\nu_1(t)\nu_2(t) = ((\nu_1)^i_a(\nu_2)^a_j)$ is a deformation with infinitesimal deformation $\zeta_1 + \zeta_1$ and that $\nu^{-1}(t)$ is a deformation with infinitesimal deformation $-\zeta$. Now let $\nu(t)(M, \varepsilon, \mathcal{G})$ and $\eta(t)(M, \varepsilon, \mathcal{G})$ be two le-paths. Unfortunately, $\nu(t)\eta(t)(M, \varepsilon, \mathcal{G})$ need not be an le-path, and therefore the infinitesimal deformation $\zeta_1 + \zeta_1$ need not be an infinitesimal le-deformation. Indeed, if $\nu(t) = I + \xi(t)$ and $\eta(t) = I + \sigma(t)$, then $\nu(t)\eta(t) = I + \sigma(t) + \xi(t) + \sigma(t)\xi(t)$ and, using (8.13), we can check that $\mathcal{D}_2(\xi(t)) = 0$ and $\mathcal{D}_2(\sigma(t)) = 0$ do not imply $\mathcal{D}_2(\sigma(t) + \xi(t) + \sigma(t)\xi(t)) = 0$. Hence, $\nu(t)\eta(t)$ need not preserve the point brackets. In fact, $\nu(t)\eta(t)(M, \varepsilon, \mathcal{G})$ need not even be an LLG for $t \neq 0$. Therefore, it is not possible to define $a - b$ and $\asymp$, and there is no obvious vector space structure on $T_{(M,\varepsilon,\mathcal{G})}O(M, \mathcal{G})$.

3. We have assumed up to now the existence of le-paths and have derived some consequences. However, we have not explicitly constructed any le-paths, and indeed it is not at all obvious that there exist any le-paths other than $\nu(t) = \mathrm{Id}$ for all t.

We will circumvent the above difficulties by restricting (9.13) to a subcomplex and using some "special" le-paths.

Let $(T^* \otimes T)_{\mathcal{E}} \subset T^* \otimes T$ denote the space of $\mathcal{E}$-invariant (equivalently, $\widetilde{\nabla}$-parallel by Proposition 5.5) sections. For $\xi \in (T^* \otimes T)_{\mathcal{E}}$, we have

$$\widetilde{\nabla}_r \xi^i_j = \frac{\partial \xi^i_j}{\partial x^r} - \Gamma^i_{ra}\xi^a_j + \Gamma^a_{rj}\xi^i_a = 0 \tag{9.16}$$

Lemma 9.9 *Let $\xi \in (T^* \otimes T)$ satisfy $\widetilde{\nabla}\xi = 0$. Then the following are equivalent:*

(i) $D_2(\xi) = 0$;

(ii) $\xi(p)$ *is a derivation of the Lie algebra* $\widetilde{\mathfrak{X}}_{\mathcal{E}}(p)$ *for all* $p \in M$.

Proof We recall that $\xi(p)$ is a derivation of the Lie algebra $\widetilde{\mathfrak{X}}_{\mathcal{E}}(p)$ if and only if the identity

$$\{\xi(p)X, Y\}^{\mathcal{E}}_p + \{X, \xi(p)Y\}^{\mathcal{E}}_p - \xi(p)\{X, Y\}^{\mathcal{E}}_p = 0 \tag{9.17}$$

holds for all $X, Y \in T_pM$. Note that $\xi(p)$ is a derivation of $\widetilde{\mathfrak{X}}_{\mathcal{E}}(p)$ if and only if it is a derivation of $\mathfrak{X}_{\widetilde{\mathcal{E}}}(p)$. Using (6.2) and choosing $X = (\partial/\partial x^k)_{|p}$ and $Y = (\partial/\partial x^j)_{|p}$, we check that (9.17) is equivalent to

$$\xi^a_k T^i_{aj} + \xi^a_j T^i_{ka} - \xi^i_a T^a_{kj} = 0 \tag{9.18}$$

where the variables are evaluated at p. By (9.5), we have

$$(D_2(\xi))^i_{kj} = \frac{\partial \xi^i_k}{\partial x^j} - \Gamma^i_{ak}\xi^a_j - \frac{\partial \xi^i_j}{\partial x^k} - \Gamma^i_{aj}\xi^a_k = 0 \tag{9.19}$$

Now substituting $\partial\xi^i_k/\partial x^j$ and $\partial\xi^i_j/\partial x^k$ from (9.16) into (9.19), a straightforward computation shows that (9.19) becomes identical with (9.18), proving the claim. ■

In view of Lemma 9.9, we now have

$$(T^* \otimes T)_{\mathcal{E}} \xrightarrow{D_2} (\wedge^2 T^*) \otimes T \tag{9.20}$$

and the cocyles in $(T^* \otimes T)_{\mathcal{E}}$ are precisely derivations.

Now, the sections of $T \to M$ $(= \mathfrak{X}(M))$ satisfying $\widetilde{\nabla}X = 0$ are precisely elements of $\widetilde{\mathfrak{X}}_{\mathcal{E}}$, by the definition of $\widetilde{\mathfrak{X}}_{\mathcal{E}}$. Therefore, $T_{\mathcal{E}} = \widetilde{\mathfrak{X}}_{\mathcal{E}}$. Obviously, $\widetilde{\nabla}(\widetilde{\mathfrak{X}}_{\mathcal{E}}) = 0$, but note that the first operator D_1 in (9.13) is ∇ and not $\widetilde{\nabla}$! This point is quite subtle and will have a rather surprising consequence in Chapter 11. Therefore, the kernel of D_1 is $\widetilde{\mathfrak{X}}_{\mathcal{E}} \cap \mathfrak{X}_{\widetilde{\mathcal{E}}}$, and it is easy to check that some X belongs to $\widetilde{\mathfrak{X}}_{\mathcal{E}} \cap \mathfrak{X}_{\widetilde{\mathcal{E}}}$, if and only if $X(p)$ belongs to the center $C(\widetilde{\mathfrak{X}}_{\mathcal{E}}(p)) = C(\mathfrak{X}_{\widetilde{\mathcal{E}}}(p))$ of the Lie algebras $\widetilde{\mathfrak{X}}_{\mathcal{E}}(p) \simeq \mathfrak{X}_{\widetilde{\mathcal{E}}}(p)$.

We now claim that $D_1(T_\varepsilon) \subset (T^* \otimes T)_\varepsilon$. Indeed, since $\widetilde{\nabla}_j X^i = 0$, (6.15) gives

$$(D_1 X)^i_j = \nabla_j X^i = -T^i_{aj} X^a \tag{9.21}$$

Therefore, $\widetilde{\nabla} D_1(X) = 0$, since $\widetilde{\nabla} X = 0$ and $\widetilde{\nabla} T = 0$, by (6.1). We also see from (9.21) that $D_1 X$ is an inner derivation defined by X. We now have the diagram

$$\begin{array}{ccccccccc} 0 & \longrightarrow & \mathfrak{X}_{\widetilde{\varepsilon}} & \longrightarrow & T & \xrightarrow{D_1} & T^* \otimes T & \xrightarrow{D_2} & (\wedge^2 T^*) \otimes T \\ & & \cup & & \cup & & \cup & & \| \\ 0 & \longrightarrow & C(\mathfrak{X}_{\widetilde{\varepsilon}}) & \longrightarrow & T_\varepsilon & \xrightarrow{D_1} & (T^* \otimes T)_\varepsilon & \xrightarrow{D_2} & (\wedge^2 T^*) \otimes T \end{array} \tag{9.22}$$

We denote the cohomology of the lower sequence in (9.22) at $(T^* \otimes T)_\varepsilon$ by $H^*_{\widetilde{\nabla}=0}(D, T)$. Thus, we have proved that

$$H^0_{\widetilde{\nabla}=0}(D, T) \simeq C(\widetilde{\mathfrak{X}}_\varepsilon), \qquad H^1_{\widetilde{\nabla}=0}(D, T) \simeq \frac{\mathrm{Der}(\widetilde{\mathfrak{X}}_\varepsilon)}{\mathrm{IDer}(\widetilde{\mathfrak{X}}_\varepsilon)} \tag{9.23}$$

where $\mathrm{IDer}(\widetilde{\mathfrak{X}}_\varepsilon) \subset \mathrm{Der}(\widetilde{\mathfrak{X}}_\varepsilon)$ is the ideal of inner derivations. Our notation in (9.23) suggests that we also have the cohomology groups $H^*_{\nabla=0}(D, T)$, $H^*_{\nabla=0,\widetilde{\nabla}=0}(D, T)$, $H^*(\widetilde{D}, T)$, $H^*_{\widetilde{\nabla}=0}(\widetilde{D}, T)$, $H^*_{\nabla=0}(\widetilde{D}, T)$, and $H^*_{\nabla=0,\widetilde{\nabla}=0}(\widetilde{D}, T)$. We will study them in Chapter 11.

We will now define some special le- and ge-paths whose infinitesimal deformations each have an obvious vector space structure, with the quotient of these vector spaces being *isomorphic* to $H^1_{\widetilde{\nabla}=0}(D, T)$.

We recall that some $\nu(p) \in \mathcal{U}_1^{p,p}$ preserves $\underset{\sim}{T(p)} \Leftrightarrow \nu(p)$ preserves the point bracket at $p \Leftrightarrow \nu(p)$ is an automorphism of the Lie algebra $\widetilde{\mathfrak{X}}_\varepsilon \simeq \mathfrak{X}_{\widetilde{\varepsilon}}$.

Definition 9.10 *A gauge transformation ν is special if $\nu(p) \in \mathrm{Aut}(\widetilde{\mathfrak{X}}_\varepsilon(p)) = \mathrm{Aut}(\mathfrak{X}_{\widetilde{\varepsilon}}(p))$ for all $p \in M$.*

Therefore, the two LLGs $(M, \varepsilon, \mathcal{G})$ and $\nu(M, \varepsilon, \mathcal{G})$ have the same torsion everywhere, but (8.16) shows that they may have quite different 1-arrows. In particular, if they both globalize, their global transformations may be quite different. Now, $\varepsilon(q,p) \circ \nu(q) \circ \varepsilon(p,q)$ is an automorphism of $\widetilde{\mathfrak{X}}_\varepsilon(q)$ that may not be equal to $\nu(q)$; that is, ν may not be ε-invariant. We will be interested in gauge transformations that are special and ε-invariant. It is very easy to construct such gauge transformations: Choose some derivation at some point, exponentiate it to some automorphism at that point, and translate this automorphism over M by ε. We let $\mathrm{LE}^{s,\varepsilon}(M, \mathcal{G})$ denote the set of special ε-invariant le-paths; that is, an le-path $\underset{\sim}{\nu(t)}(M, \varepsilon, \mathcal{G})$ belongs to $\mathrm{LE}^{s,\varepsilon}(M, \mathcal{G})$ if and only if $\nu(t,p) \in \mathrm{Aut}(\widetilde{\mathfrak{X}}_\varepsilon(p))$ for all $p \in M$ and $\widetilde{\nabla}\nu = 0$. It is also easy to construct such paths: Choose some derivation at some point, multiply it by t, exponentiate it to an automorphism depending on t at that point, and translate this automorphism over M by ε. We observe that the infinitesimal deformation ξ of this le-path is also ε-invariant and is obtained by translating the original derivation over M by ε. We similarly define $\mathrm{GE}^{s,\varepsilon}(M, \mathcal{G}) \subset \mathrm{LE}^{s,\varepsilon}(M, \mathcal{G})$. Now the infinitesimal deformations of both

$\mathrm{LE}^{s,\mathcal{E}}(M, \mathcal{G})$ and $\mathrm{GE}^{s,\mathcal{E}}(M, \mathcal{G})$ have an obvious vector space structure. It remains to make the following definition:

Definition 9.11 *Let V denote the space of infinitesimal deformations of the paths in $\mathrm{LE}^{s,\mathcal{E}}(M, \mathcal{G})$ and W denote the space of infinitesimal deformations of the paths in $\mathrm{GE}^{s,\mathcal{E}}(M, \mathcal{G})$. We define*

$$T_{[(M,\varepsilon,\mathcal{G})]}(O(M, \mathcal{G})/\sim) \stackrel{\text{def}}{=} V/W \tag{9.24}$$

Using these definitions, we can now easily fill in the details of the following proposition:

Proposition 9.12 *We have the isomorphism*

$$\mathrm{KS} : T_{[(M,\varepsilon,\mathcal{G})]}(O(M, \mathcal{G})/\sim) \longrightarrow H^*_{\widetilde{\nabla}=0}(D, T) \tag{9.25}$$

We will conclude with three remarks.

(1) Does V/W in Definition 9.11 deserve to be identified as the tangent space that we are searching for? Could there be some le-paths whose infinitesimal deformations are not taken into account by (9.24)? We will leave it to the interested reader to contemplate this deep question.

(2) Since $\mathrm{IDer}(\widetilde{\mathfrak{X}}_\varepsilon) \subset \mathrm{Der}(\widetilde{\mathfrak{X}}_\varepsilon)$ is an ideal, the tangent space of the set $O(M, \mathcal{G})/\sim$ at the base point $[(M, \varepsilon, \mathcal{G})]$ is a Lie algebra. If we choose a new base point, then the tangent space at this new base point can be identified with the tangent space at the original point. We believe that $O(M, \mathcal{G})/\sim$ has a smooth structure, these identifications are 1-arrows of some local diffeomorphisms, and the set $O(M, \mathcal{G})/\sim$ is an LLG.

(3) According to Chapter 7, (M, ε) is equivalent to (M, w), where w is the structure object and $\mathcal{R} = 0$ if and only if w satisfies certain integrability conditions (namely, (7.18)). Thus, we studied in Chapters 8-9 the set of locally equivalent structure objects modulo global equivalence. Now we may replace w by any other geometric object (for instance, a metric g, a symplectic form ω, or an almost-complex structure J) and pose the same problem. However, the pseudogroups we deal with may no longer act simply transitively. We have the nonlinear and linear Spencer sequences also in this case, and the philosophy is the same: This moduli space is related to the first cohomology group by the KS map. To our surprise, we have not come across this simple philosophy in the literature, even though Spencer sequences have been studied extensively by a number of authors ([Sp], [GS1], [P1], [P2]).

10

The de Rham Cohomology of an LLG

This purpose of this auxiliary chapter is to reformulate some well-known facts from the theory of Lie groups in our framework as preparation for the next chapter.

We recall that a differential k-form ω on M is given locally as $\omega = (\omega_{i_1,\ldots,i_k}(x))$ on (U,x), where the functions $\omega_{i_1,\ldots,i_k}(x)$ are alternating in the indices $i_1,\ldots,i_k$. We can interpret these functions as the components of ω defined by

$$\omega = \frac{1}{k!}\omega_{i_1\ldots i_k}(x)dx^{i_1} \wedge \cdots \wedge dx^{i_k} \tag{10.1}$$

Let $\wedge^k = \wedge^k(M)$ denote the space of k-forms on M, where $\wedge^0 = C(M) =$ smooth functions on M. We have the exterior derivative

$$d : \wedge^k \longrightarrow \wedge^{k+1} \tag{10.2}$$

defined by

$$\begin{aligned}(d\omega)_{i_1\ldots i_{k+1}} &\overset{\text{def}}{=} \left[\frac{\partial \omega_{i_2\ldots i_{k+1}}}{\partial x^{i_1}}\right]_{[i_1\ldots i_{k+1}]} \\ &= \frac{\partial \omega_{i_2\ldots i_{k+1}}}{\partial x^{i_1}} - \frac{\partial \omega_{i_1 i_3\ldots i_{k+1}}}{\partial x^{i_2}} - \cdots - \frac{\partial \omega_{i_2 i_3\ldots i_k i_1}}{\partial x^{i_{k+1}}}\end{aligned} \tag{10.3}$$

where we have omitted the factor $1/(k+1)$ for notational simplicity. Equivalently, we may write in brief

$$\omega = \omega_I dx^I$$

and $d\omega$ can be defined as

$$d\omega = \frac{\partial \omega_I}{\partial x^a} dx^a \wedge dx^I = (d\omega_I) \wedge dx^I$$

We have $d^2 = 0$ and the de Rham complex

$$0 \longrightarrow \mathbb{R} \xrightarrow{d} \wedge^0 \xrightarrow{d} \wedge^1 \xrightarrow{d} \cdots \xrightarrow{d} \wedge^n \tag{10.4}$$

An Alternative Approach to Lie Groups and Geometric Structures, Ercüment H. Ortaçgil (2018).

noting that the set of locally constant functions on M is $\mathbb{R}$, since M is connected. Therefore, the "representation bundle" $T \to M$ in (9.13) is replaced by the trivial bundle $M \times \mathbb{R} \to M$ in (10.4). We invite the reader to keep an eye on the question whether the arguments in this chapter remain valid if we change the representation bundle.

We denote the cohomology of (10.4) by $H^*_{\mathrm{dR}}(M, \mathbb{R})$. Now (10.4) is locally exact and $H^*_{\mathrm{dR}}(M, \mathbb{R})$ is finite-dimensional if M is compact. Furthermore, this cohomology depends only on the homotopy type of M and is therefore a rather coarse invariant (see [BT]).

Now let (M, ε) be given and $\wedge^k_\varepsilon \subset \wedge^k$ denote the space of ε-invariant k-forms, or, equivalently, k-forms ω satisfying $\widetilde{\nabla}\omega = 0$. We have

$$\widetilde{\nabla}_r \omega_{i_1 \ldots i_k} = \frac{\partial \omega_{i_1 \ldots i_k}}{\partial x^r} + \Gamma^a_{ri_1} \omega_{ai_2 \ldots i_k} + \cdots + \Gamma^a_{ri_k} \omega_{i_1 \ldots i_{k-1}a} = 0 \tag{10.5}$$

Therefore, if $\widetilde{\nabla}\omega = 0$, then

$$\begin{aligned}(d\omega)_{ri_1 \ldots i_k} &= \left[\frac{\partial \omega_{i_1 \ldots i_k}}{\partial x^r}\right]_{[ri_1 \ldots i_k]} \\ &= -\left[\Gamma^a_{ri_1} \omega_{ai_2 \ldots i_k} + \cdots + \Gamma^a_{ri_k} \omega_{i_1 \ldots i_{k-1}a}\right]_{[ri_1 \ldots i_k]}\end{aligned} \tag{10.6}$$

The expression on the right-hand side of (10.6) is well known in its modern form. For instance, let $\omega = (\omega_{jk})$ be a 2-form. Then

$$\begin{aligned}-\left[\Gamma^a_{rj}\omega_{ak} + \Gamma^a_{rk}\omega_{ja}\right]_{[rjk]} &= -\Gamma^a_{rj}\omega_{ak} - \Gamma^a_{jk}\omega_{ar} - \Gamma^a_{kr}\omega_{aj} \\ &\quad - \Gamma^a_{rk}\omega_{ja} - \Gamma^a_{kj}\omega_{ra} - \Gamma^a_{jr}\omega_{ka} \\ &= -T^a_{rj}\omega_{ak} - T^a_{jk}\omega_{ar} - T^a_{kr}\omega_{aj}\end{aligned} \tag{10.7}$$

If $X = (X^i)$ and $Y = (Y^i)$, we recall that their torsion bracket is given by $T(X, Y)^i = T^i_{ab} X^a Y^b$. Since $\omega(X, Y) = \omega_{ab} X^a Y^b$, (10.7) shows that (10.6) is equivalent to

$$d\omega(X, Y, Z) = -\omega(T(X, Y), Z) - \omega(T(Y, Z), X) - \omega(T(Z, X), Y) \tag{10.8}$$

modulo the factor $\frac{1}{3}$ that we have absorbed in our notation $[rjk]$. It is standard to write (10.8) as

$$d\omega(X_0, X_1, X_3) = \sum_{0 \le i \lneqq j \le 3} (-1)^{i+j} \omega(T(X_i, X_j), X_0, \widehat{X}_i, \widehat{X}_j, X_3) \tag{10.9}$$

and we observe the well-known formula for the exterior derivative without the "representation term." More generally for any k-form, it is not difficult to show that the coordinate-free version of (10.6) is

$$d\omega(X_0, \ldots, X_k) = \sum_{0 \le i \lneqq j \le k} (-1)^{i+j} \omega(T(X_i, X_j), X_0, \ldots, \widehat{X}_i, \ldots, \widehat{X}_j, \ldots, X_k) \tag{10.10}$$

Therefore, we have the following proposition:

Proposition 10.1 *If $\omega \in \wedge^k_\varepsilon$, then $d\omega$ is given by (10.6) or equivalently by (10.10).*

Note that we do not have $d^2 = 0$, since $T(X, Y)$ does not satisfy the Jacobi identity unless $\mathfrak{R} = 0$ (see Proposition 6.4). For the moment, what matters for us is the fact that the right-hand side of (10.6) is a sum of terms that are products of the form $T\omega$ with various contractions as in (10.7). Now assuming $\mathfrak{R} = 0$ and $\widetilde{\nabla}\omega = 0$, we apply $\widetilde{\nabla}$ to (10.6). Since $\widetilde{\nabla}T = \mathfrak{R} = 0$ and $\widetilde{\nabla}$ is a derivation commuting with contractions, the right-hand side of (10.6) vanishes and we get $\widetilde{\nabla}(d\omega) = 0$. We have thus proved the following proposition:

Proposition 10.2 *Suppose $\mathfrak{R} = 0$. Then $\widetilde{\nabla}\omega = 0$ implies $\widetilde{\nabla}(d\omega) = 0$.*

Thus, if $\mathfrak{R} = 0$, Proposition 10.2 gives the subcomplex

$$0 \longrightarrow \mathbb{R} \xrightarrow{d} \wedge^0_\varepsilon \xrightarrow{d} \wedge^1_\varepsilon \xrightarrow{d} \ldots. \xrightarrow{d} \wedge^n_\varepsilon \tag{10.11}$$

Henceforth, we assume $\mathfrak{R} = 0$, so that $T(X, Y)$ satisfies the Jacobi identity by Proposition 6.4. We recall that $T(X, Y) = [X, Y]$ for $X, Y \in \widetilde{\mathfrak{X}}_\varepsilon$. Since the Lie algebra $\widetilde{\mathfrak{X}}_\varepsilon$ localizes to $\widetilde{\mathfrak{X}}_\varepsilon(p)$ at any $p \in M$, it follows that we can compute (10.11) at any point, and therefore (10.11) is a purely algebraic object. Following our notation in Chapter 9, we denote the cohomology of (10.11) by $H^*_{\widetilde{\nabla}=0}(d, \mathbb{R})$, where $\widetilde{\nabla} = 0$ stands for the fact that the sections of all spaces in (10.11) are $\widetilde{\nabla}$-parallel. With this notation, $H^*(d, \mathbb{R}) = H^*_{\mathrm{dR}}(M, \mathbb{R})$. If $\mathfrak{R} = 0$ and $(M, \varepsilon, \mathcal{G})$ is globalizable, we can also define $\wedge^k_{\widetilde{\varepsilon}} \subset \wedge^k$, and we easily show that this gives a subcomplex like (10.11) that can be identified with (10.11) and therefore computes the same cohomology.

Since $\wedge^k_\varepsilon \subset \wedge^k$, the inclusion maps induce the commutative diagram

$$\begin{array}{ccccccc}
\wedge^0 & \xrightarrow{d} & \wedge^1 & \xrightarrow{d} & \cdots & \xrightarrow{d} & \wedge^n \\
\uparrow & & \uparrow & & \uparrow & & \uparrow \\
\wedge^0_\varepsilon & \xrightarrow{d} & \wedge^1_\varepsilon & \xrightarrow{d} & \cdots & \xrightarrow{d} & \wedge^n_\varepsilon
\end{array} \tag{10.12}$$

Now our purpose is to prove the following proposition:

Proposition 10.3 *Let $(M, \varepsilon, \mathcal{G})$ be an LLG. If M is compact, then the inclusion maps in (10.12) induce an injection in cohomology. If $(M, \varepsilon, \mathcal{G})$ is globalizable (and M is connected), then they induce an isomorphism in cohomology.*

The proof depends on the important idea of "averaging" due to Weyl, which makes essential use of compactness and the existence of a unique Haar measure on a compact Lie group. However, some of these constructions work on any parallelizable manifold (M, ε) with compact M, as we shall now briefly outline. We fix some $p \in M$ and choose a nonzero n-form $\overline{\upsilon}(p)$ at the tangent space T_pM; that is, $\overline{\upsilon}(p) \in \wedge^n(T^*_p)$. We define the n-form $\overline{\upsilon}$ on M by translating $\overline{\upsilon}(p)$ over M using ε. Since M is compact, $\int_M \overline{\upsilon} = c$ for some $c \neq 0$.

We define $\upsilon \stackrel{\text{def}}{=} (1/c)\overline{\upsilon}$ and get $\int_M \upsilon = 1$. If σ is another ε-invariant n-form satisfying $\int_M \sigma = 1$, then $\sigma(p) = c\upsilon(p)$, and we get $c = 1$ by integrating both sides over M. We call υ the unique normalized volume element of the parallelizable (M, ε). By construction, $\widetilde{\nabla}\upsilon = 0$, since υ is ε-invariant. Now, for any subset $A \subset M$ satisfying the condition $U \subset A \subset \overline{U}$ for some open set U, where $\overline{U}$ denotes its closure, the volume $\upsilon(A)$ is defined by the formula $\upsilon(A) \stackrel{\text{def}}{=} \int_U \upsilon$. We call such subsets "nice." More generally, for any smooth function $\phi : M \to \mathbb{R}$, $\int_M \phi(x)\, d\upsilon$ is defined in terms of Riemann sums as follows: We partition M into nice subsets $A_1, \ldots, A_k$, choose points $a_i \in A_i$, consider the Riemann sum $\sum_{1\leq i\leq k} \phi(a_i)\upsilon(A_i)$, and show that such Riemann sums converge to the common finite limit $\int_M \phi(x)\, d\upsilon$ independently of the choice of the points $a_i \in A_i$ as we partition "more and more finely." Note that these arguments do not need $\mathfrak{R} = 0$.

Now, if $\mathfrak{R} = 0, f \in \mathcal{G}$ and $A \subset \mathrm{Dom}(f)$, then $\upsilon(A) = \upsilon(f(A))$ because the 1-arrows of f are equal to the 1-arrows of ε and υ is ε-invariant. So the pseudogroup $\mathcal{G}$ preserves the volume. Using this fact and the above Riemann sum definition, we easily see that

$$\int_A (\phi \circ f)(x)\, d\upsilon = \int_{f(A)} \phi(x)\, d\upsilon, \qquad f(A) \subset \mathrm{Dom}(\phi), \quad f \in \mathcal{G} \tag{10.13}$$

In particular, if $(M, \varepsilon, \mathcal{G})$ is globalizable, then υ becomes the unique Haar measure on the PHS $(M, \mathcal{G})$, and (10.13) becomes

$$\int_M \phi(f(x))\, d\upsilon = \int_M \phi(x)\, d\upsilon, \qquad \phi \in C(M), \quad f \in \mathcal{G} \tag{10.14}$$

For simplicity of notation let $E = \wedge^k$. For the first part of Proposition 10.3, we need the following properties of the vector bundle $E \to M$:

(i) $\varepsilon(p, q)$ induces an isomorphism $\sharp\varepsilon(p, q) : E_p \to E_q$, where E_p is the fiber over p.

(ii) $\sharp\varepsilon(q, r) \circ \sharp\varepsilon(p, q) = \sharp\varepsilon(p, r)$.

We call a section s of $E \to M$ ε-invariant if $\sharp\varepsilon(p, q)(s(p)) = s(q)$, $p, q \in M$. Let $E_\varepsilon \subset E$ be the space of ε-invariant sections, where, as always, we denote the space of sections of $E \to M$ by E.

Now we fix a base point e and define the map

$$\begin{aligned} \mathcal{I}_e &: E \longrightarrow E_e \\ &: s \longrightarrow \int_M \sharp\varepsilon(x, e)(s(x))\, d\upsilon \end{aligned} \tag{10.15}$$

For each $x \in M$, the integrand $\sharp\varepsilon(x, e)(s(x))$ belongs to the fiber E_e, which is a vector space. Therefore, the integral in (10.15) is defined in terms of Riemann sums as above. We now define the average map

$$\begin{aligned} &\Sigma : E \longrightarrow E_\varepsilon \\ &\Sigma(s)\,(x) \stackrel{\text{def}}{=} \sharp\varepsilon(e, x)\, \mathcal{I}_p(s) \end{aligned} \tag{10.16}$$

Clearly, $\Sigma(s) \in E_{\varepsilon}$, in view of (i) and (ii), and it is easy to see that Σ does not depend on the choice of e. If $s \in E_{\varepsilon}$, then $\Sigma(s) = s$, because $\sharp\varepsilon(x, e)(s(x)) = s(e)$ for all $x \in M$. Therefore, $\mathcal{I}(s) = s(e)$, since $\int_M d\upsilon = 1$ and $\Sigma(s)(x) = s(x)$.

Applying the construction of Σ to the sections of the spaces in the top sequence of (10.12), we get the diagram

$$\begin{array}{ccccccc} \wedge^0 & \xrightarrow{d} & \wedge^1 & \xrightarrow{d} & \cdots & \xrightarrow{d} & \wedge^n \\ \Sigma \downarrow \uparrow i & & \Sigma \downarrow \uparrow i & & \Sigma \downarrow \uparrow i & & \Sigma \downarrow \uparrow i \\ \wedge^0_{\varepsilon} & \xrightarrow{d} & \wedge^1_{\varepsilon} & \xrightarrow{d} & \cdots & \xrightarrow{d} & \wedge^n_{\varepsilon} \end{array} \tag{10.17}$$

where $\Sigma \circ i = \mathrm{Id}$. Note that the construction of Σ does not need $\mathfrak{R} = 0$, but without this assumption we do not have the bottom complex in (10.17). The crucial fact now is that Σ commutes with the exterior derivative d, that is,

$$\Sigma \circ d = d \circ \Sigma \tag{10.18}$$

because

$$\begin{aligned} (d \circ \Sigma)(\omega)(x) &= d\left(\sharp\varepsilon(e, x) \int_M \sharp\varepsilon(x, e)(\omega(x))\, d\upsilon\right) \\ &= \sharp\varepsilon(e, x)\left(d \int_M \sharp\varepsilon(x, e)(\omega(x))\, d\upsilon\right) \\ &= \sharp\varepsilon(e, x) \int_M d\left(\sharp\varepsilon(x, e)(\omega(x))\right) d\upsilon \\ &= \sharp\varepsilon(e, x) \int_M \sharp\varepsilon(x, e)(d\omega)(x))\, d\upsilon \\ &= (\Sigma \circ d)(\omega)(x) \end{aligned} \tag{10.19}$$

The second and fourth equalities in (10.19) follow from the fact that the exterior derivative commutes with local diffeomorphisms and the 1-arrows of ε are induced by the elements of $\mathcal{G}$. The third equality follows from the linearity of d and the definition of the integral as the limit of Riemann sums.

Now, since i and Σ both commute with d, they induce maps $H(i) : H^*(\widetilde{\mathfrak{X}}_{\varepsilon}) \to H^*(d, \mathbb{R})$ and $H(\Sigma) : H^*(d, \mathbb{R}) \to H^*(\widetilde{\mathfrak{X}}_{\varepsilon})$. We can now prove that $H(i)$ is injective: If $\omega \in \wedge^k_{\varepsilon}$, $d\omega = 0$, and $d\xi = \omega$ for some $\xi \in \wedge^{k-1}$, then $\Sigma\xi \in \wedge^{k-1}_{\varepsilon}$ and $d(\Sigma\xi) = \Sigma(d\xi) = \Sigma\omega = \omega$. Therefore, the closed ε-invariant form ω, which is exact in the top complex, is already exact in the bottom complex of (10.17). Thus, we have proved the first part of Proposition 10.3. The surjectivity of $H(i)$ is more subtle and requires the connectedness of M and the globalizability of $(M, \varepsilon, \mathcal{G})$. Let $\omega \in \wedge^k$ with $d\omega = 0$, so that we have the cohomology class $[\omega] \in H^k(d, \mathbb{R})$. It suffices to show that

$$[\Sigma(\omega)] = [\omega] \tag{10.20}$$

that is, ω and $\Sigma(\omega)$ differ by an exact form. Since $\Sigma(\omega)$ is ε-invariant, this implies that any cohomology class in $H^k(d, \mathbb{R})$ is represented by an ε-invariant form and therefore $H(i)$ is surjective. The idea of (10.20) is simple, but it takes some work to make this idea rigorous: We recall that a diffeomorphism $f : M \to M$ induces an isomorphism $H(f) : H^*(d, \mathbb{R}) \to H^*(d, \mathbb{R})$, and $H(f) = H(g)$ if f and g are homotopic. Now let $x \in M$ and choose a path $C(t)$, $0 \leq t \leq 1$, from the basepoint e to x. Now $\varepsilon(e, C(t))$ is the 1-arrow of the unique *global* diffeomorphism $f_t^x \in \mathcal{G}$, $f_0^x = \mathrm{Id}$, $f_1^x(e) = x$. Hence, f_1 is homotopic to $f_0 = \mathrm{Id}$. Therefore, if $\omega \in \wedge^k$ and $d\omega = 0$, then $[\omega] = [(f_1^x)^* \omega]$, and $[(f_1^x)^* \omega]$ depends only on the point x but not on the path. Now, if $x_1, \ldots, x_k \in M$, we get

$$[\omega] = \left[\frac{1}{k}\left((f^{x_1})^*\omega + \cdots + (f^{x_k})^*\omega\right)\right] \tag{10.21}$$

and the idea is to prove (10.20) as a limit of (10.21) as $k \to \infty$, using the definition of Σ.

If $(M, \varepsilon, \mathcal{G})$ is globalizable, there is a complete symmetry between $\widetilde{\varepsilon}$ and ε, and all of the above arguments remain valid if we replace ε by $\widetilde{\varepsilon}$. In particular, we can construct a unique μ with $\int_M \mu = 1$ if M is compact. It turns out that $\mu = \upsilon$: (4.21) gives $\det \mathrm{Ad}(e, xy) = \det \mathrm{Ad}(e, x) \det \mathrm{Ad}(e, y)$, and therefore $\det \mathrm{Ad}(e, \cdot) : M \to \mathbb{R}$ is a homomorphism. Since M is compact and connected, we have $\det \mathrm{Ad}(e, M) \subset \{-1, 1\}$ and $\det \mathrm{Ad}(e, x) = 1$ for all $x \in M$. Therefore, the translations of some n-form at some point over M by $\widetilde{\varepsilon}$ and ε define the same n-form on M, which implies $\mu = \upsilon$. This fact allows another construction as follows: Let $\wedge^k_{\varepsilon,\widetilde{\varepsilon}} \overset{\text{def}}{=} \wedge^k_{\varepsilon} \cap \wedge^k_{\widetilde{\varepsilon}}$. Clearly, $\wedge^*_{\varepsilon,\widetilde{\varepsilon}}$ is stable under d, and we denote its cohomology by $H^*_{\widetilde{\nabla}=0,\nabla=0}(d, \mathbb{R})$. At this point, we invite the reader to go carefully over the proof of the well-known *first* isomorphism in

$$H^*_{\widetilde{\nabla}=0,\nabla=0}(d, \mathbb{R}) \simeq H^*(d, \mathbb{R}) \simeq H^*_{\nabla=0}(d, \mathbb{R}) \simeq H^*_{\widetilde{\nabla}=0}(d, \mathbb{R}) \tag{10.22}$$

which becomes *false* if we change the representation bundle, as can already be seen from (9.23): We have $H^0_{\widetilde{\nabla}=0}(D, T) = C(\widetilde{\mathfrak{X}}_\varepsilon) = H^0_{\widetilde{\nabla}=0,\nabla=0}(D, T)$, which need not be isomorphic to $H^0(D, T) = \mathfrak{X}_{\widetilde{\varepsilon}}$, even if M is compact! In the same way, $H^1_{\widetilde{\nabla}=0,\nabla=0}(D, T)$ need not be isomorphic to $H^1_{\widetilde{\nabla}=0}(D, T) \simeq \mathrm{Der}(\mathfrak{g})/\mathrm{IDer}(\mathfrak{g})$, $\mathfrak{g} = \widetilde{\mathfrak{X}}_\varepsilon$, either. To see this, let $\xi \in \mathrm{Der}(\mathfrak{g}) \subset T^* \otimes T$ satisfy $\widetilde{\nabla}\xi = \nabla\xi = 0$. Therefore, in addition to (9.16), we also have

$$\nabla_r \xi^i_j = \frac{\partial \xi^i_j}{\partial x^r} - \Gamma^i_{ar}\xi^a_j + \Gamma^a_{jr}\xi^i_a = 0 \tag{10.23}$$

Subtracting (10.23) from (9.16) gives $T^i_{ar}\xi^a_j + T^a_{rj}\xi^i_a = 0$. Interchanging r, j and adding, we deduce $T^i_{ar}\xi^a_j + T^i_{aj}\xi^a_r = 0$. Therefore, $T^a_{rj}\xi_a = 0$ by (9.18), which means that the derivation $\xi \in \mathrm{Der}(\mathfrak{g})$ vanishes on $[\mathfrak{g}, \mathfrak{g}]$. Since the inner derivations defined by D_1 vanish on the center $C(\mathfrak{g})$, we conclude that

$$H^1_{\tilde{\nabla}=0,\nabla=0}(D,T) \simeq \{\xi \in \mathrm{Der}(\mathfrak{g}) \mid \xi(x) = 0,\ x \in [\mathfrak{g}, \mathfrak{g}]\} \tag{10.24}$$

and (10.24) need not be isomorphic tc $\mathrm{Der}(\mathfrak{g})/\mathrm{IDer}(\mathfrak{g})$, even if M is compact.

11

The Linear Spencer Sequence

In this chapter, we extend (9.13) to a full complex $C^*(D, T)$ called the linear Spencer sequence of the LLG $(M, \varepsilon, \mathcal{G})$ and compute its cohomology. We also define the "dual" complex $C^*(\widetilde{D}, T)$ for *any* (M, ε) and the various subcomplexes $C^*_{\nabla=0}(D, T)$, $C^*_{\widetilde{\nabla}=0}(D, T)$, $C^*_{\widetilde{\nabla}=0,\nabla=0}(D, T)$, $C^*_{\widetilde{\nabla}=0}(\widetilde{D}, T)$, etc. We observe that these complexes can be defined if we replace $T \to M$ by any tensor bundle $T^{r,s} \to M$. In the introduction to [Ks], Kostant states that it is neither easy to compute nor easy to interpret geometrically the cohomology $H^*(\mathfrak{g}, \rho, V)$ of a Lie algebra with a representation $\rho : \mathfrak{g} \to gl(V)$ except in some well-known cases, and he makes a significant contribution to this problem in [Ks]. We hope that this chapter will shed some further light on the geometric meaning of this purely algebraic concept.

We start with the dual complex $C^*(\widetilde{D}, T)$. Let $\omega = (\omega^i_{j_1\ldots j_k}(x))$ be a section of $(\wedge^k T^*) \otimes T \to M$. We define the local operator $\widetilde{d}_r$ by the formula

$$\widetilde{d}_r \omega^i_{j_1\ldots j_k} \overset{\text{def}}{=} \frac{\partial \omega^i_{j_1\ldots j_k}}{\partial x^r} - \Gamma^i_{ra}\omega^a_{j_1\ldots j_k} \tag{11.1}$$

Note that $\widetilde{d}_r$ is "covariant differentiation" with respect to the "representation index" i, but does not touch the "form indices" $j_1, \ldots, j_k$ like $\widetilde{\nabla}_r \omega^i_{j_1\ldots j_k}$. For $k = 0$, $\widetilde{d}_r = \widetilde{\nabla}_r$. For $k \geq 1$, (11.1) is not a tensor. We define

$$\begin{aligned} \widetilde{D} : (\wedge^k T^*) \otimes T &\longrightarrow (\wedge^{k+1} T^*) \otimes T \\ : \omega^i_{j_1\ldots j_k} &\longrightarrow (\widetilde{D}\omega)^i_{rj_1\ldots j_k} \end{aligned} \tag{11.2}$$

where

$$\begin{aligned} (\widetilde{D}\omega)^i_{rj_1\ldots j_k} &\overset{\text{def}}{=} \left[\widetilde{d}_r \omega^i_{j_1\ldots j_k}\right]_{[rj_1\ldots j_k]} \\ &= \widetilde{d}_r \omega^i_{j_1 j_2\ldots j_k} - \widetilde{d}_{j_1} \omega^i_{rj_2\ldots j_k} - \widetilde{d}_{j_2} \omega^i_{j_1 r j_3\ldots j_k} - \cdots - \widetilde{d}_{j_k} \omega^i_{j_1 j_2\ldots j_{k-1} r} \end{aligned} \tag{11.3}$$

An Alternative Approach to Lie Groups and Geometric Structures, Ercüment H. Ortaçgil (2018).

Setting $r = j_0$, we can write (11.3) in the equivalent form

$$(\widetilde{D}\omega)^i_{j_0 j_1 \cdots j_k} = \sum_{0\le m\le k} (-1)^m \widetilde{d}_{j_m} \omega^i_{j_0 j_1 \cdots j_{m-1} j_{m+1} \cdots j_k} \tag{11.4}$$

We can define $\widetilde{D}$ in coordinate-free form as follows: We have the exterior derivative $d : \wedge^k T^* \to \wedge^{k+1} T^*$ and the operator $\widetilde{\nabla} : T \to T^* \otimes T$ defined by $(\mu^i) \to (\widetilde{\nabla}_r \mu^i)$. We define $\widetilde{D}$ by requiring

$$\widetilde{D}(\omega \otimes \mu) = (d\omega) \otimes \mu + (-1)^k \omega \wedge (\widetilde{\nabla}\mu) \tag{11.5}$$

If we write

$$\xi = \frac{1}{k!}\xi^a_{j_1 \cdots j_k} \frac{\partial}{\partial x^a} \otimes dx^{j_1} \wedge \cdots \wedge dx^{j_k} \tag{11.6}$$

and apply $\widetilde{D}$ to (11.6) using (11.5), we get (11.3) (modulo some constants). Thus, we get the sequence

$$0 \longrightarrow \widetilde{\mathfrak{X}}_\varepsilon \longrightarrow T \xrightarrow{\widetilde{D}_1} T^* \otimes T \xrightarrow{\widetilde{D}_2} (\wedge^2 T^*) \otimes T \xrightarrow{\widetilde{D}_2} \cdots \xrightarrow{\widetilde{D}_n} (\wedge^n T^*) \otimes T \tag{11.7}$$

Proposition 11.1 $\widetilde{D}^2 = 0$

Proof Using the definitions of $\widetilde{D}_1$ and $\widetilde{D}_2$ by (11.3), a straightforward computation gives

$$((\widetilde{D}_2 \circ \widetilde{D}_1)(\xi))^i_{jk} = -\widetilde{\mathfrak{R}}^i_{jk,a}\xi^a \tag{11.8}$$

Therefore, $\widetilde{D}_2 \circ \widetilde{D}_1 = 0$, since $\widetilde{\mathfrak{R}} = 0$ by Proposition 5.6. Now it is easy to show that (11.7) is a complex using (11.8) and (11.5). ■

We denote the complex (11.7) by $C^*(\widetilde{D}, T)$, (10.4) by $C^*(d, \wedge^0)$, and their cohomologies by $H^*(\widetilde{D}, T)$ and $H^*(d, \wedge^0) = H^*_{\text{dR}}(M, \mathbb{R})$, respectively. At first sight, it seems that we have assigned a new cohomology $H^*(\widetilde{D}, T)$ to any (M, ε). Unfortunately, this is not the case.

Proposition 11.2 $H^k(\widetilde{D}, T) \simeq \oplus_n H^k(d, \wedge^0), 0 \le k \le n.$

Proof We choose a basis $\sigma_{(k)} = (\sigma^i_{(k)})$ of $\widetilde{\mathfrak{X}}_\varepsilon$ and the dual basis $\eta^{(k)} = \left(\eta^{(k)}_i\right) \in \widetilde{\mathfrak{X}}^*_\varepsilon$, $1 \le k \le n$. Therefore, $\sigma^a_{(k)}\eta^{(m)}_a = \delta^m_k$ and $\widetilde{\nabla}\eta^{(k)} = 0$; that is

$$\widetilde{\nabla}_r \eta^{(k)}_i = \frac{\partial \eta^{(k)}_i}{\partial x^r} + \Gamma^a_{ri}\eta^{(k)}_a = 0, \qquad 1 \le k \le n \tag{11.9}$$

We define

$$\begin{aligned} \alpha &: (\wedge^s T^*) \otimes T \to \oplus_n \wedge^s T^* \\ &: (\omega^i_{j_1 \cdots j_s}) \longrightarrow (\eta^{(1)}_a \omega^a_{j_1 \cdots j_s}, \ldots, \eta^{(n)}_a \omega^a_{j_1 \cdots j_s}) \end{aligned} \tag{11.10}$$

Clearly, α is linear. It is injective because if $X(p) \in T_pM$ and $\eta^{(k)}(p)(X(p)) = 0$, then $X(p) = 0$. It is also surjective: Let $\omega = (\omega^{(1)}, \ldots, \omega^{(n)}) \in \oplus_n \wedge^s T^*$. We define $\xi \in (\wedge^s T^*) \otimes T$ by

$$\xi^i_{j_1\ldots j_s} \stackrel{\text{def}}{=} \sum_{1\leq k\leq n} \sigma^i_{(k)} \omega^{(k)}_{j_1\ldots j_s} \tag{11.11}$$

and get $\alpha(\xi) = \omega$. Now we have the diagram

$$\begin{array}{ccc} (\wedge^k T^*) \otimes T & \xrightarrow{\widetilde{D}} & (\wedge^{k+1} T^*) \otimes T \\ \alpha\big\downarrow & & \alpha\big\downarrow \\ \oplus_n \wedge^k T^* & \xrightarrow{\oplus d} & \oplus_n \wedge^{k+1} T^* \end{array} \tag{11.12}$$

where the vertical maps are isomorphisms. We claim that (11.12) commutes. Denoting the mth component of α by α_m, we compute

$$\begin{aligned} \left(\alpha_m \circ \widetilde{D}\right) \omega^i_{j_1\ldots j_k} &= \left[\eta^{(m)}_a \frac{\partial \omega^c_{j_1\ldots j_k}}{\partial x^r} - \eta^{(m)}_a \Gamma^a_{rb} \omega^b_{j_1\ldots j_k}\right]_{[rj_1\ldots j_k]} \\ &= \left[\frac{\partial \eta^{(m)}_a}{\partial x^r} \omega^a_{j_1\ldots j_k} + \eta^{(m)}_a \frac{\partial \omega^a_{j_1\ldots j_k}}{\partial x^r}\right]_{[rj_1\ldots j_k]} \\ &= \left[\frac{\partial}{\partial x^r} (\eta^{(m)}_a \omega^a_{j_1\ldots j_k})\right]_{[rj_1\ldots j_k]} \\ &= ((d \circ \alpha_m)(\omega))_{rj_1\ldots j_k} \end{aligned} \tag{11.13}$$

where we have used (11.9) in the second equality. Therefore, $\alpha \circ \widetilde{D} = d \circ \alpha$, and α induces the asserted isomorphism in cohomology. ∎

Note that the isomorphism given by Proposition 11.2 is not canonical. To summarize, (11.7) is nothing but n copies of the de Rham complex. In particular, (11.7) is locally exact.

Now let $((\wedge^s T^*) \otimes T)_\varepsilon \subset (\wedge^s T^*) \otimes T$ be the subspace of ε-invariant forms. If $\mathfrak{R} = 0$, these forms are stable under $\widetilde{D}$ because

$$\widetilde{\nabla}_r \omega^i_{j_1\ldots j_k} = \widetilde{d}_r \omega^i_{j_1\ldots j_k} + \Gamma^a_{rj_1} \omega^i_{aj_2\ldots j_k} + \cdots + \Gamma^a_{rj_k} \omega^i_{j_1 j_2\ldots a} \tag{11.14}$$

Therefore, if $\widetilde{\nabla}_r \omega^i_{j_1\ldots j_k} = 0$, then

$$(\widetilde{D}\omega)^i_{rj_1\ldots j_k} = -\left[\Gamma^a_{rj_1} \omega^i_{aj_2\ldots j_k} + \cdots + \Gamma^a_{rj_k} \omega^i_{j_1 j_2\ldots a}\right]_{[rj_1\ldots j_k]} \tag{11.15}$$

The only difference between (10.6) and (11.15) is that d is replaced with $\widetilde{D}$ and the $\mathbb{R}$-valued ω is replaced with the T-valued ω. Therefore, if we apply $\widetilde{\nabla}$ to (11.15), the right-hand side vanishes since $\widetilde{\nabla}\omega = 0$ and $\widetilde{\nabla}T = \mathfrak{R} = 0$, as in the derivation of Proposition 10.2. Thus, we obtain the following proposition:

Proposition 11.3 *Let $(M, \varepsilon, \mathcal{G})$ be an LLG. Then, we have the subcomplex*

$$T_{\varepsilon} \xrightarrow{\widetilde{D}_1} \left(T^* \otimes T\right)_{\varepsilon} \xrightarrow{\widetilde{D}_2} \left((\wedge^2 T^*) \otimes T\right)_{\varepsilon} \xrightarrow{\widetilde{D}_3} \cdots \xrightarrow{\widetilde{D}_n} \left((\wedge^n T^*) \otimes T\right)_{\varepsilon} \tag{11.16}$$

of (11.7), where $T_{\varepsilon} = \widetilde{\mathfrak{X}}_{\varepsilon}$.

We denote the cohomology of (11.16) by $H^*_{\widetilde{\nabla}=0}(\widetilde{D}, T)$. It is crucial to observe that $\widetilde{D}_1$ vanishes on T_{ε} because $\widetilde{D}_1 = \widetilde{\nabla}$ and $T_{\varepsilon} = \widetilde{\mathfrak{X}}_{\varepsilon}$. Therefore, $H^0_{\widetilde{\nabla}=0}(\widetilde{D}, T) = \widetilde{\mathfrak{X}}_{\varepsilon}$.

Now (11.16) localizes at any point $p \in M$ and is an algebraic object. Let us write its localization at p as

$$T_{\varepsilon}(p) \xrightarrow{\widetilde{D}_1(p)} \left(T^* \otimes T\right)_{\varepsilon}(p) \xrightarrow{\widetilde{D}_2(p)} \cdots \xrightarrow{\widetilde{D}_n(p)} \left((\wedge^n T^*) \otimes T\right)_{\varepsilon}(p) \tag{11.17}$$

where $T_{\varepsilon}(p) = \widetilde{\mathfrak{X}}_{\varepsilon}(p) \simeq \widetilde{\mathfrak{X}}_{\varepsilon}$. All localizations are isomorphic in the obvious sense by an isomorphism induced by ε, and they all compute $H^*_{\widetilde{\nabla}=0}(\widetilde{D}, T)$.

Now recalling the definition of the cohomology $H^*(\mathfrak{g}, \rho, V)$ of a (finite-dimensional) abstract Lie algebra $\mathfrak{g}$ with the representation $\rho : \mathfrak{g} \to gl(V)$, we have the following trivial proposition:

Proposition 11.4 (11.17) *computes the cohomology of the Lie algebra $\mathfrak{g} = \widetilde{\mathfrak{X}}_{\varepsilon}(p)$ with respect to the trivial representation $t : \mathfrak{g} \to gl(\mathfrak{g})$.*

The proof is immediate from (11.15) and (10.6)–(10.10). Indeed, since $T(X_i, X_j) = [X_i, X_j]$, (10.10) is the well known coboundary operator on $V = \widetilde{\mathfrak{X}}_{\varepsilon}(p)$-valued cochains ω. Therefore, Proposition 11.4 states

$$H^*_{\widetilde{\nabla}=0}(\widetilde{D}, T) \simeq H^*(\mathfrak{g}, t, \mathfrak{g}) = \oplus_n H^*(\mathfrak{g}, \mathbb{R}) \tag{11.18}$$

where $H^*(\mathfrak{g}, \mathbb{R})$ denotes the Lie algebra cohomology of $\mathfrak{g}$ computed with $\mathbb{R}$-valued cochains using (10.10) as usual. If M is compact, we deduce from Proposition 11.2 the extra second isomorphism in

$$\oplus_n H^k(d, \wedge^0) \simeq H^k(\widetilde{D}, T) \simeq H^*_{\widetilde{\nabla}=0}(\widetilde{D}, T) = H^*(\mathfrak{g}, t, \mathfrak{g}) = \oplus_n H^*(\mathfrak{g}, \mathbb{R}) \tag{11.19}$$

In fact, the isomorphism between the first and last terms is induced by α in the proof of Proposition 11.2.

We now come to the linear Spencer sequence. It is quite obvious how we shall proceed. We define d_r by

$$\begin{aligned} d_r \omega^i_{j_1 j_2 \ldots j_k} &\overset{\text{def}}{=} \frac{\partial \omega^i_{j_1 j_2 \ldots j_k}}{\partial x^r} - \Gamma^i_{ar} \omega^a_{j_1 j_2 \ldots j_k} \\ &= \widetilde{d}_r \omega^i_{j_1 j_2 \ldots j_k} - T^i_{ar} \omega^a_{j_1 j_2 \ldots j_k} \end{aligned} \tag{11.20}$$

Repeating the above construction with d_r instead of $\widetilde{d}_r$, we get the sequence

$$0 \longrightarrow \mathfrak{X}_{\widetilde{\varepsilon}} \longrightarrow T \xrightarrow{D_1} T^* \otimes T \xrightarrow{D_2} (\wedge^2 T^*) \otimes T \xrightarrow{D_3} \cdots \xrightarrow{D_n} (\wedge^n T^*) \otimes T \quad (11.21)$$

However this is not a complex unless $\mathfrak{R} = 0$. Now (11.8) is replaced with (9.12).

Definition 11.5 *Suppose $\mathfrak{R} = 0$. The complex $C^*(D, T)$ defined by (11.21) is the linear Spencer sequence of the LLG $(M, \varepsilon, \mathcal{G})$.*

We now assume that $(M, \varepsilon, \mathcal{G})$ is globalizable. With this assumption there is a complete symmetry between (11.7) and (11.21). In particular, the local splitting $\widetilde{\varepsilon}$ in Chapter 4 is now a global splitting, and the proof of Proposition 11.2 works through with (11.21). We denote the cohomology of (11.21) by $H^*(D, T)$.

Thus, we deduce the following corollary:

Corollary 11.6 *Let $(M, \varepsilon, \mathcal{G})$ be a globalizable LLG. Then $H^*(D, T) \simeq H^*(\widetilde{D}, T)$.*

By Corollary 6.5, if $\mathfrak{R} = 0$, then $\nabla T = 0$. Therefore, $\nabla \omega = 0$ implies $\nabla(D(\omega)) = 0$ in the same way as $\widetilde{\nabla}\omega = 0$ implies $\widetilde{\nabla}(\widetilde{D}\omega) = 0$, as shown above. Thus, we get the cohomology $H^*_{\nabla=0}(D, T)$, which is in complete symmetry with $H^*_{\widetilde{\nabla}=0}(\widetilde{D}, T)$ if $(M, \varepsilon, \mathcal{G})$ is globalizable. With this assumption, we obtain the isomorphism (11.18) with $\widetilde{D}$ and $\widetilde{\nabla}$ replaced by D and ∇, respectively, and if M is compact, we obtain the analog of (11.19).

Now we recall $\mathfrak{g} = \widetilde{\mathfrak{X}}_\varepsilon(p) \simeq \widetilde{\mathfrak{X}}_\varepsilon \simeq \mathfrak{X}_{\widetilde{\varepsilon}}$ and define the "true" representation $\rho : \mathfrak{g} \to \mathrm{gl}(\mathfrak{g})$ by $\rho(x)(y) = [x, y]$ and consider $H^*(\mathfrak{g}, \rho, \mathfrak{g})$. Now it is easy to guess from (9.22) and (9.23) that $C^*_{\widetilde{\nabla}=0}(D, T)$ is a complex with cohomology $H^*_{\widetilde{\nabla}=0}(D, T) \simeq H^*(\mathfrak{g}, \rho, \mathfrak{g})$. It is worth checking here the details of this intriguing fact. So, let $\omega \in (\wedge^k T^*) \otimes T$. Now, from (11.15) and (11.20), we deduce

$$\widetilde{\nabla}_r \omega^i_{j_1 \ldots j_k} = d_r \omega^i_{j_1 \ldots j_k} - T^i_{ra}\omega^i_{j_1 \ldots j_k} + \Gamma^a_{rj_1}\omega^i_{aj_2 \ldots j_k} + \cdots + \Gamma^a_{rj_k}\omega^i_{j_1 j_2 \ldots a} \quad (11.22)$$

Therefore, if $\widetilde{\nabla}\omega = 0$, then

$$\begin{aligned}(D_r\omega)^i_{j_1 \ldots j_k} &= \left[T^i_{ra}\omega^a_{j_1 \ldots j_k}\right]_{[rj_1 \ldots j_k]} \\ &\quad - \left[\Gamma^a_{rj_1}\omega^i_{aj_2 \ldots j_k} + \cdots + \Gamma^a_{rj_k}\omega^i_{j_1 j_2 \ldots a}\right]_{[rj_1 \ldots j_k]}\end{aligned} \quad (11.23)$$

Comparing (11.15) and (11.23), we see that the latter has an extra first term on the right-hand side. We recall that the second term on the right-hand side of (11.23) is (10.10) in coordinate-free form, with the only difference being as explained above. Applying $\widetilde{\nabla}$ to (11.23), we get the following proposition:

Proposition 11.7 *If $\widetilde{\nabla}\omega = 0$, then $\widetilde{\nabla}(D\omega) = 0$.*

Proposition 11.7 gives the subcomplex

$$T_\varepsilon \xrightarrow{D_1} \left(T^* \otimes T\right)_\varepsilon \xrightarrow{D_2} \left((\wedge^2 T^*) \otimes T\right)_\varepsilon \xrightarrow{D_3} \cdots \xrightarrow{D_n} \left((\wedge^n T^*) \otimes T\right)_\varepsilon \quad (11.24)$$

of (11.21). Clearly (11.24) is the extension of the bottom sequence of (9.22). We note again that $T_{\varepsilon} = \widetilde{\mathfrak{X}}_{\varepsilon}$ but not $\mathfrak{X}_{\widetilde{\varepsilon}}$, and $D_1 = \nabla$ but not $\widetilde{\nabla}$.

Now the first term on the right-hand side of (11.23) is another expression that we know very well in its modern form. For $k = 3$, for instance,

$$\left[T^i_{ra}\omega^a_{jms}\right]_{[rjms]} = T^i_{ra}\omega^a_{jms} - T^i_{ja}\omega^a_{rms} - T^i_{ma}\omega^a_{jrs} - T^i_{sa}\omega^a_{jmr} \tag{11.25}$$

which we can write as

$$\sum_{0\leq m\leq 3} (-1)^m T^i_{j_m a}\omega^a_{j_0\ldots\widehat{j_m}\ldots j_3} \tag{11.26}$$

or in coordinate-free form as

$$\sum_{0\leq m\leq 3} (-1)^m T(X_m, \omega(X_0, .., \widehat{X}_m, \ldots, X_3)) \tag{11.27}$$

where $T(X, Y)$ is the torsion bracket and $T(X, Y) = [X, Y]$ for $X, Y \in \widetilde{X}_{\varepsilon}$. We will leave it to the interested reader to check that the coordinate-free expression for the first term on the right-hand side of (11.23) is

$$\sum_{0\leq i\leq k} (-1)^i T(X_i, \omega(X_0, \ldots, \widehat{X}_i, \ldots, X_k)) \tag{11.28}$$

which is the "representation term" in the definition of $H^*(\mathfrak{g}, \rho, \mathfrak{g})$. Therefore, the distinction between the cohomology of the trivial and true representations of $\mathfrak{g}$ emerges from the tension between ∇ and $\widetilde{\nabla}$, that is, from left/right!

We summarize the above argument in the following proposition:

Proposition 11.8 (11.24) *is isomorphic to the deformation complex* $C^*(\mathfrak{g}, \rho, \mathfrak{g})$*, and therefore* $H^*_{\widetilde{\nabla}=0}(D, T) \simeq H^*(\mathfrak{g}, \rho, \mathfrak{g})$.

To summarize, a globalizable local Lie group $(M, \varepsilon, \mathcal{G})$ gives the diagram

$$\begin{array}{ccccccccccc}
0 & \longrightarrow & \mathfrak{X}_{\widetilde{\varepsilon}} & = & T_{\widetilde{\varepsilon}} & \xrightarrow{D_1} & (T^*\otimes T)_{\widetilde{\varepsilon}} & \xrightarrow{D_2} & \cdots & \xrightarrow{D_n} & ((\wedge^n T^*)\otimes T)_{\widetilde{\varepsilon}} \\
 & & \cap & & \cap & & \cap & & \cap & & \cap \\
0 & \longrightarrow & \mathfrak{X}_{\widetilde{\varepsilon}} & \longrightarrow & T & \xrightarrow{D_1} & (T^*\otimes T) & \xrightarrow{D_2} & \cdots & \xrightarrow{D_n} & ((\wedge^n T^*)\otimes T) \\
 & & \| & & \cup & & \cup & & \cup & & \cup \\
0 & \longrightarrow & C(\mathfrak{X}_{\widetilde{\varepsilon}}) & \longrightarrow & T_{\varepsilon} & \xrightarrow{D_1} & (T^*\otimes T)_{\varepsilon} & \xrightarrow{D_2} & \cdots & \xrightarrow{D_n} & ((\wedge^n T^*)\otimes T)_{\varepsilon}
\end{array} \tag{11.29}$$

The middle subcomplex is the linear Spencer sequence $C^*(D,T)$, the top subcomplex is $C^*_{\nabla=0}(D,T)$ and computes the trivial cohomology $H^*(\mathfrak{g},t,\mathfrak{g})$, and the bottom subcomplex is $C^*_{\widetilde{\nabla}=0}(D,T)$ and computes the deformation cohomology $H^*(\mathfrak{g},\rho,\mathfrak{g})$. We can replace the middle subcomplex by the dual complex $C^*(\widetilde{D},T)$ defined on any (M,ε) by (11.7), the top subcomplex by $C^*_{\widetilde{\nabla}=0}(\widetilde{D},T)$, and the bottom complex by $C^*_{\nabla=0}(\widetilde{D},T)$ with the assumption $\mathfrak{R}=0$. This gives a diagram isomorphic to (11.29).

It is straightforward at this stage to observe that we can replace the representation bundle $T\to M$ by any tensor bundle $T^{k,m}\to M$. We define $\widetilde{d}_r$ and d_r on $\omega=\left(\omega^{s_1\ldots s_k}_{l_1\ldots l_m,j_1\ldots j_t}\right)=\left(\omega^S_{L,J}\right)\in\wedge^t(T^*)\otimes T^{k,m}$ by differentiating covariantly only with respect to the representation indices S and L in the obvious way. All the arguments work through, and we get (11.29) where the complexes $C^*(D,T)$, $C^*_{\nabla=0}(D,T)$ and $C^*_{\widetilde{\nabla}=0}(D,T)$ are replaced by $C^*(D,T^{k,m})$, $C^*_{\nabla=0}(D,T^{k,m})$ and $C^*_{\widetilde{\nabla}=0}(D,T^{k,m})$. Now,

(i) $C^*(D,T^{k,m})$ computes $\oplus_{\dim T^{k,m}(p)}H^*(d,\wedge^0)$;

(ii) $C^*_{\nabla=0}(D,T^{k,m})$ computes $H^*(\mathfrak{g},t,T^{k,m}(\mathfrak{g})\simeq\oplus_{\dim T^{k,m}(p)}H^*(\mathfrak{g},\mathbb{R})$, where $t:\mathfrak{g}\to gl(T^{k,m}(\mathfrak{g}))$ is the trivial representation;

(iii) $C^*_{\widetilde{\nabla}=0}(D,T^{k,m})$ computes $H^*(\mathfrak{g},\rho,T^{k,m}(\mathfrak{g}))$, where $\rho:\mathfrak{g}\to gl(T^{k,m}(\mathfrak{g}))$ is induced by the "true" representation $\mathfrak{g}\to gl(\mathfrak{g})$.

We believe that, in principle, the cohomology of *any* abstract Lie algebra with respect to *any* representation (all finite-dimensional) arises in this way from tensor representations.

We will end this chapter with the remark that the linear Spencer sequence is the origin (at least from the present standpoint) of many modern and highly developed theories. For instance,

1. The theory of connections on vector bundles may be viewed as a generalization of $\widetilde{\nabla}$ and ∇, as we have already observed in Chapter 6. In particular, a flat vector bundle gives rise to a complex generalizing (11.7) and (11.21).
2. Given a sheaf Θ on M, we can define the cohomology $H^*(M,\Theta)$ using fine resolutions. Since the standard partition-of-unity argument applies to the sections of the spaces in (11.7) and (11.21), they are fine resolutions of the sheaf of Lie algebras $U\to\widetilde{X}_\varepsilon(U)$ and $U\to\mathfrak{X}_{\widetilde{\varepsilon}}(U)$.
3. The cohomology theory of Lie algebras was initiated in the fundamental paper [CE] and has grown into a vast subject. We have seen that this purely algebraic theory emerges from the linear Spencer sequence by invariance and localization.
4. The cohomology theory of Lie algebroids is another developing modern theory (see [MK1], [Cr], [CF], and [F]). The linear Spencer sequence is also a prototype here.

5. The horizontal cohomology of PDEs was initiated by [Vn] (see [V] and references therein). Now, (11.7) and (11.21) are the horizontal cohomologies of the linear PDEs $\widetilde{\nabla}\xi = 0$ and $\nabla\xi = 0$, as we will see in Chapter 14.
6. A linear system of PDEs satisfying certain conditions (such as formal integrability and involutivity) gives rise to a Janet sequence. These sequences are defined and studied in detail by Pommaret in [P1], [P2] and other works. We remark here that (11.7) and (11.21) are also the Janet sequences of the linear PDEs $\widetilde{\nabla}\xi = 0$ and $\nabla\xi = 0$.

12

The Secondary Characteristic Classes

We defined the complex $C^*(\widetilde{D}, T^{r,s})$ in Chapter 11. In this chapter, we will take a closer look at $C^*(\widetilde{D}, T^{1,1})$, which is special for many reasons.

We recall that a section ω of $\wedge^k(T^*) \otimes T^{1,1} \to M$ is of the form $\omega = \left(\omega^i_{j_1 \ldots j_k, m}(x)\right)$ on (U, x). We have the complex $C^*(\widetilde{D}, T^{1,1})$ given by

$$0 \longrightarrow T^{1,1}_{\varepsilon} \longrightarrow T^{1,1} \xrightarrow{\widetilde{D}_1} T^* \otimes T^{1,1} \xrightarrow{\widetilde{D}_2} (\wedge^2 T^*) \otimes T^{1,1} \xrightarrow{\widetilde{D}_3} \cdots \xrightarrow{\widetilde{D}_n} (\wedge^n T^*) \otimes T^{1,1} \tag{12.1}$$

where

$$\widetilde{d}_r \omega^i_{j_1 \ldots j_k, m} = \frac{\partial \omega^i_{j_1 \ldots j_k, m}}{\partial x^r} - \Gamma^i_{ra} \omega^a_{j_1 \ldots j_k, m} + \Gamma^a_{rm} \omega^i_{j_1 \ldots j_k, a} \tag{12.2}$$

and

$$(\widetilde{D}\omega)^i_{r j_1 \ldots j_k, m} = \left[\widetilde{d}_r \omega^i_{j_1 \ldots j_k, m}\right]_{[r j_1 \ldots j_k]} \tag{12.3}$$

Now, $\widetilde{D}_1 = \widetilde{\nabla}$ and (12.1) is a complex since $\widetilde{\mathfrak{R}} = 0$. We also know that $H^*(\widetilde{D}, T^{1,1}) \simeq \oplus_{n^2} H^*_{dR}(M, \mathbb{R})$. Everything stated in Chapter 11 for tensor representations holds for (12.1).

One reason why (12.1) is special is that an obvious wedge product is defined on its sections. If $\omega \in (\wedge^r T^*) \otimes T^{1,1}$ and $\varphi \in (\wedge^s T^*) \otimes T^{1,1}$, we define $\omega \wedge \varphi \in (\wedge^{r+s} T^*) \otimes T^{1,1}$ by the formula

$$(\omega \wedge \varphi)^i_{j_1 \ldots j_r k_1 \ldots k_s, m} \stackrel{\text{def}}{=} \left[\omega^i_{j_1 \ldots j_r, a} \varphi^a_{k_1 \ldots k_s, m}\right]_{[j_1 \ldots j_r k_1 \ldots k_s]} \tag{12.4}$$

where, as before, the constants are absorbed in the alternation sign. We can easily show that $\wedge$ is associative (the constants are important here). More important is the fact that

$$\widetilde{D}(\omega \wedge \varphi) = (\widetilde{D}\omega) \wedge \varphi + (-1)^{\deg \omega} \omega \wedge (\widetilde{D}\varphi) \tag{12.5}$$

This can be verified by a straightforward computation using the definitions, but its proof boils down to the fact that $\widetilde{\nabla}$ is a derivation. We will leave the rather straightforward details to the interested reader.

We can interpret the torsion $T = (T^i_{jk})$ in two ways:

1. As a section of $\wedge^2(T^*) \otimes T \to M$ and therefore as a section of the fourth space in (11.7) and (11.21). With this interpretation, the torsion T is a T-valued 2-form where jk are the 2-form indices and i is the T-index. We can easily check that $D_2 T = \widetilde{D}_2 T = 0$ (we always assume $\mathfrak{R} = 0$ when we mention D, so that (11.21) is a complex), but unfortunately T is exact in those complexes, with the primitives $\pm I = \left(\pm\delta^i_j\right)$, as the reader can easily check. Therefore, it is not possible to define an interesting cohomology class in this way using T.
2. As a $T^{1,1}$-valued 1-form where j is the 1-form index and i, k are $T^{1,1}$-indices. With this new interpretation, we write $\left(T^i_{jk}\right)$ with a comma as $\left(T^i_{j,k}\right)$, and T becomes a section of the third space in (12.1). Note that $\mathfrak{R} = \left(\mathfrak{R}^i_{jk,m}\right)$ is a section of the fourth space in (12.1).

Now,

$$\begin{aligned}\mathfrak{R}^i_{jk,r} &= \widetilde{\nabla}_r T^i_{j,k} = \frac{\partial T^i_{j,k}}{\partial x^r} - \Gamma^i_{ra} T^a_{j,k} + \Gamma^a_{rj} T^i_{a,k} + \Gamma^a_{rk} T^i_{j,a} \\ &= \widetilde{d}_r T^i_{j,k} + \Gamma^a_{rj} T^i_{a,k}\end{aligned} \tag{12.6}$$

Alternating r, j in (12.6) gives

$$(\widetilde{D}T)^i_{rj,k} = \mathfrak{R}^i_{jk,r} - \mathfrak{R}^i_{rk,j} - T^a_{r,j} T^i_{a,k} \tag{12.7}$$

By Proposition 6.4, we have

$$\mathfrak{R}^i_{jk,r} - \mathfrak{R}^i_{rk,j} = -\mathfrak{R}^i_{rj,k} - T^i_{a,r} T^a_{k,j} - T^i_{a,j} T^a_{r,k} - T^i_{a,k} T^a_{j,r} \tag{12.8}$$

Substituting (12.8) into (12.7), we get

$$\begin{aligned}(\widetilde{D}_2 T)^i_{rj,k} &= -\mathfrak{R}^i_{rj,k} - T^i_{a,r} T^a_{k,j} - T^i_{a,j} T^a_{r,k} - T^i_{a,k} T^a_{j,r} - T^a_{r,j} T^i_{a,k} \\ &= -\mathfrak{R}^i_{rj,k} - T^i_{r,a} T^a_{j,k} + T^i_{j,a} T^a_{r,k} \\ &= -\mathfrak{R}^i_{rj,k} - \left[T^i_{r,a} T^a_{j,k}\right]_{[rj]}\end{aligned} \tag{12.9}$$

Rewriting (12.9) using (12.4), we deduce the following proposition:

Proposition 12.1 *For (M, ε), we have*

$$-\mathfrak{R} = \widetilde{D}_2(T) + T \wedge T \tag{12.10}$$

At this moment, we pause a little and inspect (12.10) carefully. Note that the expression on its right-hand side of (12.10) looks very much like the left-hand side of (7.15). Indeed, let A be a 1-form on a smooth manifold M with values in some Lie algebra $\mathfrak{h}$. It is standard to define the curvature $\mathbf{R}$ of A by the formula

$$\mathbf{R} \stackrel{\text{def}}{=} dA + A \wedge A \tag{12.11}$$

so that A is identified with T in (12.10). With this identification, however, note that $\mathfrak{h}$ cannot be equal to $\widetilde{X}_\varepsilon$, whose definition requires $\mathfrak{R} = 0$.

Now we apply $\widetilde{D}_3$ to (12.10) using (12.5) and $\widetilde{D}_3 \circ \widetilde{D}_2 = 0$. Writing $\widetilde{D}$ for $\widetilde{D}_k$, we get

$$\begin{aligned}\widetilde{D}(\mathfrak{R}) &= -\widetilde{D}(T) \wedge T + T \wedge \widetilde{D}(T) \\ &= -(-\mathfrak{R} - T \wedge T) \wedge T + T \wedge (-\mathfrak{R} - T \wedge T) \\ &= \mathfrak{R} \wedge T - T \wedge \mathfrak{R}\end{aligned} \tag{12.12}$$

Using (12.12), we can now prove the following proposition:

Proposition 12.2 (The Second Bianchi Identity) *For (M, ε), we have*

$$D_3(\mathfrak{R}) = 0 \tag{12.13}$$

where $D_3 : (\wedge^2 T^*) \otimes T^{1,1} \to (\wedge^3 T^*) \otimes T^{1,1}$ is the operator defined by using d_r in (12.2).

Note that we do not have the Spencer complex $C^*(D, T^{1,1})$ in Proposition 12.2, since we do not assume $\mathfrak{R} = 0$, and with this assumption (12.13) becomes the tautology $0 = 0$!

Proof We have

$$d_r \mathfrak{R}^i_{jk,m} = \widetilde{d}_r \mathfrak{R}^i_{jk,m} - T^i_{ar} \mathfrak{R}^a_{jk,m} + T^a_{mr} \mathfrak{R}^i_{jk,a} \tag{12.14}$$

Therefore,

$$\begin{aligned} D(\mathfrak{R})^i_{rjk,m} &= \widetilde{D}(\mathfrak{R})^i_{rjk,m} + \left[T^i_{ra} \mathfrak{R}^a_{jk,m} - \mathfrak{R}^i_{jk,a} T^a_{rm}\right]_{[rjk]} \\ &= \widetilde{D}(\mathfrak{R})^i_{rjk,m} + \left[T^i_{ra} \mathfrak{R}^a_{jk,m} - \mathfrak{R}^i_{rj,a} T^a_{km}\right]_{[rjk]} \\ &= \widetilde{D}(\mathfrak{R})^i_{rjk,m} + (T \wedge \mathfrak{R} - \mathfrak{R} \wedge T)^i_{rjk,m} \\ &= 0 \qquad \blacksquare \end{aligned} \tag{12.15}$$

Now, for $\omega \in (\wedge^k T^*) \otimes T^{1,1}$, we define its trace $\mathrm{Tr}(\omega) \in \wedge^k$ by

$$\mathrm{Tr}(\omega)_{j_1 \ldots j_k} \stackrel{\text{def}}{=} \omega^a_{j_1 \ldots j_k, a} \tag{12.16}$$

From this, we deduce

$$\begin{aligned}
\operatorname{Tr}(\widetilde{D}(\omega))_{rj_1\ldots j_k} &= \left[\frac{\partial \omega^a_{j_1\ldots j_k,a}}{\partial x^r} - \Gamma^b_{ra}\omega^a_{j_1\ldots j_k,b} + \Gamma^a_{rb}\omega^b_{j_1\ldots j_k,a}\right]_{[rj_1\ldots j_k]} \\
&= \left[\frac{\partial \omega^a_{j_1\ldots j_k,a}}{\partial x^r}\right]_{[rj_1\ldots j_k]} \\
&= d(\operatorname{Tr}(\omega))_{rj_1\ldots j_k}
\end{aligned} \tag{12.17}$$

that is,

$$\operatorname{Tr}\circ\widetilde{D} = d\circ\operatorname{Tr} \tag{12.18}$$

We can easily check also that

$$\operatorname{Tr}(\omega\wedge\xi) = (-1)^{\deg\omega\deg\xi}\operatorname{Tr}(\xi\wedge\omega) \tag{12.19}$$

We now define

$$\mathfrak{R}^{(k)} \stackrel{\text{def}}{=} \mathfrak{R}\wedge\mathfrak{R}\wedge\cdots\wedge\mathfrak{R} \qquad (k\text{ copies}) \tag{12.20}$$

$$T^{(k)} \stackrel{\text{def}}{=} T\wedge T\wedge\cdots\wedge T \qquad (k\text{ copies}) \tag{12.21}$$

So, we have

$$\operatorname{Tr}(\mathfrak{R}^{(k)}) \in \wedge^{2k} \tag{12.22}$$

$$\operatorname{Tr}(T^{(k)}) \in \wedge^{k} \tag{12.23}$$

Each of (12.22) and (12.23) *has a story of its own* to tell, but they are tightly interwoven. We start with (12.22).

Proposition 12.3 $d\operatorname{Tr}(\mathfrak{R}^{(k)}) = 0, k \geq 1.$

Proof We claim

$$d\operatorname{Tr}\left(\mathfrak{R}^{(k)}\right) = \operatorname{Tr}\big(\widetilde{D}(\mathfrak{R}^{(k-j)})\wedge\mathfrak{R}^{(j)}\big), \qquad 0\leq j\leq k \tag{12.24}$$

We proceed by induction on j. Clearly, (12.24) holds for $j = 0$, where we set $\mathfrak{R}^{(0)} = I$. Omitting $\wedge$, we compute

$$\begin{aligned}
\operatorname{Tr}\big(\widetilde{D}(\mathfrak{R}^{(k-j)})\mathfrak{R}^{(j)}\big) &= \operatorname{Tr}\left(\widetilde{D}(\mathfrak{R}^{(k-j-1)}\mathfrak{R})\mathfrak{R}^{(j)}\right) \\
&= \operatorname{Tr}\left(\widetilde{D}(\mathfrak{R}^{(k-j-1)})\mathfrak{R}^{(j+1)}\right) \\
&\quad + \operatorname{Tr}\left(\mathfrak{R}^{(k-j-1)}\widetilde{D}(\mathfrak{R})\mathfrak{R}^{(j)}\right)
\end{aligned}$$

$$\begin{aligned}&= \mathrm{Tr}\left(\widetilde{D}(\mathfrak{R}^{(k-j-1)})\mathfrak{R}^{(j+1)}\right)\\&\quad + \mathrm{Tr}\left(\mathfrak{R}^{(k-j)}T\mathfrak{R}^{(j)}\right)\\&\quad - \mathrm{Tr}\left(\mathfrak{R}^{(k-j-1)}T\mathfrak{R}^{(j+1)}\right)\\&= \mathrm{Tr}\left(\widetilde{D}(\mathfrak{R}^{(k-j-1)})\mathfrak{R}^{(j+1)}\right)\end{aligned} \tag{12.25}$$

which completes the inductive step. Setting $j = k-1$ in (12.24) gives

$$\begin{aligned}d\mathrm{Tr}(\mathfrak{R}^{(k)}) &= \mathrm{Tr}\big(\widetilde{D}(\mathfrak{R})\mathfrak{R}^{(k-1)}\big)\\&= \mathrm{Tr}(\mathfrak{R}T\mathfrak{R}^{(k-1)}) - \mathrm{Tr}(T\mathfrak{R}^{(k)})\\&= 0 \qquad \blacksquare\end{aligned} \tag{12.26}$$

Note that we can also prove Proposition 12.3 by showing (12.5) and (12.18) for D and then using Proposition 12.2, but (12.24) is of some interest by itself.

Unfortunately, all the closed forms $\mathrm{Tr}(\mathfrak{R}^{(k)})$ are exact in the de Rham complex of M. The reason is that the connection ∇ on the tangent bundle $T \to M$ defines closed forms by substituting its curvature $\mathfrak{R}$ into the invariant polynomials $\mathrm{Tr}(A_1, \ldots, A_k) = \mathrm{Tr}(A_1A_2 \ldots A_k)$, $A_i \in M_{n\times n}(\mathbb{R})$. According to the general theory, these cohomology classes are the Pontryagin classes of $T \to M$. However, $T \to M$ is a trivial vector bundle trivialized by ε, and therefore these classes vanish. Note also that $\widetilde{\nabla}$ is another connection on $T \to M$ with curvature $\widetilde{\mathfrak{R}} = 0$, and these classes are independent of the connection.

Our purpose is now to find some explicit primitives of the closed forms (12.22). We need the following lemma:

Lemma 12.4 $\mathrm{Tr}(T^{(2k)}) = 0, k \geq 1$.

Proof Let K denote the set of k-tuples $(\sigma(1), \ldots, \sigma(k))$, where σ is a permutation. We define an equivalence relation on K: $(\sigma(1), \ldots, \sigma(k)) \sim (\tau(1), \ldots, \tau(k))$ if $(\sigma(1), \ldots, \sigma(k)) = (\tau(i), \tau(i+1), \ldots, \tau(k), \tau(1), \ldots, \tau(i-1))$ for some i. So, two k-tuples are equivalent if they differ by a cyclic permutation. This equivalence relation is imposed by $\mathrm{Tr}(AB) = \mathrm{Tr}(BA)$. Now consider the formal sum

$$\sum_{\sigma \in S_k} \mathrm{sgn}(\sigma)(\sigma(1), \ldots, \sigma(k)) \tag{12.27}$$

The proof reduces to the following combinatorial statement: If k is even and we identify the equivalent k-tuples in (12.27), then (12.27) vanishes. To prove this, we choose two odd integers a, b with $k = a + b$. For some fixed σ, we have

$$\begin{aligned}&\mathrm{sgn}(\sigma)(\sigma(1), \ldots, \sigma(a), \sigma(a+1), \ldots, \sigma(a+b))\\&\qquad = (-1)^{ab}\mathrm{sgn}(\sigma)(\sigma(a+1), \ldots, \sigma(a+b), \sigma(1), \ldots, \sigma(a))\end{aligned} \tag{12.28}$$

and all terms cancel in pairs. $\blacksquare$

Taking the trace of (12.10) and using (12.18) and Lemma 12.4, we get

$$\mathrm{Tr}(\mathfrak{R}) = d(-\mathrm{Tr}(T)) \tag{12.29}$$

Therefore, $c_1 \stackrel{\text{def}}{=} -\mathrm{Tr}(T)$ is a primitive for $\mathrm{Tr}(\mathfrak{R})$. Now,

$$\begin{aligned}\widetilde{D}\left(T^{(3)}\right) &= \widetilde{D}\left(TT^{(2)}\right) = \widetilde{D}\,(T)\,T^{(2)} - T\widetilde{D}\left(T^{(2)}\right)\\ &= \widetilde{D}(T)T^{(2)} - T(\widetilde{D}(T)T - T\widetilde{D}(T))\\ &= \widetilde{D}(T)T^{(2)} - T\widetilde{D}(T)T + T^{(2)}\widetilde{D}(T)\end{aligned} \tag{12.30}$$

Taking the trace of (12.30) and using (12.19), we get

$$\mathrm{Tr}\big(\widetilde{D}\left(T^{(3)}\right)\big) = 3\,\mathrm{Tr}\big(\widetilde{D}(T)T^{(2)}\big) = 3\,\mathrm{Tr}\left(\mathfrak{R}T^{(2)} - T^{(4)}\right) = 3\,\mathrm{Tr}\left(\mathfrak{R}T^{(2)}\right) \tag{12.31}$$

Now,

$$\begin{aligned}\widetilde{D}(\mathfrak{R}T) &= (\mathfrak{R}T - T\mathfrak{R})T + \mathfrak{R}\left(-\mathfrak{R} - T^{(2)}\right)\\ &= \mathfrak{R}T^{(2)} - T\mathfrak{R}T - \mathfrak{R}^{(2)} - \mathfrak{R}T^{(2)}\end{aligned} \tag{12.32}$$

Therefore,

$$\mathrm{Tr}\big(\widetilde{D}(\mathfrak{R}T)\big) = -\mathrm{Tr}\,\mathfrak{R}^{(2)} + \mathrm{Tr}\left(\mathfrak{R}T^{(2)}\right) = -\mathrm{Tr}\,\mathfrak{R}^{(2)} + \tfrac{1}{3}\mathrm{Tr}\left(\widetilde{D}(T^{(3)})\right) \tag{12.33}$$

and

$$c_3 \stackrel{\text{def}}{=} -\mathrm{Tr}\big(\mathfrak{R}T - \tfrac{1}{3}T^{(3)}\big) \tag{12.34}$$

is a primitive for $\mathrm{Tr}\,\mathfrak{R}^{(2)}$. We observe that c_3 is the Chern–Simons 3-form modulo the sign (see [CSi]). The method is now clear: To find the primitive of $\mathrm{Tr}\,\mathfrak{R}^{(3)}$, we compute $\widetilde{D}\left(T^{(5)}\right)$, $\widetilde{D}(\mathfrak{R}T^{(3)})$, and $\widetilde{D}(\mathfrak{R}^{(2)}T)$, take traces, and exhibit a primitive of the form

$$c_5 = \mathrm{Tr}(\alpha\mathfrak{R}^{(2)}T + \beta\mathfrak{R}T^{(3)} + \gamma T^{(5)}) \tag{12.35}$$

for some nonzero scalars α, β, γ. We invite the interested reader to show that this algorithm works and produces all the primitives. The general conclusion is

$$\mathrm{Tr}\,\mathfrak{R}^{(k)} = d\mathrm{Tr}\left(\cdots + \gamma T^{(2k-1)}\right), \qquad k \geq 1 \tag{12.36}$$

where the dots "$\cdots$" here denote a linear combination of terms, each being a wedge product of T's and $\mathfrak{R}$'s and of total degree $2k-1$. Furthermore, each term has $\mathfrak{R}$ as a factor and $\gamma \neq 0$.

Definition 12.5 *The primitives given by* (12.36) *are the Chern–Simons forms* c_{2k-1} *of* (M, ε).

Now suppose $\mathfrak{R} = 0$ and (M, ε) is the LLG $(M, \varepsilon, \mathcal{G})$. Thus, we obtain from (12.36) the following important proposition:

Proposition 12.6 *For an LLG* $(M, \varepsilon, \mathcal{G})$*, we have*

$$d\mathrm{Tr}\, T^{(2k+1)} = 0 \tag{12.37}$$

where $\mathrm{Tr}\, T^{(2k+1)} = c_{2k+1}$ modulo a nonzero constant factor.

Definition 12.7 *The cohomology classes* $[c_{2k+1}] \in H^{2k+1}_{\mathrm{dR}}(M, \mathbb{R})$ *are the Chern–Simons classes of the LLG* $(M, \varepsilon, \mathcal{G})$.

Note that it is easier to show (12.37) directly using Lemma 12.4: An easy induction using $\mathfrak{R} = 0$ and (12.10) shows that $d\mathrm{Tr}\, T^{(2k+1)} = -\mathrm{Tr}\, T^{(2k+2)} = 0$. Also note that $\widetilde{\nabla}\left(\mathrm{Tr}\, T^{(2k+1)}\right) = \nabla\left(\mathrm{Tr}\, T^{(2k+1)}\right) = 0$, and $\mathrm{Tr}\, T^{(2k+1)}$ lives in the subcomplexes $C^*_{\widetilde{\nabla}=0}(d, \wedge^0)$ and $C^*_{\nabla=0}(d, \wedge^0)$ of the de Rham complex $C^*(d, \wedge^0)$. All these complexes compute the same cohomology if M is compact and $(M, \varepsilon, \mathcal{G})$ is globalizable.

Henceforth in this chapter, we assume $\mathfrak{R} = 0$ and deal with the LLG $(M, \varepsilon, \mathcal{G})$, concentrating on the following question:

Q: What is the geometric meaning of the Chern–Simons classes?

We will give a fairly satisfactory answer for $[c_1]$, and $[c_3]$ will turn out to be quite familiar. Surprisingly, we will find out that these classes were already known to Pontryagin in 1939 in a special case, and have been studied extensively by various authors since then, but from a different standpoint.

Now (12.4) and (12.21) show that

$$\left(\mathrm{Tr}\, T^{(2k+1)}\right)_{j_1 \cdots j_{2k+1}} = \left[T^b_{j_1, a_1} T^{a_1}_{j_2, a_2} T^{a_2}_{j_3, a_3} \cdots T^{a_k}_{j_{2k+1}, b}\right]_{[j_1 \cdots j_{2k+1}]}, \qquad k \geq 0 \tag{12.38}$$

To write this in coordinate-free form, we observe that for $\xi \in \widetilde{X}_\varepsilon$, $T(\xi)$ is the linear map $\widetilde{X}_\varepsilon \to \widetilde{X}_\varepsilon$ defined by $T(\xi)(\eta) = T(\xi, \eta) = [\xi, \eta]$; that is, $T(\xi) = \mathrm{ad}(\xi)$. Therefore, (12.38) becomes

$$(\mathrm{Tr}\, T^{(2k+1)})(\xi_1, \ldots, \xi_{2k+1}) = \sum_{\sigma \in S_{2k+1}} \mathrm{sgn}(\sigma) \mathrm{Tr}\left(\mathrm{ad}(\xi_1) \circ \mathrm{ad}(\xi_2) \circ \cdots \circ \mathrm{ad}(\xi_{2k+1})\right) \tag{12.39}$$

Here, $\xi_1, \ldots, \xi_{2k+1} \in \widetilde{X}_\varepsilon \simeq X_{\widetilde{\varepsilon}}$, but we can see that we can define $\mathrm{Tr}\, T^{(2k+1)}$ in any abstract Lie algebra $\mathfrak{h}$ using the right-hand side of (12.39). In fact, we can replace ad in (12.39) by any representation $\rho : \mathfrak{h} \to V$ and define odd-degree closed forms in the much more general framework of algebroids (see [Cr] and [CF]). The reader who is familiar with the algebraic approach to the cohomology of classical compact Lie groups will notice the so called "primitive elements" in (12.39) arising from the image of the Cartan map (see VI, 6 of [GHV]). We will comment more on this issue below.

Now

$$(\operatorname{Tr} T)_j = T^a_{j,a} \tag{12.40}$$

We refer to [Wn1] for a generalization of (12.40). Our purpose is to construct an explicit local primitive for (12.40), that is, a function f with $df = \operatorname{Tr} T$ on some connected $U \subset M$. We recall the adjoint map Ad_e based at $e \in M$ in Chapter 4: For $e, x \in U$, we have the 1-arrow $\operatorname{Ad}_\varepsilon(x) = \varepsilon(x,e)\widetilde{\varepsilon}(e,x)$ with source and target at e. So $g(x) \overset{\text{def}}{=} \det \operatorname{Ad}_\varepsilon(x)$ is a function defined on U if U is sufficiently small. If $(M, \varepsilon, \mathcal{G})$ is globalizable, then g is defined on M. Since $g(e) = 1$, g is continuous, and U is connected, $g(x) \gneqq 0$ on U and therefore $\log(g(x))$ is defined on U. We call a globalizable $(M, \varepsilon, \mathcal{G})$ unimodular if $g(x) = 1$ on M.

Proposition 12.8 *$-\log(g(x))$ is a primitive for* $\operatorname{Tr} T$ *on U. In particular, if $(M, \varepsilon, \mathcal{G})$ is globalizable, then $[\operatorname{Tr} T] = 0$ in $H^1_{\mathrm{dR}}(M, \mathbb{R})$. If $(M, \varepsilon, \mathcal{G})$ is also unimodular, then* $\operatorname{Tr} T = 0$.

The second and third assertions are clear. To show the first assertion, we need the following well-known lemma:

Lemma 12.9 *Let $A = \left(a^i_j(x)\right)$ be an invertible $n \times n$ matrix with smooth entries on $U \subset \mathbb{R}^n$. Let $B = \left(b^i_j(x)\right)$ be the inverse of A. Then*

$$(\det A)^{-1} \frac{\partial \det A}{\partial x^j} = \frac{\partial a^s_t}{\partial x^j} b^t_s = \operatorname{Tr}\left(\frac{\partial A}{\partial x^j} A^{-1}\right) \tag{12.41}$$

The proof follows easily from the cofactor expansion of $\det A$. Note that if $\det A$ is positive on U, then the left-hand side of (12.41) is equal to

$$\frac{\partial \log(\det A)}{\partial x^j} \tag{12.42}$$

We first compute $\operatorname{Ad}_\varepsilon(x)$ in coordinates. We recall from Chapter 4 that $h(y) = g(p,q,y)$ is the local solution of (2.4) in the variable y satisfying the initial condition $h(p) = q$ and

$$\widetilde{\varepsilon}^i_j(p,q) = \left[\frac{\partial g^i(p,y,q)}{\partial y^j}\right]_{y=p}, \qquad y \in U \tag{12.43}$$

Therefore,

$$\begin{aligned} \operatorname{Ad}_\varepsilon(x)^i_j &= \varepsilon^i_a(x,e)\widetilde{\varepsilon}^a_j(e,x) \\ &= \varepsilon^i_a(x,e)\left[\frac{\partial g^a(e,y,x)}{\partial y^j}\right]_{y=e} \end{aligned} \tag{12.44}$$

We first show that $-d\left(\log(\det \mathrm{Ad}_e)\right)$ and $\mathrm{Tr}\, T$ have the same value at $x = e$. Now,

$$
\begin{aligned}
-\left[d(\log(\det \mathrm{Ad}_e))(e)\right]_i &= -\left[\frac{\partial}{\partial x^i} \log\det(\mathrm{Ad}_\varepsilon(x))\right]_{x=e} \\
&= -\left[\frac{\partial}{\partial x^i}\left(\varepsilon^j_a(x,e)\frac{\partial g^a(e,y,x)}{\partial y^k}\right)\right]_{x=e,y=e} \delta^k_j \\
&= -\left[\frac{\partial}{\partial x^i}\left(\varepsilon^b_a(x,e)\frac{\partial g^a(e,y,x)}{\partial y^b}\right)\right]_{x=e,y=e}
\end{aligned}
\tag{12.45}
$$

where we have used Lemma 12.9 and $\mathrm{Ad}_e(e,e)^i_j = \delta^i_j$. The right-hand side of (12.45) is equal to

$$
\begin{aligned}
&-\left[\frac{\partial \varepsilon^b_a(x,e)}{\partial x^i}\frac{\partial g^a(e,x,y)}{\partial y^b} + \varepsilon^b_a(x,e)\frac{\partial^2 g^a(e,y,x)}{\partial x^i \partial y^b}\right]_{x=e,y=e} \\
&= -\left[\frac{\partial \varepsilon^b_a(x,e)}{\partial x^i}\right]_{x=e} \delta^a_b - \delta^b_a \frac{\partial}{\partial y^b}\left[\frac{\partial g^a(e,y,x)}{\partial x^i}\right]_{x=e,y=e} \\
&= \Gamma^a_{ia}(e) - \frac{\partial}{\partial y^a}\left[\frac{\partial g^a(e,y,x)}{\partial x^i}\right]_{x=e,y=e} \\
&= \Gamma^a_{ia}(e) - \left[\frac{\partial \varepsilon^a_i(e,y)}{\partial y^a}\right]_{y=e} \\
&= \Gamma^a_{ia}(e) - \Gamma^a_{ai}(e) \\
&= (\mathrm{Tr}\, T(e))_i
\end{aligned}
\tag{12.46}
$$

as claimed. Now let $e' \in U$ *be another basepoint.* We have

$$
\log\det \mathrm{Ad}_e(x) = \log\det \mathrm{Ad}_{e'}(x) + \log\det \mathrm{Ad}_e(e') \tag{12.47}
$$

Differentiating this at $x = e'$ and using (12.45) and (12.46), we get

$$
-d\left(\log\det \mathrm{Ad}_e\right)(e') = c_1(e') \tag{12.48}
$$

which finishes the proof of Proposition 12.8, since e' is arbitrary.

The next proposition states that the converse of the second assertion of Proposition 12.8 is also true:

Proposition 12.10 *Suppose* $[\mathrm{Tr}\, T] = 0$ *in* $H^1_{\mathrm{dR}}(M, \mathbb{R})$. *Then* $(M, \varepsilon, \mathcal{G})$ *is globalizable.*

Proof Suppose $df = \mathrm{Tr}\,T$ for some $f : M \to \mathbb{R}$. We fix $e \in M$, and choose an arbitrary $x \in M$ and a path C from e to x. Adding a constant to $f_{|U}$ in a neighborhood of e (and still calling it $f_{|U}$), we can choose points $e_0 = e, e_1, \ldots, e_n = x$ on C and open sets U_i such that

(i) $e_i \in U_i$;

(ii) $\{U_i\}$ covers C;

(iii) $-\log\det \mathrm{Ad}(e_i, x) = f(x)$ on U_i.

We define $\widetilde{\varepsilon}(e,x) \stackrel{\text{def}}{=} \widetilde{\varepsilon}(e_{n-1},x) \circ \cdots \circ \widetilde{\varepsilon}(e_1,e_2) \circ \widetilde{\varepsilon}(e,e_1)$. Since f is globally defined, $\widetilde{\varepsilon}(e,x)$ is independent of the path, and therefore $(M, \varepsilon, \mathcal{G})$ is globalizable. With this definition of $\widetilde{\varepsilon}$, it follows that $\mathrm{Ad}(e,x) = \mathrm{Ad}(e_{n-1},x) \circ \cdots \circ \mathrm{Ad}(e,e_1)$ and $-\log\det \mathrm{Ad}(e,x) = f(x)$ on M. ■

Corollary 12.11 $[\mathrm{Tr}\,T] \in H^1_{\mathrm{dR}}(M,\mathbb{R})$ *is an obstruction to globalizability:* $(M, \varepsilon, \mathcal{G})$ *is globalizable if and only* $[\mathrm{Tr}\,T] = 0$.

Now let C be a loop at e. We integrate $\mathrm{Tr}\,T$ along C. Since $d(\mathrm{Tr}\,T) = 0$, the value of this integral depends only on the homotopy class of C. This gives a homomorphism

$$[\mathrm{Tr}\,T]_* : \Pi(M,e) \to \mathbb{R} \tag{12.49}$$

Clearly, $[\mathrm{Tr}\,T]_* = 0$ if and only if $[\mathrm{Tr}\,T] = 0$. Letting K be the kernel of (12.49), we construct the covering $\pi : \mathcal{M} \to M$ for K and lift $(M, \varepsilon, \mathcal{G})$ to $\pi^{-1}(M, \varepsilon, \mathcal{G})$ on $\mathcal{M}$. If $(M, \varepsilon, \mathcal{G})$ is complete, it is easy to see that $\pi^{-1}(M, \varepsilon, \mathcal{G})$ is globalizable and is the "smallest" such covering. We also note that $[\mathrm{Tr}\,T]$ behaves functorially with respect to the covering maps.

We now come to $\mathrm{Tr}\,T^{(3)}$, which is given by

$$\begin{aligned}
\mathrm{Tr}\,T^{(3)}(x,y,z) &= 3\,\mathrm{Tr}\big(\mathrm{ad}(x)\circ\mathrm{ad}(y)\circ\mathrm{ad}(z) - \mathrm{ad}(x)\circ\mathrm{ad}(z)\circ\mathrm{ad}(y)\big)\\
&= 3\,\mathrm{Tr}\big(\mathrm{ad}(x)\circ\big(\mathrm{ad}(y)\circ\mathrm{ad}(z) - \mathrm{ad}(z)\circ\mathrm{ad}(y)\big)\big)\\
&= 3\,\mathrm{Tr}\big(\mathrm{ad}(x)\circ\mathrm{ad}[y,z]\big)\\
&= 3\kappa(x,[y,z]), \qquad x,y,z \in \widetilde{X}_\varepsilon
\end{aligned} \tag{12.50}$$

where κ is the Killing form. The following proposition is well known (see [CE]):

Proposition 12.12

(i) $[\mathrm{Tr}\,T^{(3)}] \neq 0$ *in* $H^3(\mathfrak{g},\mathbb{R})$ *if* $\mathfrak{g} = \widetilde{X}_\varepsilon$ *is semisimple.*

(ii) $\mathrm{Tr}\,T^{(3)} = 0$ *if and only if* $\widetilde{X}_\varepsilon$ *is solvable.*

Note that the conditions of Proposition 12.12 are algebraic. We recall that the concept of solvability emerged from Galois' work on the solvability of the quintic by radicals, which is a concrete problem. Later, this concept was generalized to other algebraic structures, such as Lie algebras. However, $\widetilde{X}_\varepsilon$ is a Lie algebra of vector fields on M and not an abstract Lie algebra. What does solvability of $\widetilde{X}_\varepsilon$ really mean—that is, what is it that we want to solve? From this standpoint, Proposition 12.12 is clearly not very satisfactory. We refer the interested reader to the classical book [O2] for some interesting relations between algebraic solvability and solvability of differential equations.

Around 1930, the important question arose as to how the structure of the cohomology ring of a compact Lie group could be determined. We refer to the excellent survey [Sa] for the history of this fascinating problem. In 1939 Pontryagin showed in [Pn] that, for certain classical compact Lie groups, this cohomology ring is generated by certain odd-degree classes later called the primitive elements. Today, it is known that this is true for any connected compact Lie group. Note that all these classes may be concentrated in $H^1_{\mathrm{dR}}(M,\mathbb{R})$, as the example of a torus shows. This cohomology ring is the prototype of what is now called a Hopf algebra. Since the cohomology of a compact Lie group G is determined by its Lie algebra cohomology $H^*(\mathfrak{g},\mathbb{R})$, the problem was to express the primitive elements purely algebraically. We refer to [GHV] for this algebraic approach, together with an extensive list of references.

Note that $[\mathrm{Tr}\, T]=0$ in $H^1_{\mathrm{dR}}(M,\mathbb{R})$ for any Lie group by Corollary 12.11. We will state here the following proposition, which follows from (12.39) and the general theory (see [GHV]):.

Proposition 12.13 *Suppose $(M,\varepsilon,\mathcal{G})$ is globalizable, so that M is a connected Lie group according to Proposition 4.4. If M is compact, then the nontrivial classes among the Chern–Simons classes are primitive classes. If M is also simple, then they coincide with the set of all primitive classes.*

There also seems to be an intriguing relation between the Chern–Simons classes and the Godbillon–Vey classes of foliations. We recall that a codimension-k foliation $\mathcal{F}_k$ on a smooth M determines a characteristic class $\mathrm{GV}(\mathcal{F}_k)\in H^{2k+1}_{\mathrm{dR}}(M,\mathbb{R})$. The first nontrivial example is due to Poenaru: Let $G=\mathrm{SL}(2,\mathbb{R})$, let $\Delta\subset G$ be a cocompact discrete subgroup, and let $\mathcal{F}_1$ be the codimension-1 foliation on G/Δ defined by the left-invariant vector fields $X=\begin{bmatrix}0&1\\0&0\end{bmatrix}$ and $H=\begin{bmatrix}1&0\\0&-1\end{bmatrix}$. Using (12.50), it is then easy to check that $\mathrm{GV}(\mathcal{F}_1)=[\mathrm{Tr}\, T^{(3)}]$. This may not seem very exciting, because in this example $[\mathrm{Tr}\, T^{(3)}]\neq 0$ by Proposition 12.12, $\mathrm{GV}(\mathcal{F}_1)\neq 0$, as shown by Poenaru, and $\dim H^1_{\mathrm{dR}}(G/\Delta,\mathbb{R})=1$. However, we are naturally lead to ask the following question:

Q: Let $(M,\varepsilon,\mathcal{G})$ be a globalizable LLG and H be *any* connected subgroup of the Lie group M of codimension k. What is the relation between $\mathrm{GV}(\mathcal{F}_k)$ and $[\mathrm{Tr}\, T^{(2k+1)}]$ in $H^{2k+1}_{\mathrm{dR}}(M,\mathbb{R})$?

Note that $[\mathrm{Tr}\, T^{(2k+1)}]$ is defined without any reference to any foliation on M.

13

The Homogeneous Flow

We recall that an orientable 3-manifold is parallelizable. In particular this is true for a simply connected 3-manifold. By Proposition 1.2, M is parallelizable if and only if it admits a splitting, and by Lemma 8.4, the gauge group acts transitively on the set of all splittings. The question asked at the end of Chapter 3 as to whether (M, ε) admits a splitting ε' with $\mathfrak{R}(\varepsilon') = 0$ is very subtle, as the following proposition shows:

Proposition 13.1 *The following are equivalent.*

(i) *A compact and simply connected 3-manifold is diffeomorphic to S^3 (the Poincaré Conjecture).*

(ii) *A compact and simply connected 3-manifold admits a splitting ε with $\mathfrak{R}(\varepsilon) = 0$.*

Proof (i) $\Longrightarrow$ (ii) is clear since S^3 is a Lie group and therefore a globalizable LLG.

(ii) $\Longrightarrow$ (i): Since $\mathfrak{R}(\varepsilon) = 0$, M is an LLG. Since M is compact, $(M, \varepsilon, \mathcal{G})$ is complete by Proposition 3.5 and therefore globalizable, since M is also simply connected. It follows from Proposition 4.4 that M is a Lie group. However, S^3 is the only compact and simply connected Lie group in dimension 3, and therefore M is diffeomorphic to S^3. ∎

Therefore, the following question is quite relevant:

Q: Suppose M admits a splitting. Is there a flow that starts with any splitting ε_0 as the initial condition and flows in the direction of a splitting ε satisfying $\mathfrak{R}(\varepsilon) = 0$?

The answer turns out to be in the affirmative, and the purpose of this chapter is to give a detailed construction of such a flow. We will also prove the existence of unique short-time solutions of this flow, which, like the Ricci flow, is a second-order nonlinear parabolic evolution equation. As we proceed, it will become apparent that this flow is much more elementary than the Ricci flow, and we believe that it will simplify the present proof of the Poincaré Conjecture.

An Alternative Approach to Lie Groups and Geometric Structures, Ercüment H. Ortaçgil (2018).

We start by recalling some facts from Chapter 7. We fix a basepoint $e \in M$ and a coordinate system around e once and for all. We rewrite (6.32) in the form

$$\widetilde{\nabla}_r \varepsilon^i_j(e,x) = \frac{\partial \varepsilon^i_j(e,x)}{\partial x^r} - \Gamma^i_{ra}(x)\varepsilon^a_j(e,x) = 0 \tag{13.1}$$

We recall that $\varepsilon^i_j(e,x)$ transforms only in the index i and $\widetilde{\nabla}$ acts only on this index in (13.1). Similarly, we have $\widetilde{\nabla}_r \varepsilon^i_j(x,e) = 0$. We now define a tensor g by defining its components $g_{ij}(x)$ on (U,x) by

$$g_{ij}(x) \stackrel{\text{def}}{=} \sum_{1 \le a \le \dim M} \varepsilon^a_i(x,e)\varepsilon^a_j(x,e) \tag{13.2}$$

We write (13.2) in brief as $g_{ij}(x) = \varepsilon^a_i(x,e)\varepsilon^a_j(x,e)$. Clearly, g_{ij} is symmetric. It is also positive-definite, since the matrix $\varepsilon^i_j(e,x)$ is invertible.

Definition 13.2 *g is the canonical metric of (M,ε).*

The canonical metric is unique modulo the choice of basepoint. We recall that we can remove this arbitrariness by identifying e with the origin o of $\mathbb{R}^n$ and considering 1-arrows with source at o and target in M, but this point is not important. From (13.1) and (13.2), we deduce

$$\widetilde{\nabla} g = 0 \tag{13.3}$$

Since g is determined by ε, all the invariants of g (such as the Christoffel symbols, the Riemann curvature tensor, and the Ricci tensor) can be expressed in terms of ε (see [YB] for a detailed study of metric connections on a Riemannian manifold).

Now let t be any tensor field on M, for instance $t = \left(t^{ij}_{km}(x)\right)$. We define

$$t^{ij}_{k(m)}(x) \stackrel{\text{def}}{=} \varepsilon^a_m(e,x)t^{ij}_{ka}(x) \tag{13.4}$$

Now, $t^{ij}_{k(m)}(x)$ does not transform in the index m. We say that $t^{ij}_{k(m)}$ is obtained from t^{ij}_{km} by moving the index m to the basepoint e, or, in other words, the active index m is made passive. Similarly, we can move the index k to the basepoint e. Similarly, we define, for instance

$$t^{(i)j}_{km}(x) \stackrel{\text{def}}{=} \varepsilon^i_a(x,e)t^{aj}_{km}(x) \tag{13.5}$$

and $t^{(i)j}_{km}$ does not transform in the index i. In the same way as we make an active index passive, we can make a passive index active, as in (13.11) below.

We now define

$$\begin{aligned} H^i_j(\varepsilon) \stackrel{\text{def}}{=} -\widetilde{\nabla}_a\left(\varepsilon^b_j(e,x)g^{ac}T^i_{bc}\right) &= -\varepsilon^b_j(e,x)\widetilde{\nabla}_a\left(g^{ac}T^i_{bc}\right) \\ &= -\varepsilon^b_j(e,x)g^{ac}\widetilde{\nabla}_a T^i_{bc} \\ &= -\varepsilon^b_j(e,x)g^{ac}\mathfrak{R}^i_{bc,a} \\ &= -g^{ac}\mathfrak{R}^i_{(b)c,a} \end{aligned} \tag{13.6}$$

Definition 13.3 *$H = (H^i_j(\varepsilon))$ is the homogeneous tensor of (M, ε).*

Note that $H^i_j(\varepsilon)$ transforms only in the index i like $\varepsilon^i_j(e,x)$.

We now assume that $\varepsilon(e,x,t)$ evolves with time for $t \in (-\epsilon, \epsilon)$. So, for any t, $\varepsilon(e,x,t)$ assigns a 1-arrow with source at e and target at x, and this assignment is smooth in both t and x, as in Chapter 9. Equivalently, $\varepsilon(e,x,t)$ defines a global cross section of the principal bundle $\mathcal{U}^{e,\bullet}_1 \to M$ that evolves with time. We set $\varepsilon(e,x,0) = \varepsilon_0(e,x)$ and call ε_0 the initial condition. We observe that $\varepsilon(\epsilon,x,t)$ determines a deformation starting from $\varepsilon_0(e,x)$ at $t = 0$ according to Definition 9.1, and, conversely, any deformation determines $\varepsilon(e,x,t)$. However, to formulate the problem equivalently in terms of deformations will not simplify the problem.

Definition 13.4 *The homogeneous flow of the parallelizable manifold (M, ε_0) is the second-order nonlinear evolution equation*

$$\frac{\partial \varepsilon^i_j(e,x,t)}{\partial t} = H^i_j(\varepsilon(e,x,t)) \tag{13.7}$$

with initial condition $\varepsilon(e,x,0) = \varepsilon_0(e,x)$.

For simplicity of notation below, we will omit the arguments of our functions, except for ε since the notation ε does not distinguish between $\varepsilon(e,x)$ and $\varepsilon(x,e)$, which will be crucial.

Proposition 13.5 *If M is compact, (13.7) admits a unique short-time solution with any initial condition ε_0.*

Proof We will first compute the symbol of the linearization of H. So, we set

$$\frac{\partial \varepsilon^i_j(\epsilon,x,t)}{\partial t} = h^i_j \tag{13.8}$$

and compute the terms that depend on the second-order derivatives of h^i_j with respect to x in the expression

$$\frac{\partial H^i_j(\varepsilon(e,x,t))}{\partial t} \tag{13.9}$$

According to (13.1), $\widetilde{\nabla}_r \varepsilon^i_j(e,x,t) = 0$ for all t, where $\widetilde{\nabla} = \widetilde{\nabla}(t)$ is the operator defined by $\varepsilon^i_j(e,x,t)$. Therefore,

$$\begin{aligned}
0 &= \frac{\partial}{\partial t}\widetilde{\nabla}_r \varepsilon^i_j(e,x,t) \\
&= \frac{\partial}{\partial t}\left(\frac{\partial \varepsilon^i_j(e,x,t)}{\partial x^r} - \Gamma^i_{ra}\varepsilon^a_j(e,x,t)\right) \\
&= \frac{\partial}{\partial x^r}\left(\frac{\partial \varepsilon^i_j(e,x,t)}{\partial t}\right) - \Gamma^i_{ra}\frac{\partial \varepsilon^a_j(e,x,t)}{\partial t} - \frac{\partial \Gamma^i_{ra}}{\partial t}\varepsilon^a_j(e,x,t) \\
&= \frac{\partial h^i_j}{\partial x^r} - \Gamma^i_{ra}h^a_j - \frac{\partial \Gamma^i_{ra}}{\partial t}\varepsilon^a_j(e,x,t) \\
&= \widetilde{\nabla}_r h^i_j - \frac{\partial \Gamma^i_{ra}}{\partial t}\varepsilon^a_j(e,x,t)
\end{aligned} \tag{13.10}$$

which gives

$$\begin{aligned}
\frac{\partial \Gamma^i_{rk}}{\partial t} &= \varepsilon^a_k(x,e,t)\widetilde{\nabla}_r h^i_a \\
&= \widetilde{\nabla}_r\left(\varepsilon^a_k(x,e,t)h^i_a\right) \\
&= \widetilde{\nabla}_r h^i_{(k)}
\end{aligned} \tag{13.11}$$

where we have moved the index k at the basepoint e to x. Note the simplicity of the variation (13.11) compared with the variation of the Christoffel symbols of the metric in the Ricci flow. From (13.11), we deduce

$$\frac{\partial T^i_{rk}}{\partial t} = \widetilde{\nabla}_r h^i_{(k)} - \widetilde{\nabla}_k h^i_{(r)} \tag{13.12}$$

Now,

$$\begin{aligned}
\frac{\partial H^i_j}{\partial t} = &- \frac{\partial \varepsilon^a_j(e,x,t)}{\partial t}g^{bc}\widetilde{\nabla}_b T^i_{ac} - \varepsilon^a_j(e,x,t)\frac{\partial g^{bc}}{\partial t}\widetilde{\nabla}_b T^i_{ac} \\
&- \varepsilon^a_j(e,x,t)g^{bc}\frac{\partial}{\partial t}\widetilde{\nabla}_b T^i_{ac}
\end{aligned} \tag{13.13}$$

It is only the last term in (13.13) that contains the second-order derivatives of h^i_j. Further,

$$\frac{\partial}{\partial t}\widetilde{\nabla}_b T^i_{ac} = \widetilde{\nabla}_b \frac{\partial T^i_{ac}}{\partial t} + \text{lower-order terms} \tag{13.14}$$

Substituting (13.14) into (13.13), we get

$$
\begin{aligned}
-\varepsilon_j^a(e,x,t)g^{bc}\widetilde{\nabla}_b\frac{\partial T^i_{ac}}{\partial t} &= -\varepsilon_j^a(e,x,t)g^{bc}\widetilde{\nabla}_b\big(\varepsilon_c^d(x,e,t)\widetilde{\nabla}_a h^i_d - \varepsilon_a^d(x,e,t)\widetilde{\nabla}_c h^i_d\big) \\
&= -\varepsilon_j^c(e,x,t)\varepsilon_c^d(x,e,t)g^{bc}\widetilde{\nabla}_b\widetilde{\nabla}_a h^i_d + g^{bc}\widetilde{\nabla}_b\widetilde{\nabla}_c h^i_j \qquad (13.15)
\end{aligned}
$$

and the second "elliptic term" on the right-hand side of this equation shows that (13.7) is weakly parabolic and also explains our choice of the minus sign in (13.6).

The idea now, due to DeTurck (see [D]), is to transform (13.15)into an equivalent strongly parabolic equation by canceling its first "bad term." For this purpose, we now choose an arbitrary "connection" $\overline{\Gamma}^i_{jk}$ and define a time-dependent vector field $W(x,t)$ by

$$W^i \stackrel{\text{def}}{=} g^{ab}(\Gamma^i_{ab} - \overline{\Gamma}^i_{ab}) \qquad (13.16)$$

The key fact in this definition is that $\Gamma - \overline{\Gamma}$ is a tensor and $\overline{\Gamma}$ does not depend on t. We now define a second-order nonlinear operator M by the formula

$$M^i_j \stackrel{\text{def}}{=} \varepsilon_j^a(e,x,t)\widetilde{\nabla}_a W^i = \widetilde{\nabla}_{(j)}W^i \qquad (13.17)$$

We compute

$$
\begin{aligned}
\frac{\partial M^i_j}{\partial t} &= \varepsilon_j^c(e,x,t)g^{ab}\widetilde{\nabla}_c\frac{\partial \Gamma^i_{ab}}{\partial t} + \cdots \\
&= \varepsilon_j^c(e,x,t)g^{ab}\widetilde{\nabla}_c\big(\varepsilon_b^d(x,e,t)\widetilde{\nabla}_a h^i_d\big) + \cdots \\
&= \varepsilon_j^c(e,x,t)g^{ab}\varepsilon_b^d(x,e,t)\widetilde{\nabla}_c\widetilde{\nabla}_a h^i_d + \cdots \\
&= \varepsilon_j^c(e,x,t)g^{cb}\varepsilon_b^d(x,e,t)\widetilde{\nabla}_a\widetilde{\nabla}_c h^i_d + \cdots \qquad (13.18)
\end{aligned}
$$

From (13.15) and (13.18), we obtain

$$\frac{\partial M^i_j}{\partial t} - \frac{\partial H^i_j}{\partial t} = g^{bc}\widetilde{\nabla}_b\widetilde{\nabla}_c h^i_j + \cdots \qquad (13.19)$$

Thus, we conclude that the evolution equation

$$\frac{\partial \varepsilon^i_j(e,x,t)}{\partial t} = M^i_j + H^i_j \qquad (13.20)$$

is strongly parabolic. Since M is compact, the general theory (see [LSU]) tells us that (13.20) admits a unique short-time solution for any initial condition. Now let

$\varepsilon(e, x, t) = \varepsilon_t$ be the unique short-time solution of (13.20) starting from ε_0. Let φ_t be the unique short-time solution of the ODE

$$\frac{\partial \varphi_t}{\partial t} = W \tag{13.21}$$

so that φ_t is a family of diffeomorphisms of M with $\varphi_0 = \mathrm{Id}$. We can now easily show following [D] that $\varphi_t^* \varepsilon_t$ solves (13.7) and that this solution is unique, finishing the proof. ■

By techniques in the study of the Ricci flow that have now become standard, we can now prove the following proposition:

Proposition 13.6 *Suppose there exists some a (possibly ∞) such that the solutions of* (13.7) *on* $[0, a)$ *converge to some splitting ε as $t \to a$. Then the limit splitting ε is smooth and* $H(\varepsilon) = 0$.

It is worth noting the following: From (13.2), we get

$$\begin{aligned} \frac{\partial g_{ij}}{\partial t} &= \frac{\partial \varepsilon_i^a}{\partial t} \varepsilon_j^a + \frac{\partial \varepsilon_j^a}{\partial t} \varepsilon_i^a \\ &= \varepsilon_j^a H_i^a + \varepsilon_i^a H_j^a \end{aligned} \tag{13.22}$$

However, (13.22) is not the Ricci flow of g, because its right-hand side cannot be expressed solely in terms of g. Therefore, the unique short-time solutions of (13.22) and the Ricci flow

$$\frac{\partial g_{ij}}{\partial t} = -2\, \mathrm{Ric}_{ij}(g) \tag{13.23}$$

starting from the same initial condition are different. For instance, (13.23) is contracting on S^n, constant on $\mathbb{R}^n$, and expanding on $\mathbb{H}^n$, whereas (13.22) flows in the direction of a locally homogeneous metric. Therefore, we believe that many technical difficulties with (13.23) do not arise with (13.7). However, note that (13.22) is defined only on (M, ε), whereas (13.23) is defined on any Riemannian manifold.

Clearly, $\mathfrak{R} = 0$ implies $H = 0$. Conversely, we will show that $H = 0$ implies $\mathfrak{R} = 0$ if $\dim M = 3$. First, we will prove the following proposition:

Proposition 13.7 *The number of independent components of $\mathfrak{R}$ is $\frac{1}{12}n^2(n^2-1)$, $n = \dim M$, which is equal to the number of independent components of the Riemann curvature tensor R of the metric* (13.2).

Proof We recall that R is determined by the symmetries

$$\begin{aligned} &\text{(i)} \quad R_{ij,km} = -R_{ji,km} \\ &\text{(ii)} \quad R_{ij,km} = -R_{ij,mk} \\ &\text{(iii)} \quad R_{ij,km} + R_{jk,im} + R_{ki,jm} = 0 \end{aligned} \tag{13.24}$$

Therefore, the number of constraints on the components of $\mathfrak{R}$ imposed by $\mathfrak{R}^i_{jk,r} = -\mathfrak{R}^i_{kj,r}$ and the First Bianchi Identity (Proposition 6.4) is equal to the number of constraints on the components of R imposed by (i) and (iii). We will now derive the analog of (ii), which will prove the assertion. We define

$$\begin{aligned} \mathfrak{R}_{jk,rm} &\stackrel{\text{def}}{=} \mathfrak{R}^a_{jk,r}\, g_{am} \\ T_{jk,m} &\stackrel{\text{def}}{=} T^a_{jk}\, g_{am} \end{aligned} \tag{13.25}$$

Therefore,

$$\widetilde{\nabla}_r T_{jk,m} = \widetilde{\nabla}_r(T^a_{jk} g_{am}) = \widetilde{\nabla}_r(T^a_{jk}) g_{am} = \mathfrak{R}^a_{jk,r} g_{am} = \mathfrak{R}_{jk,rm} \tag{13.26}$$

Now,

$$\begin{aligned} \nabla_r g_{ij} &= \widetilde{\nabla}_r g_{ij} + T^a_{ri} g_{aj} + T^a_{rj} g_{ai} \\ &= T_{ri,j} + T_{rj,i} \end{aligned} \tag{13.27}$$

Therefore,

$$\begin{aligned} \nabla_m \nabla_r g_{ij} &= \nabla_m (T_{ri,j} + T_{rj,i}) \\ &= \widetilde{\nabla}_m (T_{ri,j} + T_{rj,i}) + T^b_{mr}(T_{bi,j} + T_{bj,i}) \\ &\quad + T^b_{mi}(T^a_{rb,j} + T^a_{rj,b}) + T^b_{mj}(T_{ri,b} + T_{rb,i}) \\ &= \mathfrak{R}_{ri,mj} - \mathfrak{R}_{rj,mi} + T^b_{mr} T_{bi,j} + T^b_{mr} T_{bj,i} \\ &\quad + T^b_{mi} T_{rb,j} + T^b_{mi} T_{rj,b} + T^b_{mj} T_{ri,b} + T^b_{mj} T_{rb,i} \end{aligned} \tag{13.28}$$

Like (6.55), we have

$$\begin{aligned} \nabla_m \nabla_r g_{ij} - \nabla_r \nabla_m g_{ij} &= \mathfrak{R}^a_{mr,i} g_{aj} + \mathfrak{R}^a_{mr,j} g_{ia} - T^a_{mr} \nabla_a g_{ij} \\ &= \mathfrak{R}_{mr,ij} + \mathfrak{R}_{mr,ij} - T^a_{mr} T_{ai,j} - T^a_{mr} T_{aj,i} \end{aligned} \tag{13.29}$$

We alternate m, r in (13.28) and substitute from (13.29), which gives

$$\mathfrak{R}^a_{mr,ij} + \mathfrak{R}^a_{mr,ij} - \mathfrak{R}_{ri,mj} - \mathfrak{R}_{rj,mi} + \mathfrak{R}_{mi,rj} + \mathfrak{R}_{mj,ri} = A_{mrij} \tag{13.30}$$

where A is an expression depending only on T and g. Note that the left-hand side of (13.30) is skew-symmetric in m, r and symmetric in i, j.

We now rewrite the Bianchi identity (6.18) in two equivalent forms:

$$\mathfrak{R}_{mj,ri} + \mathfrak{R}_{rm,ji} + \mathfrak{R}_{jr,mi} = T_{am,i} T^a_{jr} + T_{aj,i} T^a_{rm} + T_{ar,i} T^a_{mj} \tag{13.31}$$

and

$$\mathfrak{R}_{mi,rj} + \mathfrak{R}_{ir,mj} + \mathfrak{R}_{rm,ij} = T_{am,j}T^a_{ir} + T_{ai,j}T^a_{rm} + T_{ar,j}T^a_{mi} \tag{13.32}$$

Subtracting (13.31) and (13.32) from (13.30), we get

$$2\mathfrak{R}_{mr,ij} + 2\mathfrak{R}_{mr,ji} = B_{mrij} \tag{13.33}$$

for some expression B depending only on T and g. This finishes the proof. ■

Corollary 13.8 *If* $\dim M = 3$, *then* H *and* $\mathfrak{R}$ *have six independent components. Therefore,* $H = 0$ *implies* $\mathfrak{R} = 0$.

Proof The independent components of $\mathfrak{R}$ are

$$\mathfrak{R}_{12,12}, \quad \mathfrak{R}_{13,12}, \quad \mathfrak{R}_{12,23}, \quad \mathfrak{R}_{13,13}, \quad \mathfrak{R}_{23,23}, \quad \mathfrak{R}_{13,23} \tag{13.34}$$

We define

$$H_{ij} \stackrel{\text{def}}{=} H^a_i g_{aj} \tag{13.35}$$

Clearly, H^i_j and H_{ij} have the same number of independent components. From (13.6), we deduce

$$\begin{aligned} H_{jk} = & - g^{11}\mathfrak{R}_{j1,1k} - g^{21}\mathfrak{R}_{j2,1k} - g^{31}\mathfrak{R}_{j3,1k} \\ & - g^{12}\mathfrak{R}_{j1,2k} - g^{22}\mathfrak{R}_{j2,2k} - g^{32}\mathfrak{R}_{j3,2k} \\ & - g^{13}\mathfrak{R}_{j1,3k} - g^{23}\mathfrak{R}_{j2,3k} - g^{33}\mathfrak{R}_{j3,3k} \end{aligned} \tag{13.36}$$

For simplicity, we assume that we make the count at a point where $g_{ij} = \delta_{ij}$. Therefore,

$$H_{jk} = -\mathfrak{R}_{j1,1k} - \mathfrak{R}_{j2,2k} - \mathfrak{R}_{j3,3k} \tag{13.37}$$

and

$$\begin{aligned} H_{11} &= -\mathfrak{R}_{12,21} - \mathfrak{R}_{13,31} \\ H_{12} &= -\mathfrak{R}_{13,32} = H_{21} \\ H_{13} &= -\mathfrak{R}_{12,23} = H_{31} \\ H_{22} &= -\mathfrak{R}_{21,12} - \mathfrak{R}_{23,32} \\ H_{23} &= -\mathfrak{R}_{21,13} = H_{32} \\ H_{33} &= -\mathfrak{R}_{31,13} - \mathfrak{R}_{32,23} \end{aligned} \tag{13.38}$$

From (13.34) and (13.38), we conclude that H_{ij} has the six independent components $H_{11}, H_{12}, H_{13}, H_{22}, H_{23}$, and H_{33}. ■

Clearly, Corollary 13.8 is not very satisfactory: It is desirable to derive a decomposition of $\mathfrak{R}$ into a direct sum of an H-component and a trace-free component in any dimension so that Corollary 13.8 will assert that the trace-free component vanishes if $\dim M = 3$.

Finally, a serious issue is that of whether the solution of (13.7) converges (in finite or infinite time) on a compact and simply connected M with $\dim M = 3$. An affirmative answer will imply the Poincaré Conjecture by Propositions 13.1 and 13.6 and Corollary 13.8 *without surgeries.*

We hope to have presented sufficient evidence to convince the reader that it is worth studying the behavior of (13.7) on particular examples.

14

The Van Est Theorem

As we remarked at the end of Chapter 11, the linear Spencer sequence (11.21) is also the horizontal complex of the linear PDE $\nabla\xi = 0$. In this chapter, we will first make this statement precise and then apply this idea to the nonlinear PDE (2.4). This will give another nonlinear analog of (11.21) other than the nonlinear Spencer sequence in (8.14). Our purpose in this chapter is to construct this nonlinear horizontal complex. Once this is done, this complex extends naturally to a double complex and gives an easy proof of the Van Est theorem in [Es].

We start by taking a glance at the main idea of horizontal cohomology. Consider k-forms of the form

$$\omega = \frac{1}{k!}\omega_{i_1\ldots i_k}(x,\xi)dx^{i_1} \wedge \cdots \wedge dx^{i_k} \tag{14.1}$$

where ξ denotes an element in the fiber $\pi^{-1}(x)$ of the tangent bundle $\pi : T \to M$. As in Chapter 11, we write (14.1) in brief as $\omega = (\omega_{i_1\ldots i_k}(x,\xi)) = (\omega_I(x,\xi))$. We define the total differentiation operator $\widetilde{\delta}_r$ by the formula

$$\left(\widetilde{\delta}_r\omega_{i_1\ldots i_k}\right)(x,\xi) \stackrel{\text{def}}{=} \frac{\partial\omega_{i_1\ldots i_k}(x,\xi)}{\partial x^r} + \frac{\partial\omega_{i_1\ldots i_k}(x,\xi)}{\partial\xi^a}\Gamma^a_{rb}\xi^b \tag{14.2}$$

Note that we regard ξ as a section $\xi(x)$ of $T \to M$ in (14.2), differentiate with respect to x^r, and substitute $\partial\xi^a/\partial x^r$ formally from (5.9). Denoting the space of k-forms (14.1) by $\wedge^k(T,\cdot)$, we define

$$\begin{aligned}\widetilde{D} &: \wedge^k(T,\cdot) \longrightarrow \wedge^{k+1}(T,\cdot)\\ &: \omega_{i_1\ldots i_k}(x,\xi) \longrightarrow \left[\delta_r\omega_{i_1\ldots i_k}(x,\xi)\right]_{[ri_1\ldots i_k]}\end{aligned} \tag{14.3}$$

Using $\widetilde{\mathfrak{R}} = 0$, we can easily check that $\widetilde{D}^2 = 0$. Similarly, if $\mathfrak{R} = 0$, we can define δ_r accordingly by formally substituting from $\nabla\xi = 0$ and get $D^2 = 0$. The idea is now clear: We can define forms $\omega = (\omega_{i_1\ldots i_k}(x,\xi))$, where $\xi \in \pi^{-1}(x)$, $\pi : T^{r,s} \to M$, define $\widetilde{\delta}_r$ (or δ_r)

An Alternative Approach to Lie Groups and Geometric Structures, Ercüment H. Ortaçgil (2018).

by formally substituting from $\widetilde{\nabla}\xi = 0$ (or $\nabla\xi = 0$), and get $\widetilde{D} : \wedge^k(T^{r,s}, \cdot) \longrightarrow \wedge^{k+1}(T^{r,s}, \cdot)$, with $\widetilde{D}^2 = 0$ (or $D^2 = 0$, assuming $\mathfrak{R} = 0$).

Now, since $T \to M$ is a vector bundle, it is reasonable to assume that ω in (14.1) depends linearly on ξ; that is, $\omega_I(x, \xi) = \omega_{I,a}(x)\xi^a$. With this assumption, the complex defined by (14.3) is called the horizontal complex of the linear PDE $\widetilde{\nabla}\xi = 0$. It is now easy to see that the operator $D : \wedge^k(T, \cdot) \longrightarrow \wedge^{k+1}(T, \cdot)$ becomes the same operator $D : \wedge^k(T^*) \otimes T^* \longrightarrow \wedge^{k+1}(T^*) \otimes T^*$ in the linear Spencer sequence with the representation $T^* \to M$ as defined in Chapter 11, and the horizontal operator $D : \wedge^k(T^*, \cdot) \longrightarrow \wedge^{k+1}(T^*, \cdot)$ becomes $D : \wedge^k(T^*) \otimes T \longrightarrow \wedge^{k+1}(T^*) \otimes T$ with the representation $T \to M$. More generally, the horizontal complex $(D, \wedge^*(T^{r,s}, \cdot))$ is the linear Spencer complex with the representation $(T^{r,s})^* = T^{s,r} \to M$.

The above simple derivation of the linear Spencer complexes with tensorial representations in Chapter 11 in a unified manner shows that the formal substitution from a PDE due to [Vn] (see [V]) is a fundamental idea, and now we want to apply this idea to the nonlinear PDE (2.4). So, we consider k-forms of the form

$$\omega_{j_1 \ldots j_k}(x, y) \tag{14.4}$$

where $(x, y) \in M \times M$. Without worrying much about the meaning of the local expression (14.4) for the moment, we define δ_r formally by

$$\left(\delta_r \omega_{j_1 \ldots j_k}\right)(x, y) \stackrel{\text{def}}{=} \frac{\partial \omega_{j_1 \ldots j_k}(x, y)}{\partial x^r} + \frac{\partial \omega_{j_1 \ldots j_k}(x, y)}{\partial y^a} \varepsilon^a_r(x, y) \tag{14.5}$$

As in (14.2), note that we regard y as a solution $y = y(x)$ of (2.4), differentiate with respect to x^r, and substitute formally from (2.4). Denoting the space of k-forms (14.4) by $\wedge^k(M, M)$, we define

$$\begin{aligned} D : \wedge^k(M, M) &\longrightarrow \wedge^{k+1}(M, M) \\ : \omega_{j_1 \ldots j_k}(x, y) &\longrightarrow \left[\delta_r \omega_{j_1 \ldots j_k}(x, y)\right]_{[r j_1 \ldots j_k]} \end{aligned} \tag{14.6}$$

If $\mathcal{R} = 0$, we can easily check that $D^2 = 0$ (see below), and we obtain the complex

$$\wedge^0(M, M) \longrightarrow \wedge^1(M, M) \longrightarrow \ldots \longrightarrow \wedge^n(M, M) \tag{14.7}$$

Coming back to (14.4), we recall that $\wedge^k_\varepsilon$ denotes the space of ε-invariant k-forms on M. Now, ω assigns to any $(p, q) \in M \times M$, an element of $\wedge^k_\varepsilon$. However, there is an ambiguity with the notation of (14.4): Choosing coordinates $p \in (U, x)$, $q \in (V, y)$, it is not clear from (14.4) whether the form indices $I = (j_1, \ldots, j_k)$ in the coordinate expression of $\omega(p, q) \in \wedge^k_\varepsilon$ refer to (U, x) or (V, y) or any other $(W, z) \subset M$. We agree that (14.4) is the local expression of $\omega(x, y)$ on the neighborhood U of the first argument x, keeping in mind that we can always translate $\omega(x, y)_{|U}$ by ε without changing its value. In fact, we can interpret $\omega(x, y)$ as a k-form at the tangent space at x, since we can translate this k-form to M, obtaining an element of $\wedge^k_\varepsilon$.

Clearly, $\wedge^0(M, M) = C(M \times M) =$ the space of smooth functions on $M \times M$.
Now we want to understand the kernel of the first operator in (14.7), which is defined by

$$(D\varphi)(x,y) = \frac{\partial\varphi(x,y)}{\partial x^r} + \frac{\partial\varphi(x,y)}{\partial y^a}\varepsilon_r^a(x,y) \tag{14.8}$$

Henceforth, we assume that $(M, \varepsilon, \mathcal{G})$ is globalizable, so that we have the PHS $(M, \mathcal{G})$. For $f \in \mathcal{G}$, we define its graph $\mathrm{gr}(f) \overset{\text{def}}{=} \{(x, f(x)) \mid x \in M\} \subset M \times M$.

Lemma 14.1 *The following are equivalent:*

(i) $D(\varphi) = 0, \varphi \in C(M \times M)$;

(ii) φ *is constant on* $\mathrm{gr}(f)$ *for all* $f \in \mathcal{G}$.

Proof (ii) $\Longrightarrow$ (i): Fix some $(x_0, y_0) \in M \times M$. There exists a unique $f \in \mathcal{G}$ with $f(x_0) = y_0$ and solves

$$\frac{\partial f^i(x)}{\partial x^r} = \varepsilon_r^i(x, f(x)) \tag{14.9}$$

on M. Since $\varphi(x, f(x)) = c$, $x \in M$, (14.8) gives

$$\begin{aligned}(D\varphi)(x_0,y_0) &= \frac{\partial\varphi(x_0,y_0)}{\partial x^r} + \frac{\partial\varphi(x_0,y_0)}{\partial y^a}\varepsilon_r^a(x_0,y_0)\\ &= \left[\frac{\partial\varphi(x,f(x))}{\partial x^r} + \frac{\partial\varphi(x,f(x))}{\partial y^a}\varepsilon_r^a(x,f(x))\right]_{x=x_0}\\ &= \left[\frac{\partial\varphi(x,f(x))}{\partial x^r} + \frac{\partial\varphi(x,f(x))}{\partial y^a}\frac{\partial f^i(x)}{\partial x^r}\right]_{x=x_0}\\ &= [d\varphi(x,f(x))_r]_{x=x_0}\\ &= 0\end{aligned} \tag{14.10}$$

(i) $\Longrightarrow$ (ii): This follows from (14.10) by reversing the steps. ■

Now let $\mathcal{C}(\mathcal{G})$ denote the space of smooth real valued functions on $\mathcal{G}$ (not the centralizer $C(\mathcal{G})$ in Chapter 4!). Since a choice of basepoint gives a 1–1 correspondence between $\mathcal{G}$ and M, smoothness has a well-defined meaning. Since $f \in \mathcal{G}$ is determined on M (globalizable!) by $(x_0, f(x_0))$ for any $x_0 \in M$ and φ is constant on $gr(f)$, we assign this constant value to f, which identifies the kernel of the first operator in (14.7) with $\mathcal{C}(\mathcal{G})$. So, we get

$$0 \longrightarrow \mathcal{C}(\mathcal{G}) \longrightarrow \wedge^0(M,M) \longrightarrow \ldots \longrightarrow \wedge^n(M,M) \tag{14.11}$$

Definition 14.2 *For a globalizable LLG* $(M, \varepsilon, \mathcal{G})$, (14.11) *is the horizontal complex of the PHS* $(M, \mathcal{G})$.

If $(M, \varepsilon, \mathcal{G})$ is not globalizable, (14.7) is still defined and $\mathcal{C}(\mathcal{G})$ denotes the sheaf of "locally constant functions" on the pseudogroup $\mathcal{G}$.

We now repeat the construction of (14.11) with M instead of $M \times M$. So, a k-form ω assigns to $p \in M$ an element of $\wedge_\varepsilon^k$. However, this is the same as assigning to p a k-form at the tangent space at p that is the value $\omega(p)$ of $\omega \in \wedge_\varepsilon^k$ at p. Therefore, any k-form in this sense defines an ordinary k-form w. Conversely, any ordinary k-form w assigns to p an element of $\wedge_\varepsilon^k$ obtained by translating the value $w(p)$ to M. Thus, we can regard ω as an ordinary k-form on M. With this identification, the second term on the right-hand side of (14.5) drops out and D becomes the exterior derivative. Thus, we get the de Rham complex

$$0 \longrightarrow \mathbb{R} \longrightarrow \wedge^0(M) \longrightarrow \wedge^1(M) \longrightarrow \ldots \longrightarrow \wedge^n(M) \tag{14.12}$$

Finally, we repeat the construction of (14.11) with no M. This gives the complex (10.11), which we now write as

$$0 \longrightarrow \mathbb{R} \longrightarrow \wedge_\varepsilon^1(\varnothing) \longrightarrow \wedge_\varepsilon^2(\varnothing) \ldots \longrightarrow \wedge_\varepsilon^n(\varnothing) \tag{14.13}$$

We now change our notation, setting $\wedge_\varepsilon^k(\varnothing) = \wedge^{k,0}$, $\wedge^k(M) = \wedge^{k,1}$ and $\wedge^k(M, M) = \wedge^{k,2}$. We claim that we have the commutative diagram

$$\begin{array}{ccccccccccc}
0 & \longrightarrow & C(\mathcal{G}) & \longrightarrow & \wedge^{0,2} & \longrightarrow & \wedge^{1,2} & \longrightarrow & \cdots & \longrightarrow & \wedge^{n,2} \\
 & & \uparrow & & \uparrow & & \uparrow & & \uparrow & & \uparrow \\
0 & \longrightarrow & \mathbb{R} & \longrightarrow & \wedge^{0,1} & \longrightarrow & \wedge^{1,1} & \longrightarrow & \cdots & \longrightarrow & \wedge^{n,1} \\
 & & & & \uparrow & & \uparrow & & \uparrow & & \uparrow \\
 & & & & \mathbb{R} & \longrightarrow & \wedge^{1,0} & \longrightarrow & \cdots & \longrightarrow & \wedge^{n,0} \\
 & & & & \uparrow & & \uparrow & & \uparrow & & \uparrow \\
 & & & & 0 & & 0 & & 0 & & 0
\end{array} \tag{14.14}$$

The vertical maps $\wedge^{k,0} \rightarrow \wedge^{k,1}$ are injections given by (10.12), and we know that the bottom squares in (14.14) commute. The map $\mathbb{R} \rightarrow C(\mathcal{G})$ is the obvious inclusion. We need to define the top vertical maps in (14.14) and show that the top squares also commute. For $0 \leq k \leq n$, we define

$$\begin{aligned} &\wedge^{k,1} \longrightarrow \wedge^{k,2} \\ &\omega \longrightarrow \overline{\omega}, \qquad \overline{\omega}(p,q) \stackrel{\text{def}}{=} \omega(p) - \omega(q) \end{aligned} \tag{14.15}$$

We recall that ω is a k-form with values in $\wedge_\varepsilon^k$ and the subtraction in (14.15) is done in $\wedge_\varepsilon^k$. Hence, the compositions of the vertical maps in (14.14) vanish and we have exactness at the middle terms. The first top square obviously commutes. Let us check the second top square, since its commutativity implies the commutativity of the other squares by the definitions. So, let $\omega : M \to \mathbb{R}$. Now, $(d\omega)_r = \partial\omega(x)/\partial x^r$ and

$$((\delta \circ d)\omega)\,(x,y) = \frac{\partial\omega(x)}{\partial x^r} - \frac{\partial\omega(y)}{\partial y^a}\varepsilon_r^a(x,y) \tag{14.16}$$

We observe that in (14.16), we have translated the value $(d\omega)_j(y) = \partial\omega(y)/\partial y^j$ on (V,y) to (U,x) in order to compute the difference on (U,x) according to our agreement. Now $(\delta\omega)(x,y) = \omega(x) - \omega(y)$ according to (14.15), and therefore $((D \circ \delta)\omega)\,(x,y)$ is equal to (14.16) in view of (14.8). To show the commutativity of the other squares, we apply this argument to the functions ω_I and then alternate.

The diagram (14.14) and the formula (14.15) show the way in which we should proceed.

Let A be any set and V be any vector space. Let F^k denote the vector space of functions $f : A \times A \times \cdots \times A \to V$ (k-factors). We set $F^0 = V$. We define $\delta : F^k \to F^{k+1}$ by the well-known formula

$$(\delta f)(x_1, \ldots, x_{k+1}) \stackrel{\text{def}}{=} \sum_{1 \le i \le k+1} (-1)^{i+1} f(x_1, \ldots, \widehat{x_i}, ..x_{k+1}) \tag{14.17}$$

and get the complex

$$0 \longrightarrow V \longrightarrow F^1 \longrightarrow F^2 \longrightarrow F^3 \longrightarrow \cdots \tag{14.18}$$

Since an arbitrary set should not have any nontrivial cohomology, intuition tells us that the cohomology H^k of (14.18) at F^k vanishes for $k \geq 2$. This is indeed true. For instance, suppose $f \in F^3$ and $\delta f = 0$. Now

$$\begin{aligned}(\delta f)(x_1, x_2, x_3, x_4) &= f(x_2, x_3, x_4) - f(x_1, x_3, x_4) + f(x_1, x_2, x_4) - f(x_1, x_2, x_3)\\ &= 0\end{aligned} \tag{14.19}$$

We define $g \in F^2$ by

$$g(a, b) \stackrel{\text{def}}{=} f(a, b, x_4) \tag{14.20}$$

and compute

$$\begin{aligned}(\delta g)(x_1, x_2, x_3) &= g(x_2, x_3) - g(x_1, x_3) + g(x_1, x_2)\\ &= f(x_2, x_3, x_4) - f(x_1, x_3, x_4) + f(x_1, x_2, x_4)\\ &= f(x_1, x_2, x_3)\end{aligned} \tag{14.21}$$

Therefore, $\delta g = f$ and $H^3 = 0$. Similarly, $H^k = 0$ for all $k \geq 2$. If A is a topological space, V is finite-dimensional, and F^k consists of continuous maps, then again $H^* = 0$ because the primitives defined by (14.20) are also continuous. The same conclusion holds if $A = M$ is a smooth manifold and F^k consists of smooth maps.

We now fix k, $1 \leq k \leq n =$ dim M and consider the space $\wedge^{k,m}$ of smooth maps

$$f : M \times M \times \cdots \times M \longrightarrow \wedge^k_\varepsilon \qquad (m \text{ factors}) \tag{14.22}$$

The above construction gives the complex

$$0 \longrightarrow \wedge^{k,0} \longrightarrow \wedge^{k,1} \longrightarrow \wedge^{k,2} \longrightarrow \wedge^{k,3} \longrightarrow \tag{14.23}$$

with vanishing cohomology $H^{k,m}_{(v)} = 0$ for $m \geq 2$, $1 \leq k \leq n$, where (v) denotes vertical. Thus, we extended the vertical columns of (14.14) indefinitely to exact sequences, obtaining the diagram

$$\begin{array}{ccccccccccccc}
 & & & & \cdots & & \cdots & & \cdots & & \cdots & & \cdots \\
 & & & & \uparrow & & \uparrow & & \uparrow & & \uparrow & & \uparrow \\
 & & & & \wedge^{0,4} & & \wedge^{1,4} & & \wedge^{2,4} & & \cdots & & \wedge^{n,4} \\
 & & & & \uparrow & & \uparrow & & \uparrow & & \uparrow & & \uparrow \\
 & & & & \wedge^{0,3} & & \wedge^{1,3} & & \wedge^{2,3} & & \cdots & & \wedge^{n,3} \\
 & & & & \uparrow & & \uparrow & & \uparrow & & \uparrow & & \uparrow \\
0 & \longrightarrow & C(\mathcal{G}) & \longrightarrow & \wedge^{0,2} & \longrightarrow & \wedge^{1,2} & \longrightarrow & \wedge^{2,2} & \longrightarrow & \cdots & \longrightarrow & \wedge^{n,2} \\
 & & \uparrow & & \uparrow & & \uparrow & & \uparrow & & \uparrow & & \uparrow \\
0 & \longrightarrow & \mathbb{R} & \longrightarrow & \wedge^{0,1} & \longrightarrow & \wedge^{1,1} & \longrightarrow & \wedge^{2,1} & \longrightarrow & \cdots & \longrightarrow & \wedge^{n,1} \\
 & & & & \uparrow & & \uparrow & & \uparrow & & \uparrow & & \uparrow \\
 & & & & \mathbb{R} & \longrightarrow & \wedge^{1,0} & \longrightarrow & \wedge^{2,0} & \longrightarrow & \cdots & \longrightarrow & \wedge^{n,0} \\
 & & & & \uparrow & & \uparrow & & \uparrow & & \uparrow & & \uparrow \\
 & & & & 0 & & 0 & & 0 & & 0 & & 0
\end{array} \tag{14.24}$$

It remains to define the horizontal operators in the sequences

$$\wedge^{0,m} \longrightarrow \wedge^{1,m} \longrightarrow \wedge^{2,m} \longrightarrow \wedge^{3,m} \longrightarrow \ldots \longrightarrow \wedge^{n,m} \tag{14.25}$$

for $m \geq 3$, show the commutativity of the resulting diagram, and determine the kernels of the first operators in (14.25). The formula (14.8) for $m = 2$ tells us how we should proceed. We first define the first operator in (14.25). For simplicity of notation, take $m = 3$, since the general case is similar. We recall that $\wedge^{0,3}$ is the space of functions $M \times M \times M \longrightarrow \mathbb{R}$. We define

$$\delta : \wedge^{0,3} \longrightarrow \wedge^{1,3} \tag{14.26}$$

by the formula

$$\left(\delta_r f\right)(x,y,z) \stackrel{\text{def}}{=} \frac{\partial f(x,y,z)}{\partial x^r} + \frac{\partial f(x,y,z)}{\partial y^a}\varepsilon^a_r(x,y) + \frac{\partial f(x,y,z)}{\partial z^a}\varepsilon^a_r(x,z) \tag{14.27}$$

We extend δ to an operator $D\colon \wedge^{k,3} \longrightarrow \wedge^{k+1,3}$ by defining it on the k-forms $\omega = \omega_I(x,y,z)dx^I$ by

$$(D\omega)(x,y,z) \stackrel{\text{def}}{=} \left(\delta_r \omega_I(x,y,z)\right) dx^r \wedge dx^I \tag{14.28}$$

Equivalently, if $\omega = (\omega_{j_1\ldots j_k}(x,y,z))$, then

$$(D\omega)_{rj_1\ldots j_k}(x,y,z) = \left[\delta_r \omega_{j_1\ldots j_k}(x,y,z)\right]_{[rj_1\ldots j_k]} \tag{14.29}$$

We claim that $D^2 = 0$. To show this, it suffices to check that $\delta_m\delta_r f(x,y,z)$ is symmetric in m, r for a function f in view of (14.28). However, this is clear from the definition of δ_r, since (14.9) has unique solutions with arbitrary initial conditions: For instance, all we need to do is to substitute (14.9) into (14.27) to see that $\delta_m\delta_r f(x,y,z)$ is symmetric in m, r. This argument generalizes to an arbitrary number of products of M in the obvious way. In fact, applying δ_m to (14.27) and rearranging terms, a tedious computation gives the formula

$$\begin{aligned}\delta_m\delta_r f(x,y,z) = S_{mr}(x,y,z) &+ \frac{\partial f(x,y,z)}{\partial y^a}\left(\frac{\partial \varepsilon^a_r(x,y)}{\partial x^m} + \frac{\partial \varepsilon^a_r(x,y)}{\partial y^c}\varepsilon^c_m(x,y)\right)\\ &+ \frac{\partial f(x,y,z)}{\partial z^a}\left(\frac{\partial \varepsilon^a_r(x,z)}{\partial x^m} + \frac{\partial \varepsilon^a_r(x,z)}{\partial z^b}\varepsilon^b_m(x,z)\right)\end{aligned} \tag{14.30}$$

for some symmetric expression S_{mr}. Since $\mathcal{R}(x,y) = 0$, the second term on the right-hand side of (14.31) is also symmetric in m, r. Similarly, the third term is symmetric since $\mathcal{R}(x,z) = 0$.

Let $C(\mathcal{G} \times \mathcal{G}) = C(\mathcal{G}^{(2)})$ be the space of smooth real-valued functions on $\mathcal{G} \times \mathcal{G}$. We claim that the kernel of the first operator in (14.25) is $C(\mathcal{G} \times \mathcal{G})$ for $m = 3$. For $f, g \in \mathcal{G}$, let $\mathrm{gr}(g \circ f) \stackrel{\text{def}}{=} \{(x,y,z) \mid y = f(x), z = g(y)\} \subset M \times M \times M$. As in Lemma 14.1, we can easily show that $D\varphi = 0$ if and only if φ is constant on $\mathrm{gr}(g \circ f)$ for all $f, g \in \mathcal{G}$, and the claim follows. The general case is similar, and the commutativity is checked as in the case of (14.16). The final diagram is

$$
\begin{array}{ccccccccccccc}
0 & \longrightarrow & \cdots & \longrightarrow & \cdots & \longrightarrow & \cdots & \longrightarrow & \cdots & \longrightarrow & \cdots & \longrightarrow & \cdots \\
 & & \uparrow & & \uparrow & & \uparrow & & \uparrow & & \uparrow & & \uparrow \\
0 & \longrightarrow & C(\mathcal{G}^3) & \longrightarrow & \wedge^{0,4} & \longrightarrow & \wedge^{1,4} & \longrightarrow & \wedge^{2,4} & \longrightarrow & \cdots & \longrightarrow & \wedge^{n,4} \\
 & & \uparrow & & \uparrow & & \uparrow & & \uparrow & & \uparrow & & \uparrow \\
0 & \longrightarrow & C(\mathcal{G}^2) & \longrightarrow & \wedge^{0,3} & \longrightarrow & \wedge^{1,3} & \longrightarrow & \wedge^{2,3} & \longrightarrow & \cdots & \longrightarrow & \wedge^{n,3} \\
 & & \uparrow & & \uparrow & & \uparrow & & \uparrow & & \uparrow & & \uparrow \\
0 & \longrightarrow & C(\mathcal{G}) & \longrightarrow & \wedge^{0,2} & \longrightarrow & \wedge^{1,2} & \longrightarrow & \wedge^{2,2} & \longrightarrow & \cdots & \longrightarrow & \wedge^{n,2} \\
 & & \uparrow & & \uparrow & & \uparrow & & \uparrow & & \uparrow & & \uparrow \\
0 & \longrightarrow & \mathbb{R} & \longrightarrow & \wedge^{0,1} & \longrightarrow & \wedge^{1,1} & \longrightarrow & \wedge^{2,1} & \longrightarrow & \cdots & \longrightarrow & \wedge^{n,1} \\
 & & & & \uparrow & & \uparrow & & \uparrow & & \uparrow & & \uparrow \\
 & & & & \mathbb{R} & \longrightarrow & \wedge^{1,0} & \longrightarrow & \wedge^{2,0} & \longrightarrow & \cdots & \longrightarrow & \wedge^{n,0} \\
 & & & & \uparrow & & \uparrow & & \uparrow & & \uparrow & & \uparrow \\
 & & & & 0 & & 0 & & 0 & & 0 & & 0
\end{array}
\tag{14.31}
$$

Definition 14.3 (14.31) *is the Van Est double complex of the PHS* $(M, \mathcal{G})$.

We note the crucial fact that the boundary operator (14.17) is the same as (1.5) in [EM] but the cochains in the second column of (14.31) are arbitrary and do not satisfy the homogeneity condition (1.4) in [EM]. However, in view of the above identification, the cochains in the first column of (14.31) are the same as the nonhomogeneous cochains in [EM], and the operators in the first column of (14.31) are given by the usual boundary operator (2.2) in [EM]. Therefore, the first column of (14.31) computes the group cohomology $H^*(\mathcal{G}, \mathbb{R})$ of $\mathcal{G}$ with smooth cochains. As we did in Chapter 10, we denote the cohomology of the bottom horizontal sequence by $H^*(\mathfrak{g}, \mathbb{R})$, where $\mathfrak{g} = \widetilde{\mathfrak{X}}_{\varepsilon}$. By construction, the vertical cohomology $H^{i,j}_{(v)}$ of (14.31) at $\wedge^{i,j}$ vanishes for $i \geq 0, j \geq 1$. Let $H^{i,j}_{(h)}$ denote its horizontal cohomology at $\wedge^{i,j}$. Clearly, $H^{i,1}_{(h)} = H_{\mathrm{dR}}(M, \mathbb{R})$ and $H^{i,0}_{(h)} = H^i(\mathfrak{g}, \mathbb{R})$.

Since the vertical cohomology of (14.31) vanishes except for the first column, the standard "zig-zag" argument (see [BT] for details) gives a homomorphism

$$
H^i(\mathcal{G}, \mathbb{R}) \longrightarrow H^i(\mathfrak{g}, \mathbb{R}) \tag{14.32}
$$

which is defined as follows: Let a be a representative of some cohomology class in $H^i(\mathcal{G}, \mathbb{R})$. Since the square with the left bottom vertex at a commutes, a maps to the right to some $b \in \wedge^{0,i+1}$ that maps to zero at the top. Since $H^{0,i+1}_{(v)} = 0$, b is the vertical image of some

$c \in \wedge^{1,i}$. Now c maps to the right to some $d \in \wedge^{2,i}$ that maps to zero vertically, and we start again using $H^{2,i}_{(v)} = 0 \ldots$ finally getting some $e \in \wedge^{i+1,0}$ that maps to zero to the right, defining an element of $H^i(\mathfrak{g}, \mathbb{R})$. We can easily check that the choices made along the way are immaterial, and (14.32) is well-defined. Note that this argument uses only the vanishing of the vertical cohomology of (14.31) starting with the second column. If the horizontal cohomology of (14.31) vanishes too, then we can change the roles of the rows and columns using the symmetry of (14.31) and establish an inverse to (14.32), proving that (14.32) is an isomorphism. The surprising fact here is that the exactness of the second row of (14.31), which computes the de Rham cohomology of M, implies the exactness of all the rows above it.

Proposition 14.4 *If $H^{k,1}_{(h)} = H^k_{\mathrm{dR}}(M, \mathbb{R}) = 0$ for $k \geq 1$, then $H^i(\mathcal{G}, \mathbb{R}) \simeq H^i(\mathfrak{g}, \mathbb{R})$ for $i \geq 1$. In particular, $H^i(\mathcal{G}, \mathbb{R}) = 0$ for $i \geq n+1$.*

We will sketch the idea of the proof, leaving the technical details to the interested reader. Let $\wedge^{*,m}$ denote the complex (14.25) with cohomology $H^{*,m}_{(h)}$. Let $\overline{\wedge}^{k,m}$ denote the space of ordinary k-forms on $M \times M \times \cdots \times M = M^n$. The idea is to observe an inclusion map $i : \wedge^{*,m} \to \overline{\wedge}^{*,m}$ and a surjection $\alpha : \overline{\wedge}^{*,m} \to \wedge^{*,m}$ with $\alpha \circ i = \mathrm{Id}$. This implies that $H^{i,m}_{(h)}$ injects into $\overline{H}^i_{\mathrm{dR}}(M^m, \mathbb{R})$. By the Künneth formula, $\overline{H}^i_{\mathrm{dR}}(M^m, \mathbb{R}) = \bigoplus_{r+s=i} H^r_{\mathrm{dR}}(M, \mathbb{R}) \otimes H^s_{\mathrm{dR}}(M, \mathbb{R}) = 0$ for $i \geq 1$ by our assumption. It follows that all the rows of (14.31) are exact except the bottom one, and the zig-zag argument finishes the proof as explained above.

15

The Symmetry Group

In principle this chapter belongs to Part III, since it deals with the first example of a transitive transformation group that does not act simply transitively. However, the concept of the symmetry group greatly clarifies some constructions of Chapters 9 and 11, and also serves as a good motivation for Part III.

Let $(M, \varepsilon, \mathcal{G})$ be an LLG, as we will always assume in this chapter. Suppose some 1-arrow $j_1(f)^{p,q} = (x^i, y^i, f^i_j) \in \mathcal{U}_1^{p,q}$ from p to q (written briefly as (f^i_j) if the source and target are understood) maps the torsion $T(p)$ to $T(q)$, or, equivalently,

$$T^i_{ab}(q) f^a_j f^b_k = f^i_a T^a_{jk}(p) \tag{15.1}$$

If $j_1(g)^{p,q} = (g^i_j)$ is another such 1-arrow, then $(j_1(f)^{p,q})^{-1} \circ j_1(g)^{p,q} = j_1(f^{-1} \circ g)^{p,p} \in \operatorname{Aut}(\mathfrak{g}(p))$, where $\mathfrak{g} = \widetilde{\mathfrak{X}}_\varepsilon$. We define

$$S_1^{p,q} \overset{\text{def}}{=} \{j_1(f)^{p,q} \in \mathcal{U}_1^{p,q} \mid j_1(f)^{p,q}(T(p)) = T(q)\} \subset \mathcal{U}_1^{p,q} \tag{15.2}$$

and $S_1 \overset{\text{def}}{=} \cup_{p,q} S_1^{p,q} \subset \mathcal{U}_1$. The obvious composition and inversion maps $S_1^{q,r} \times S_1^{p,q} \longrightarrow S_1^{q,r}$ and $S_1^{p,q} \rightarrow S_1^{p,q}$ endow $S_1 \subset \mathcal{U}_1$ with the structure of a subgroupoid. It is easy to check that $S_1 \subset \mathcal{U}_1$ is an embedded submanifold and that the above operations are smooth.

Now, $S_1 \subset \mathcal{U}_1$ is a very special subgroupoid: Above any 1-arrow of S_1, there is a naturally defined unique 2-arrow. To see this, we first recall the definition of a 2-arrow $j_2(f)^{p,q} = (x^i, y^i, f^i_j, f^i_{jk})$ from p to q given in Chapter 1. We let $\mathcal{U}_2^{p,q}$ denote the set of all 2-arrows from p to q and define $\mathcal{U}_2 \overset{\text{def}}{=} \cup_{p,q \in M} \mathcal{U}_2^{p,q}$. Now the groupoid $\mathcal{U}_2$ has a natural differentiable structure and is called the universal groupoid on M (of order 2). Using the chain rule, it is easy to check that the composition of $\mathcal{U}_2$ is given by

$$\left(y^i, z^i, f^i_j, f^i_{jk}\right) \circ \left(x^i, y^i, g^i_j, g^i_{jk}\right) = \left(x^i, z^i, f^i_a g^a_j, f^i_a g^a_{jk} + f^i_{ab} g^a_j g^b_k\right) \tag{15.3}$$

in coordinates. We have the projection homomorphism

$$\begin{aligned} \pi : \mathcal{U}_2 &\longrightarrow \mathcal{U}_1 \\ j_2(f)^{p,q} &\longrightarrow j_1(f)^{p,q} \\ \left(x^i, y^i, f^i_j, f^i_{jk}\right) &\longrightarrow \left(x^i, y^i, f^i_j\right) \end{aligned} \tag{15.4}$$

which restricts to a homomorphism $\mathcal{U}_2^{p,p} \longrightarrow \mathcal{U}_1^{p,p}$ of groups. We will study the structure of $\mathcal{U}_2^{p,p}$ and (15.4) in more detail in Chapter 17. The abovementioned special property of S_1 now reduces to the existence of a natural splitting

$$\delta : S_1 \longrightarrow \mathcal{U}_2, \qquad \pi \circ \delta = \mathrm{Id}_{S_1} \tag{15.5}$$

which preserves the composition and inversion of arrows by the definition of a splitting.

In order to define δ, let $\Gamma^i_{jk}(p)$ be some numerical values given at p and not necessarily equal to (5.1). Given some 2-arrow $j_2(f)^{p,q} = \left(x^i, y^i, f^i_j, f^i_{jk}\right) \in \mathcal{U}_2^{p,q}$, we define $\overline{\Gamma}^i_{jk}(q)$ using the transformation rule (5.2) by

$$\overline{\Gamma}^i_{jk}(q) \stackrel{\text{def}}{=} f^i_{ca}\, g^c_k g^a_j + f^i_a \Gamma^a_{bc}(p) g^c_j g^b_k \tag{15.6}$$

where $g^i_a f^a_j = f^i_a g^a_j = \delta^i_j$. We write (15.6) in brief as $j_2(f)^{p,q}\Gamma(p) = \overline{\Gamma}(q)$. Now, if $\Gamma(p)$ is defined by (5.1), then $\overline{\Gamma}(q)$ need not be equal to $\Gamma(q)$; that is, $j_2(f)^{p,q}$ need not preserve Γ that is defined on M by (5.1). We now define

$$S_2^{p,q} \stackrel{\text{def}}{=} \{ j_2(f)^{p,q} \in \mathcal{U}_2^{p,q} \mid j_2(f)^{p,q}\Gamma(p) = \Gamma(q) \} \tag{15.7}$$

The set (15.7) is not empty: If we set $\overline{\Gamma} = \Gamma$ in (15.6) and skew-symmetrize j, k, we get (15.1). This gives the splitting δ that we are seeking: Choose some $j_1(f) = \left(f^i_j\right) \in S_1^{p,q}$, substitute into (15.6), and solve $\left(f^i_{jk}\right)$ uniquely from (15.6) so that $\left(f^i_j, f^i_{jk}\right)$ is the unique 2-arrow above $\left(f^i_j\right)$; that is, $\delta\left(f^i_j\right) = \left(f^i_j, f^i_{jk}\right)$. Thus, we obtain the subgroupoid $S_2 \subset \mathcal{U}_2$, and the projection (15.4) maps S_2 isomorphically onto S_1. Equivalently,

$$\delta(S_1) = S_2 \subset \mathcal{U}_2 \tag{15.8}$$

where δ is defined by (15.1) and (15.6). We observe the similarity between (15.8) and $\varepsilon(\mathcal{U}_0) = \varepsilon(M \times M) \subset \mathcal{U}_1$. Note that Γ is the "structure object" of S_2 in the same way as w in Chapter 7 is the structure object of $\varepsilon(\mathcal{U}_0)$.

We now claim that $\varepsilon(\mathcal{U}_0)$ can be canonically identified with a subset of S_2. This follows from the following proposition:

Proposition 15.1 *The 2-arrows of the elements of $\mathcal{G}$ and $C(\mathcal{G})$ (= the centralizer of $\mathcal{G}$ as defined in Chapter 4) preserve Γ.*

Proof Let $(p,q) \in M \times M$ and $f \in \mathcal{G}$ be the unique solution of (2.4) satisfying $f(p) = q$. Now,

$$\frac{\partial f^i}{\partial x^j} = \varepsilon^i_j(x, f(x)) = \varepsilon^i_a(q, f(x))\varepsilon^a_b(p,q)\varepsilon^b_j(x,p) \tag{15.9}$$

Differentiating (15.9) with respect to x^r at $x = p$, we deduce

$$\frac{\partial^2 f^i(p)}{\partial x^r \partial x^j} = \left[\frac{\partial \varepsilon^i_a(q, f(x))}{\partial y^c}\frac{\partial f^c}{\partial x^r}\varepsilon^a_b(p,q)\varepsilon^b_j(x,p) + \varepsilon^i_a(q,f(x))\varepsilon^a_b(p,q)\frac{\partial \varepsilon^b_j(x,p)}{\partial x^r}\right]_{x=p}$$

$$= \Gamma^i_{ca}(q)\frac{\partial f^c(p)}{\partial x^r}\frac{\partial f^a(p)}{\partial x^j} - \frac{\partial f^i}{\partial x^b}\Gamma^b_{rj}(p) \tag{15.10}$$

which shows that $j_2(f)^{p,q}$ satisfies (15.6). For $C(\mathcal{G})$, we assume that p, q, x are close to each other (unless $\mathcal{G}$ is globalizable), replace ε in (15.9) by $\widetilde{\varepsilon}$, and use Proposition 6.10 when we differentiate at $x = p$. Skew-symmetrizing r, j in (15.10), we get

$$T^i_{ca}(q)\frac{\partial f^c(p)}{\partial x^r}\frac{\partial f^a(p)}{\partial x^j} = \frac{\partial f^i(p)}{\partial x^b}T^b_{rj}(p) \tag{15.11}$$

which shows that the 1-arrows of $\mathcal{G}$ and $C(\mathcal{G})$ preserve T, as we already know by (6.1) and Corollary 6.5. Therefore Proposition 15.1 is a strengthening of $\widetilde{\nabla} T = \nabla T = 0$. ■

Proposition 15.1 now gives a splitting

$$\overline{\varepsilon} : \varepsilon(M \times M) \longrightarrow S_2 \tag{15.12}$$

by "prolonging" ε, which maps $j_1(f)^{p,q}$, where f is the unique solution of (2.4), to $j_2(f)^{p,q}$. Henceforth, we will denote $\overline{\varepsilon}$ also by ε and identify $\varepsilon(M \times M)$ with its image in S_2, writing $\varepsilon(M \times M) \subset S_2$.

We know that $\varepsilon(M \times M) \subset S_2$ integrates to the pseudogroup $\mathcal{G}$ with the assumption $\mathfrak{R} = 0$. The question is now whether S_2 integrates to some pseudogroup. The answer is affirmative, with no assumption other than $\mathfrak{R} = 0$.

Proposition 15.2 *Let $(M, \varepsilon, \mathcal{G})$ be an LLG. For any 2-arrow of $\delta(S_1) = S_2$ from p to q, there exists a unique local diffeomorphism f with $f(p) = q$ such that $j_2(f)^{x,f(x)} \in S_2$ for all x near p.*

Proof We want to solve for the local diffeomorphism $y = y(x)$ in the transformation rule (5.2). We reduce (5.2) to the first-order system

$$\frac{\partial f^i(x)}{\partial x^j} = f^i_j(x) \tag{15.13}$$

$$\frac{\partial f^i_j(x)}{\partial x^k} = -f^i_a(x)\Gamma^a_{kj}(x) + \Gamma^i_{ab}(f(x))f^a_k(x)f^b_j(x) \tag{15.14}$$

The integrability conditions for (15.13) and (15.14) are given by

$$\left[\frac{\partial^2 f^i_j(x)}{\partial x^r \partial x^k}\right]_{[rk]} = 0 \tag{15.15}$$

$$\left[\frac{\partial^2 f^i(x)}{\partial x^r \partial x^j}\right]_{[rj]} = 0 \tag{15.16}$$

To check (15.15) we differentiate (15.14) with respect to x^r, substitute back from (15.14), and alternate r, k. Using $\mathfrak{R} = 0$, a straightforward computation shows that (15.15) is identically satisfied. To check (15.16), we alternate k, j in (15.14) and get (15.1), which is identically satisfied by the definition of $\mathcal{S}_1$. Therefore we can integrate (15.13)and (15.14) locally with arbitrary initial conditions, finishing the proof. ∎

Now Proposition 15.2 gives a transitive pseudogroup on M denoted by $\mathcal{S}_2$ that integrates the 2-arrows of the groupoid S_2 and contains both $\mathcal{G}$ and $C(\mathcal{G})$ by the definition of S_2 and Proposition 15.1.

Definition 15.3 *$\mathcal{S}_2$ is the symmetry pseudogroup of $\mathcal{G}$.*

Neither compactness of M nor globalizability of $\mathcal{G}$ implies globalizability of $\mathcal{S}$ (as shown by a counterexample communicated to us by Blaom). However, if both M and $S_1^{p,p} \simeq \text{Aut}(\mathfrak{g}(p))$, $\mathfrak{g} = \widetilde{\mathfrak{X}}_\varepsilon$, are compact, then $\mathcal{S}_2$ is globalizable. The proof is similar to the proof of Proposition 3.5. Henceforth, we will denote $S_2^{p,q}$ by $\mathcal{S}_2^{p,q}$.

We recall that Diff_l denotes the universal pseudogroup of all local diffeomorphisms on M.

Proposition 15.4 $\mathcal{S}_2 = \{f \in \text{Diff}_l \mid f \circ \varphi \circ f^{-1} \in \mathcal{G}, \varphi \in \mathcal{G}\}$.

Thus, $\mathcal{S}_2$ is the normalizer of $\mathcal{G}$ inside Diff_l. If both $\mathcal{G}$ and $\mathcal{S}_2$ globalize, then $\mathcal{S}_2$ becomes the normalizer of $\mathcal{G}$ inside $\text{Diff}(M)$. The reason why we prefer the term "symmetry group" for $\mathcal{S}_2$ is that its elements permute the solutions of the nonlinear PDE (2.4) according to Proposition 15.4, and the concept of a symmetry of a differential equation due to Lie predates the more modern group-theoretic term "normalizer."

Proof The proof of Proposition 15.4 is almost identical to the proof of Proposition 5.1. Let $\varphi \in \mathcal{G}$ and $f \in \text{Diff}(M)$ such that $f \circ \varphi \circ f^{-1} \in \mathcal{G}$. Suppose $f^{-1} : (U, x) \to (V, y)$, $\varphi : (V, y) \to (Z, z)$ and $f : (Z, z) \to (W, w)$. By the chain rule,

$$\frac{\partial w^i}{\partial x^j} = \frac{\partial w^i}{\partial z^a}\frac{\partial z^a}{\partial y^b}\frac{\partial y^b}{\partial x^j} \tag{15.17}$$

Since $\varphi \in \mathcal{G}$ and $f \circ \varphi \circ f^{-1} \in \mathcal{G}$, (15.17) gives

$$\varepsilon^i_a(x, w)\frac{\partial f^a(y)}{\partial y^j} = \frac{\partial f^i(z)}{\partial z^a}\varepsilon^a_j(y, z) \tag{15.18}$$

We now fix $x = x_0$, $y = y_0$ and change z in (15.18). For any z, there exists a unique $\varphi \in \mathcal{G}$ with $\varphi(x_0) = z$. Thus, φ depends on z. However, (15.18) continues to hold, because $f \circ \varphi \circ f^{-1} \in \mathcal{G}$ *for all* $\varphi \in \mathcal{G}$. Thus, we may differentiate (15.18) with respect to z at $z = y_0$. This gives

$$\left[\frac{\partial \varepsilon^i_a(x_0, w)}{\partial w^b}\frac{\partial w^b}{\partial z^r}\frac{\partial f^a(y)}{\partial y^j}\right]_{z=y_0} = \left[\frac{\partial^2 f^i(z)}{\partial z^r \partial z^a}\varepsilon^a_j(y,z) + \frac{\partial f^i(z)}{\partial z^a}\frac{\partial \varepsilon^a_j(y,z)}{\partial z^r}\right]_{z=y_0} \tag{15.19}$$

or, equivalently,

$$\Gamma^i_{ba}(x_0)\frac{\partial x^b}{\partial y^r}\frac{\partial x^a}{\partial y^j} = \frac{\partial^2 x^i}{\partial y^r \partial y^j} + \frac{\partial x^i}{\partial y^a}\Gamma^a_{rj}(y_0) \tag{15.20}$$

Therefore, $\{f \in \mathrm{Diff}_l \mid f \circ \varphi \circ f^{-1} \in \mathcal{G}, \varphi \in \mathcal{G}\} \subset \mathcal{S}_2$. To show the reverse inclusion, we need to show that (15.20) implies (15.18). Indeed, if (15.20) holds identically, then both sides of (15.18) have the same derivatives for all z and not only for $z = y_0$. Since they are equal for $z = y_0$, they are equal locally. ■

We can now iterate the above construction. We define the transitive pseudogroups

$$\mathcal{S}_{k+1} \stackrel{\text{def}}{=} \{f \in \mathrm{Diff}_l \mid f \circ \varphi \circ f^{-1} \in \mathcal{S}_k, \ \varphi \in \mathcal{S}_k\} \tag{15.21}$$

where we set $\mathcal{S}_1 = \mathcal{G}$. Clearly,

$$\mathcal{S}_1 \subset \mathcal{S}_2 \subset \mathcal{S}_3 \subset \ldots \subset \mathrm{Diff}_l \tag{15.22}$$

If all the pseudogroups in (15.22) globalize, then we get the ascending normal tower of global transformation groups on M. Now, $\mathcal{S}_1$ is defined by the first-order nonlinear PDE (2.4), and $\mathcal{S}_2$ is defined by the second-order nonlinear PDE (5.2). It is natural to expect that $\mathcal{S}_k$ is the solution space of a nonlinear PDE of order k on M. However, the local computations become quite tedious already for $k = 3$. Also, the tower (15.22) may stabilize at some m so that the defining PDEs of $\mathcal{S}_{m+j}$ and $\mathcal{S}_m$ have the same solutions. This means that the order $m + j$ of the PDE of $\mathcal{S}_{m+j}$ may be reduced to m without changing the solutions. This reduction of the order of a differential equation through knowledge of its symmetries is an important topic in the study of differential equations ([O2]).

Our purpose in the rest of this chapter is to outline the construction of a remarkable *purely algebraic* prolongation theory of Lie algebras that arises from the linearization of (15.22). We assume that $(M, \varepsilon, \mathcal{G})$ is globalizable, set $\mathfrak{g}_1 \stackrel{\text{def}}{=} \mathfrak{X}_{\widetilde{\varepsilon}}$ that integrates to $\mathcal{S}_1 = \mathcal{G}$, and define

$$\mathfrak{g}_{k+1} \stackrel{\text{def}}{=} \{\xi \in \mathfrak{X}(M) \mid [\xi, x] \in \mathfrak{g}_k, \ x \in \mathfrak{g}_k\} \tag{15.23}$$

Thus, we obtain an ascending normal tower of Lie algebra sheaves

$$\mathfrak{g}_1 \subset \mathfrak{g}_2 \subset \mathfrak{g}_3 \subset \ldots \subset \mathfrak{X}(M) \tag{15.24}$$

Now, $\mathfrak{g}_k$ is defined by a linear PDE of order k, and the sections of $\mathfrak{g}_k(U)$ integrate to the pseudogroup $\mathcal{S}_k(U)$. For simplicity, let us assume that $\mathfrak{g}_k$ globalizes and is a Lie algebra of global vector fields on M. Now, any vector field in $\mathfrak{g}_k$ is determined on M by its k-jet at any point $p \in M$. Fixing a base point e, the evaluation maps

$$j_k(\cdot) : X \in \mathfrak{g}_k \longrightarrow j_k(X)^e \tag{15.25}$$

give the diagram

$$\begin{array}{ccccccccc} \mathfrak{g}_1 & \subset & \mathfrak{g}_2 & \subset & \mathfrak{g}_3 & \subset & \cdots & \mathfrak{X}(M) \\ \downarrow & & \downarrow & & \downarrow & & \downarrow & \\ \mathfrak{g}_1(e) & \subset & \mathfrak{g}_2(e) & \subset & \mathfrak{g}_3(e) & \subset & \cdots & \mathfrak{X}_\infty(e) \end{array} \tag{15.26}$$

where the Lie algebra structure of $\mathfrak{g}_k(e)$ is induced by the usual bracket of vector fields in $\mathfrak{g}_k$, the vertical maps are isomorphisms of Lie algebras induced by the evaluation maps in (15.25), and $\mathfrak{X}_\infty(e)$ is the Lie algebra of formal vector fields at e. Note that the evaluation map $j_\infty(\cdot) : \mathfrak{X}(M) \to \mathfrak{X}_\infty(e)$ does not induce an isomorphism in the smooth category. Clearly, $\mathfrak{g}_k(e)$ is determined by $\mathfrak{g}_1(e) = \mathfrak{X}_{\widetilde{\mathcal{E}}}(e) \simeq \mathfrak{X}_{\widetilde{\mathcal{E}}}$ for all $k \geq 1$, and it is natural to ask the following question:

Q: Given $\mathfrak{g}_1(e)$, how do we construct the bottom chain of (15.26) algebraically without using (15.22) and the upper chain in (15.26)?

As a start to answering **Q**, we first understand how $\mathfrak{g}_2(e)$ is determined by $\mathfrak{g}_2(e)$ using the above detailed analysis of $\mathcal{S}_2$. We assume that both $\mathcal{G} = \mathcal{S}_1$ and $\mathcal{S}_2$ globalize and let $f \in \mathcal{S}_2$. By Proposition 15.2, f is determined on M by its 2-arrow $j_2(f)^{p,e}$ from $p = f(e)$ to e for any $p \in M$. Now, there exists a unique $\varphi \in \mathcal{S}_1$ with $\varphi(p) = e$ and $j_2(\varphi)^{p,e} \in (\delta(\mathcal{S}_1))^{p,e} = \mathcal{S}_2^{p,e} = \mathcal{S}_2^{p,e}$. Therefore, $j_2(f)^{p,e} \circ j_2\left(\varphi^{-1}\right)^{e,p} = j_2\left(f \circ \varphi^{-1}\right)^{e,e} \in \mathcal{S}_2^{e,e} \simeq \mathcal{S}_1^{e,e} = \mathrm{Aut}(\mathfrak{g}_1(e))$. It follows that $f = g \circ \varphi$ for some *unique* $g \in \mathrm{Aut}(\mathfrak{g}_1(e))$ and

$$\mathcal{S}_2 = \mathcal{S}_1 \times \mathcal{S}_2^{e,e} = \mathcal{S}_1 \times \mathrm{Aut}(\mathfrak{g}_1(e)) \tag{15.27}$$

as a set. However, $\mathcal{S}_1$ is normal inside $\mathcal{S}_2$ by Proposition 15.4, and $\mathcal{S}$ acts on $\mathcal{S}_1$ by conjugation. Therefore, the set product (15.27) is actually the semidirect product

$$\begin{aligned} \mathcal{S}_2 &= \mathcal{S}_1 \rtimes \mathcal{S}_2^{e,e} \\ &= \mathcal{S}_1 \rtimes \mathrm{Aut}(\mathfrak{g}_1(e)) \\ &= \mathcal{S}_1 \rtimes \mathrm{Aut}(\mathfrak{X}_{\widetilde{\mathcal{E}}}) \end{aligned} \tag{15.28}$$

From (15.28), we conclude

$$\mathfrak{g}_2(e) = \mathrm{Der}(\mathfrak{g}_1(e)) \ltimes \mathfrak{g}_1(e) \tag{15.29}$$

The abstract derivation of (15.29) from (15.28) is not very satisfactory, since it does not realize the Lie algebra $\mathfrak{g}_2(e) \simeq \mathfrak{g}_2$ as a Lie algebra of *vector fields* on M that *integrates* to $\mathcal{S}_2$. To do this, we should use (15.23) for $k=1$ and construct $\mathfrak{g}_2$ explicitly, as we did $\mathcal{S}_2$. We will leave the details of this amusing computation to the interested reader.

Now, the bracket of $\mathrm{Der}(\mathfrak{g}_1) \ltimes \mathfrak{g}_1$ (omitting e) is given by

$$[(d_1, x_1), (d_2, x_2)] = \big([d_1, d_2],\ d_1(x_2) - d_2(x_1) + [x_1, x_2]\big) \tag{15.30}$$

where $d_1, d_2 \in \mathrm{Der}(\mathfrak{g})$ and $x_1, x_2 \in \mathfrak{g}$. We identify $\mathrm{Der}(\mathfrak{g})$ with the subalgebra $(\mathrm{Der}(\mathfrak{g}), 0) \subset \mathrm{Der}(\mathfrak{g}) \ltimes \mathfrak{g}$ and $\mathfrak{g}$ with the ideal $(0, \mathfrak{g}) \subset \mathrm{Der}(\mathfrak{g}) \ltimes \mathfrak{g}$ Unfortunately, (15.29) does not give much hint about what $\mathfrak{g}_k$ should be.

To make progress, let $\overline{G}$ be an abstract group that acts transitively on a set M, so we have $\overline{G}/\overline{H} = M$, where $\overline{H}$ is the stabilizer at some point. Suppose some subgroup $G \subset \overline{G}$ also acts transitively on M, so we also have $G/H = M$, where H is the stabilizer at the same point. Therefore, $H = G \cap \overline{H}$. Now $G/\left(G \cap \overline{H}\right) = \overline{G}/\overline{H}$, where "$=$" means that any coset of $\overline{H}$ in $\overline{G}$ contains exactly one coset of $G \cap \overline{H}$ in G. Therefore, if $\overline{g} \in \overline{G}$, then $\overline{g}\overline{H} \supset g(G \cap \overline{H})$ for some $g \in G$ and $\overline{g}\overline{H} = g\overline{H}$, which implies $\overline{g} = g\overline{h}$ for some $\overline{h} \in \overline{H}$; that is, $\overline{G} = G\overline{H}$. Now assuming all groups to be Lie groups, differentiation of $\overline{G} = G\overline{H}$ at the identity gives $\overline{\mathfrak{g}} = \mathfrak{g} + \overline{\mathfrak{h}}$, and we also have $\mathfrak{h} = \mathfrak{g} \cap \overline{\mathfrak{h}}$. This argument motivates the following definition given in [On]:

Definition 15.5 *Let $\overline{\mathfrak{h}} \subset \overline{\mathfrak{g}}$ and $\mathfrak{h} \subset \mathfrak{g}$ be Lie algebras. Then the pair $(\overline{\mathfrak{h}}, \overline{\mathfrak{g}})$ is an extension of $(\mathfrak{h}, \mathfrak{g})$ if*

(i) $\mathfrak{g} \subset \overline{\mathfrak{g}}$ *and* $\overline{\mathfrak{g}} = \mathfrak{g} + \overline{\mathfrak{h}}$;

(ii) $\mathfrak{h} = \mathfrak{g} \cap \overline{\mathfrak{h}}$

We write an extension as $(\mathfrak{h}, \mathfrak{g}) \preceq (\overline{\mathfrak{h}}, \overline{\mathfrak{g}})$. Note that $\dim \overline{\mathfrak{g}} - \dim \overline{\mathfrak{h}} = \dim \mathfrak{g} - \dim \mathfrak{h}$. We imagine $\mathfrak{g}$ as the Lie algebra of the infinitesimal generators of a transitive transformation group $\mathcal{G}$ on a smooth manifold M of $\dim M = \dim \mathfrak{g} - \dim \mathfrak{h}$, where $\mathfrak{h}$ is the Lie algebra of the stabilizer at some point. Therefore, an extension $(\mathfrak{h}, \mathfrak{g}) \preceq (\overline{\mathfrak{h}}, \overline{\mathfrak{g}})$ represents infinitesimally a bigger transformation group $\overline{\mathcal{G}}$ with stabilizer $\overline{H}$. Note that the inclusion $\mathfrak{h} \subset \mathfrak{g}$ is proper if and only if the inclusion $\overline{\mathfrak{h}} \subset \overline{\mathfrak{g}}$ is proper. In this case, we call the extension $(\mathfrak{h}, \mathfrak{g}) \preceq (\overline{\mathfrak{h}}, \overline{\mathfrak{g}})$ proper. We call $(\mathfrak{h}, \mathfrak{g}) \preceq (\overline{\mathfrak{h}}, \overline{\mathfrak{g}})$ normal if $\mathfrak{g} \subset \overline{\mathfrak{g}}$ is an ideal and call $(\mathfrak{h}, \mathfrak{g}) \preceq (\overline{\mathfrak{h}}, \overline{\mathfrak{g}})$ maximal if any extension of $(\mathfrak{h}, \mathfrak{g})$ injects into $(\overline{\mathfrak{h}}, \overline{\mathfrak{g}})$; that is, if $(\mathfrak{h}, \mathfrak{g}) \preceq (\mathfrak{h}', \mathfrak{g}')$ is an extension, then there exists an injection of Lie algebras $i : \mathfrak{g}' \to \overline{\mathfrak{g}}$ such that $i(\mathfrak{h}) \subset \overline{\mathfrak{h}}$. For the last definition, we require that the Lie algebras be finite-dimensional, so that a maximal extension is unique up to isomorphism defined in the obvious way. Henceforth, to be consistent with the above geometric interpretation of the pair $(\mathfrak{h}, \mathfrak{g})$, we always assume that the pairs in question are effective; that is, $\mathfrak{h}$ contains no nontrivial ideals of $\mathfrak{g}$.

Now we easily check the proper normal extension

$$(0, \mathfrak{g}_1) \preceq (\mathrm{Der}(\mathfrak{g}_1), \mathrm{Der}(\mathfrak{g}_1) \ltimes \mathfrak{g}_1) = (\mathrm{Der}(\mathfrak{g}_1), \mathfrak{g}_2) \tag{15.31}$$

Proposition 15.6 (15.31) *is the maximal normal extension.*

Proof Let $(0, \mathfrak{g}) \preceq (\mathfrak{h}', \mathfrak{g}')$ be a normal extension. Thus, $\mathfrak{g} \subset \mathfrak{g}'$ is an ideal and $\mathfrak{g} \cap \mathfrak{h}' = 0$. Therefore, $\mathfrak{g}' = \mathfrak{g} \oplus \mathfrak{h}'$. We define $i : \mathfrak{g}' \to \mathrm{Der}(\mathfrak{g}) \ltimes \mathfrak{g}$ by $i(g, h') \stackrel{\text{def}}{=} (\mathrm{ad}(h')_{|\mathfrak{g}}, g)$. Using (15.30), we check that i is a Lie algebra homomorphism. Clearly, $i(\mathfrak{h}') \subset \mathrm{Der}(\mathfrak{g})$ and $\mathrm{Ker}(i) \subset \mathfrak{h}'$. Since i is a homomorphism, $\mathrm{Ker}(i) \subset \mathfrak{g}'$ is an ideal. Therefore, $\mathrm{Ker}(i) = 0$, since $(\mathfrak{h}', \mathfrak{g}')$ is effective, proving the claim. ∎

We now set $\mathfrak{h}_1 = 0$, $\mathfrak{h}_2 = \mathrm{Der}(\mathfrak{g})$, $\mathfrak{g}_2 = \mathrm{Der}(\mathfrak{g}) \ltimes \mathfrak{g}$, and consider the chain of maximal normal extensions

$$(0, \mathfrak{g}_1) \preceq (\mathfrak{h}_2, \mathfrak{g}_2) \preceq (\mathfrak{h}_3, \mathfrak{g}_3) \preceq (\mathfrak{h}_4, \mathfrak{g}_4) \preceq \cdots \tag{15.32}$$

The reader may have noticed our trick of using the same notation $\mathfrak{g}_k$ in (15.24) and (15.32) without any proof for $k \geq 3$, which is far from being trivial! If we assume $\mathfrak{g}_k$ to be finite-dimensional in (15.32), it is not even obvious why $\mathfrak{g}_{k+1}$ should be finite-dimensional, whereas this is clear in (15.26)!

Proposition 15.7 *If* $\mathfrak{g}_1$ *is semisimple, then* (15.24) *stabilizes at* $k = 2$.

Proof Let $a \in \mathfrak{g}_3$ and $x, y \in \mathfrak{g}_1$. Now $[a, [x, y]] = [y, [x, a]] + [x, [a, y]]$. Since $[x, a] \in \mathfrak{g}_2$, $[y, [x, a]] \in \mathfrak{g}_2$. Similarly, $[x, [a, y]] \in \mathfrak{g}_2$. Therefore, $[a, [\mathfrak{g}_1, \mathfrak{g}_1]] \subset \mathfrak{g}_1$. Since $[\mathfrak{g}_1, \mathfrak{g}_1] = \mathfrak{g}_1$, we conclude that $[a, \mathfrak{g}_1] \subset \mathfrak{g}_1$, which implies $a \in \mathfrak{g}_2$, and therefore $\mathfrak{g}_3 \subset \mathfrak{g}_2$. ∎

Since we started with a simply transitive action in this chapter, we have $\mathfrak{h}_1 = 0$ in (15.32), but clearly this prolongation theory applies to arbitrary transitive actions as in Part III.

We will conclude with four questions:

1. Show that $\mathfrak{g}_k$ in (15.26) and (15.32) are the same.
2. Express $(\mathfrak{h}_k, \mathfrak{g}_k)$ explicitly in terms of the *inner structure* of $\mathfrak{g}_1$ like (15.31).
3. For any integer m (including $m = \infty$) give examples of (finite-dimensional) Lie algebras $\mathfrak{g}_1$ such that (15.24) stabilizes at m. (An example is given in [P1], page 346 for (15.22) and $m = \infty$.)
4. Clarify the relation between the above prolongation theory and the one developed by Guillemin–Singer–Sternberg for a single linear Lie algebra (see [GS2], [SS], and [Kb], and references therein).

PART III
How to Generalize?

If light is in your heart, you will find your way home

Rumi

Introduction to Part III

Our purpose in Part III is to define a geometric structure $\mathcal{P}$ and its curvature $\mathfrak{R}$ on M in such a way that $\mathfrak{R} = 0$ if and only if $\mathcal{P}_{|U}$ can be identified with $\mathcal{P}_{|V}$ by some local diffeomorphism $f : U \rightarrow V$, giving rise to a pseudogroup $\mathcal{G}$ on M. Naturally, we expect affine, Riemannian, projective, conformal, and more generally parabolic structures to be special cases. We cannot expect this generalization of Klein's Erlangen Program to include, for instance, symplectic structures, because in general a symplectic diffeomorphism is not determined locally by its k-arrows for any k unless the symplectic manifold is also homogeneous. From our experience in Chapter 15, it is natural to expect that such a structure will be a transitive subgroupoid $\mathcal{P}_{m+1} \subset \mathcal{U}_{m+1}$, where $\mathcal{U}_{m+1}$ is the universal groupoid of all $(m+1)$-arrows on M, satisfying $\mathcal{P}_{m+1} \simeq \mathcal{P}_m$, with m as the smallest such integer, called the order of $\mathcal{P}$. Equivalently, we have $\mathcal{P}_{m+1} = \varepsilon(\mathcal{P}_m)$ for some splitting ε. The $(m+1)$-arrows of $\mathcal{P}_{m+1}$ will integrate *uniquely* to local diffeomorphisms if and only if $\mathfrak{R} = 0$ on M. If all local diffeomorphisms of the emerging pseudogroup $\mathcal{G}$ globalize (necessarily uniquely), we will get a global transformation group $\mathcal{G}^{\text{ex}}$ that acts transitively on M, turning M into a homogeneous space $\mathcal{G}^{\text{ex}}/H_p$, where H_p is the stabilizer at some $p \in M$. We will fix beforehand only the "isomorphism class" of the stabilizer H_p and $\dim M = \dim \mathcal{G}^{\text{ex}} - \dim H_p$. Since some $h \in H_p$ is determined globally by $j_{m+1}(h)^{p,p} \in \mathcal{U}^{p,p}_{m+1} \simeq G_{m+1}(n) =$ the jet group of order $m+1$ in n variables, $n = \dim M$, we obtain the inclusion map

$$j_{m+1}(h) : H_p \longrightarrow \mathcal{U}^{p,p}_{m+1}$$

which allows us to identify H_p with a subgroup of $G_{m+1}(n)$ (modulo conjugation). Now the filtration of jets inside $G_{m+1}(n)$ induces a filtration inside H_p, and the problem is to detect this filtration group theoretically. This dictionary between jets and group theory is a subtle issue and will be completely clarified in Chapters 16–18, finally rendering a natural definition of a geometric structure that we call a prehomogeneous geometry (PHG). However, even though we know the meaning of the curvature from the outset, the technical definition turns out to be quite problematic. Nevertheless, this program works quite well for affine and Riemannian structures, although it runs into an unexpected difficulty with projective structures: The technical assumption that we need to define curvature forces a projective structure to be flat, and we are unable to get around this difficulty in its full generality.

The crucial observation now is that the above technical assumption also forces the total space $\mathcal{P}_m^{e,\bullet}$ of the principal bundle $\mathcal{P}_m^{e,\bullet} \to M$ to be parallelizable. The filtration by jets determines inside $\mathcal{P}_m^{e,\bullet}$ a chain of foliations. The final idea that crystallizes in Chapter 20, and was inspired by an idea communicated to us by Blaom, is to start from the top and not from the bottom: We start with some (N, ε) as in Part I, keeping in mind that N is a substitute for $\mathcal{P}_m^{e,\bullet}$, and define a foliation $\mathcal{F}_0$ on N that is "compatible" with ε. Now $\mathcal{F}_0$ iterates to a chain of foliations $\mathcal{F}_k$ inside N, and the curvatures we seek are the nonlinear and linear curvatures $\mathcal{R}$ and $\mathfrak{R}$ that we already know! However, the compatibility condition with $\mathcal{F}_k$ makes these curvatures more subtle objects. The general philosophy is now that anything done so far with (N, ε) and $(N, \varepsilon, \mathcal{G})$ can also be done for a geometric structure $\mathcal{P}$, as long as this construction is compatible with $\mathcal{F}_k$. Unfortunately, we will not have the occasion to pursue this philosophy further in this book.

Many deep questions emerge on the way. For instance, do geometric structures of arbitrarily large order m exist on compact manifolds? Note that such structures are defined by PDEs of order $m + 1$ that are *globally defined on M*. As we will discover, this problem reduces in principle to a purely algebraic question about nilpotent Lie algebras.

16

Klein Geometries

We start with a Lie algebra pair $(\mathfrak{h}, \mathfrak{g})$ where $\mathfrak{g}$ is finite-dimensional and $\mathfrak{h} \subset \mathfrak{g}$ is a proper subalgebra. As remarked in Chapter 15, we imagine $\mathfrak{g}$ to be the Lie algebra of the infinitesimal generators of a transitive transformation group $\mathcal{G}$ acting on a manifold M with $\dim M = \dim \mathfrak{g}/\mathfrak{h}$, where $\mathfrak{h}$ is the Lie algebra of the stabilizer at some point. Therefore, $\mathfrak{g}$ is a Lie algebra of vector fields and $\mathfrak{h}$ is the isotropy subalgebra at this point, namely, the subalgebra composed of all the vector fields whose values vanish at this point. Now, a transformation cannot fix all the points and still be different from the identity; that is, a transformation group always acts effectively. Therefore, we always assume that the pair $(\mathfrak{h}, \mathfrak{g})$ is effective, which means that $\mathfrak{h}$ has no nonzero ideals that are also ideals of $\mathfrak{g}$.

Definition 16.1 *An infinitesimal Klein geometry is a (effective) pair* $(\mathfrak{h}, \mathfrak{g})$.

We next recall the algebraic normal tower (15.32) and the normal PDE tower (15.24), which coincide for $k = 1, 2$, as was shown in Chapter 15. Let us assume for the moment that they coincide for all k and suppose further that (15.32) stabilizes at $(\mathfrak{h}_m, \mathfrak{g}_m)$, m being the smallest such integer. As we remarked in Chapter 15, the elements of $\mathfrak{g}_k$ are solutions of a linear PDE of order $k+1$, and this situation naturally gives rise to the following important question:

> **Q:** How can we detect the integer m algebraically by working solely with the pair $(\mathfrak{h}, \mathfrak{g}) = (\mathfrak{h}_m, \mathfrak{g}_m)$ in (15.32)?

We observe that the answer to this question will lead to the formulation of certain purely algebraic concepts in the theory of Lie algebras in terms of the derivatives appearing in the Taylor expansions (i.e., jets) of the infinitesimal generators.

A complete answer to **Q** will be given in Chapter 18. To begin with, let us set $\mathfrak{g}_0 = \mathfrak{h}$ in $(\mathfrak{h}, \mathfrak{g})$ and define

$$\mathfrak{g}_{k+1} \overset{\text{def}}{=} \{x \in \mathfrak{g}_k \mid [x, \mathfrak{g}] \subset \mathfrak{g}_k,\ k \geq 0\} = \{x \in \mathfrak{g}_0 \mid [x, \mathfrak{g}] \subset \mathfrak{g}_k,\ k \geq 0\} \tag{16.1}$$

An Alternative Approach to Lie Groups and Geometric Structures. Ercüment H. Ortaçgil (2018).

The equality of the two sets in (16.1) follows easily, and we thus obtain the following decreasing filtration of subspaces:

$$\cdots \subset \mathfrak{g}_2 \subset \mathfrak{g}_1 \subset \mathfrak{g}_0 \subset \mathfrak{g} \tag{16.2}$$

We learned (16.2) from [Sh]. It is also in [Kb] and [Ly], and, to the best of our knowledge, originally in [GS2].

We set $\mathfrak{r} \stackrel{\text{def}}{=} \cap_{i\geq 0}\mathfrak{g}_i$ in (16.2). Clearly, $[\mathfrak{r}, \mathfrak{g}] \subset \mathfrak{r}$, and therefore $\mathfrak{r} = 0$, since $\mathfrak{r} \subset \mathfrak{h}_0$ and $(\mathfrak{g}_0, \mathfrak{g})$ is effective. We infer that $\mathfrak{g}_k = 0$ for some integer k, since $\mathfrak{g}$ is finite-dimensional.

Definition 16.2 *The smallest integer m such that $\mathfrak{g}_m = 0$ is the order of* $(\mathfrak{g}_0, \mathfrak{g})$.

We denote the order m of $(\mathfrak{g}_0, \mathfrak{g})$ by $o(\mathfrak{g}_0, \mathfrak{g})$, this order vanishing if and only if $\mathfrak{g}_0 = 0$.

Lemma 16.3

$$[\mathfrak{g}_i, \mathfrak{g}_j] \subset \mathfrak{g}_{i+j}, \qquad i,j \geq 0 \tag{16.3}$$

Proof We proceed by induction on $i + j = k$. For $k = 0$, (16.3) holds since $\mathfrak{g}_0$ is a subalgebra. Suppose (16.3) holds for $k - 1$ and let $x \in \mathfrak{g}_i, y \in \mathfrak{g}_j, a \in \mathfrak{g}$. Now, $[[x,y],a] = [[x,a],y] + [[a,y],x] \in [\mathfrak{g}_{i-1}, \mathfrak{g}_j] + [\mathfrak{g}_{j-1}, \mathfrak{g}_i] \subset \mathfrak{g}_{i+j-1} + \mathfrak{g}_{i+j-1} \subset \mathfrak{g}_{i+j-1}$. Hence, $[[\mathfrak{g}_i, \mathfrak{g}_j], \mathfrak{g}] \subset \mathfrak{g}_{i+j-1}$ and $[\mathfrak{g}_i, \mathfrak{g}_j] \subset \mathfrak{g}_{i+j}$, by the definition of $\mathfrak{g}_{i+j}$. ■

In particular, $\mathfrak{g}_k \subset \mathfrak{g}_0$ is an ideal and (16.2) is a descending chain of ideals starting with $\mathfrak{g}_1$.

We recall that a Lie algebra $\mathfrak{h}$ is nilpotent if $\mathfrak{h}^k = 0$ for some k, where $\mathfrak{h}^0 = \mathfrak{h}$, $\mathfrak{h}^1 = [\mathfrak{h}, \mathfrak{h}]$ and $\mathfrak{h}^{m+1} = [\mathfrak{h}, \mathfrak{h}^m]$. As in the proof of Lemma 16.3, we can easily show $[\mathfrak{h}^i, \mathfrak{h}^j] \subset \mathfrak{h}^{i+j}$. It is also easy to see that the sum of two nilpotent ideals is nilpotent. Therefore, $\mathfrak{h}$ has a maximal nilpotent ideal $\mathrm{Nil}(\mathfrak{h})$ called the nilradical of $\mathfrak{h}$.

Proposition 16.4 *For a Klein geometry* $(\mathfrak{g}_0, \mathfrak{g})$, $\mathfrak{g}_1 \subset \mathrm{Nil}(\mathfrak{g}_0)$.

Proof Since $\mathfrak{g}_1 \subset \mathfrak{g}_0$ is an ideal, it suffices to show that $\mathfrak{g}_1$ is nilpotent. We claim that $\mathfrak{g}_1^k \subset \mathfrak{g}_{k+1}$, which is true for $k = 0$. Now $\mathfrak{g}_1^{k+1} = [\mathfrak{g}_1^k, \mathfrak{g}_1] \subset [\mathfrak{g}_{k+1}, \mathfrak{g}_1] \subset \mathfrak{g}_{k+2}$ by (16.3). Therefore, $\mathfrak{g}_1^k = 0$ for $k = o(\mathfrak{g}_0, \mathfrak{g})$, and $\mathfrak{g}_1$ is nilpotent. ■

Proposition 16.4 states that it is $\mathrm{Nil}(\mathfrak{g}_0)$ that is responsible for the possible high order. For instance, if $\mathrm{Nil}(\mathfrak{g}_0) = 0$, which is the case if $\mathfrak{g}_0$ is semisimple, then $o(\mathfrak{g}_0, \mathfrak{g}) \leq 1$. We will see in Chapter 18 that there is also a topological condition that implies $o(\mathfrak{g}_0, \mathfrak{g}) \leq 1$. Therefore, if we deal with higher-order jets, we will be in the relatively poorly known territory of nilpotent Lie algebras.

To summarize, an effective Lie algebra pair $(\mathfrak{g}_0, \mathfrak{g})$ of order $o(\mathfrak{g}_0, \mathfrak{g})$ determines the filtration (16.2) inside $\mathrm{Nil}(\mathfrak{g}_0)$ satisfying (16.3). The length of this filtration is $o(\mathfrak{g}_0, \mathfrak{g})$ including $\mathfrak{g}_0$.

We now want to show that for any integer $m \geq 0$, there exists an (effective) Klein pair $(\mathfrak{h}, \mathfrak{g})$ with $o(\mathfrak{h}, \mathfrak{g}) = m$. For this purpose, consider the Lie algebra $\mathfrak{F}(n)$ of formal polynomial vector fields in n variables. To recall, let $\mathfrak{F}_i(n)$ be the linear span of the vector fields of the form $P(x^1, \ldots, x^n)\partial_{x^j}$, where P is a homogeneous polynomial of degree $i + 1$ and $1 \leq j \leq n$. We define

$$\mathfrak{F}(n) \stackrel{\text{def}}{=} \mathfrak{F}_{-1}(n) \oplus \mathfrak{F}_0(n) \oplus \mathfrak{F}_1(n) \oplus \mathfrak{F}_2(n) \oplus \cdots \tag{16.4}$$

with the usual bracket of vector fields and $\mathfrak{F}(n)$ is an infinite-dimensional Lie algebra. Note that $[\mathfrak{F}_i(n), \mathfrak{F}_j(n)] \subset \mathfrak{F}_{i+j}(n), i, j \geq -1$ and $[\mathfrak{F}_{-1}(n), \mathfrak{F}_{-1}(n)] = 0$.

Now let $n = 2$. We fix some $k \geq 1$ and define $\widetilde{\mathfrak{g}}_{-1} \stackrel{\text{def}}{=} \text{Sp}\{\partial_x, \partial_y\} = \mathfrak{F}_{-1}(2)$, $\widetilde{\mathfrak{g}}_0 \stackrel{\text{def}}{=} \text{Sp}\{x\partial_x, x\partial_y, y\partial_y\} \subset \mathfrak{F}_0(2)$, $\widetilde{\mathfrak{g}}_i \stackrel{\text{def}}{=} \text{Sp}\{x^{i+1}\partial_y\} \subset \mathfrak{F}_i(2), 1 \leq i \leq k$, $\widetilde{\mathfrak{g}}_i \stackrel{\text{def}}{=} 0, i \geq k+1$. We define

$$\widetilde{\mathfrak{g}} \stackrel{\text{def}}{=} \widetilde{\mathfrak{g}}_{-1} \oplus \widetilde{\mathfrak{g}}_0 \oplus \widetilde{\mathfrak{g}}_1 \oplus \cdots \oplus \widetilde{\mathfrak{g}}_k \tag{16.5}$$

and easily check that $[\widetilde{\mathfrak{g}}_{-1}, \widetilde{\mathfrak{g}}_i] \subset \widetilde{\mathfrak{g}}_{i-1}$, $[\widetilde{\mathfrak{g}}_0, \widetilde{\mathfrak{g}}_i] \subset \widetilde{\mathfrak{g}}_i$ for $1 \leq i \leq k$, and $[\widetilde{\mathfrak{g}}_i, \widetilde{\mathfrak{g}}_j] = 0$ for $1 \leq i, j \leq k$. We now define

$$\begin{aligned}
\mathfrak{g} &\stackrel{\text{def}}{=} \widetilde{\mathfrak{g}}_{-1} \oplus \widetilde{\mathfrak{g}}_0 \oplus \widetilde{\mathfrak{g}}_1 \oplus \cdots \oplus \widetilde{\mathfrak{g}}_k \\
\mathfrak{g}_0 &\stackrel{\text{def}}{=} \widetilde{\mathfrak{g}}_0 \oplus \widetilde{\mathfrak{g}}_1 \oplus \cdots \oplus \widetilde{\mathfrak{g}}_k \\
\mathfrak{g}_1 &\stackrel{\text{def}}{=} \widetilde{\mathfrak{g}}_1 \oplus \cdots \oplus \widetilde{\mathfrak{g}}_k \\
&\vdots \\
\mathfrak{g}_k &\stackrel{\text{def}}{=} \widetilde{\mathfrak{g}}_k
\end{aligned} \tag{16.6}$$

and get the descending filtration

$$\mathfrak{g}_k \subset \cdots \subset \mathfrak{g}_1 \subset \mathfrak{g}_0 \subset \mathfrak{g} \tag{16.7}$$

Since $[\widetilde{\mathfrak{g}}_{-1}, \widetilde{\mathfrak{g}}_i] \subsetneq \widetilde{\mathfrak{g}}_i$, the pair $(\mathfrak{g}_0, \mathfrak{g})$ is effective and (16.1) holds by construction. Therefore, $o(\mathfrak{g}_0, \mathfrak{g}) = k + 1$. It is worth singling out this important fact as the following proposition:

Proposition 16.5 *For any integer $m \geq 0$, there exists an effective Klein geometry $(\mathfrak{g}_0, \mathfrak{g})$ with $o(\mathfrak{g}_0, \mathfrak{g}) = m$. In fact, we can choose* $\dim \mathfrak{g}/\mathfrak{g}_0 = 2$.

Proposition 16.5 has a weakness: The above example $(\mathfrak{g}_0, \mathfrak{g})$ with $o(\mathfrak{g}_0, \mathfrak{g}) = k + 1$ does not have any geometric significance. We will pose here **FP1**, which we will call *the first fundamental problem of higher-order geometry and jet theory:*

FP1: Construct examples of Klein geometries $(\mathfrak{g}_0, \mathfrak{g})$ of high order having geometric (or algebraic) significance.

Such a construction is attempted in [AO1], but the pair (27) in [AO1] is not effective and Proposition 4.2 there is incorrect. In fact, if $\dim \mathfrak{g}/\mathfrak{h} = 1$, then there are exactly three isomorphism classes of $(\mathfrak{h}, \mathfrak{g})$ as shown in Appendix C of [Sh]. As an attempt to correct this mistake, we make the following conjecture:

C1: For any integer k, there exists a semisimple Lie algebra $\mathfrak{g}$ and a parabolic subalgebra $\mathfrak{b} \subset \mathfrak{p} \subset \mathfrak{g}$, where $\mathfrak{b}$ is the maximal solvable subalgebra (= Borel subalgebra) such that $o(\mathfrak{b}, \mathfrak{p}) \geq k$.

The Lie algebra $\mathfrak{F}(n)$ and its subalgebras are studied in [Mo], [Ps], and [Dr]. It turns out that $\mathfrak{F}(n)$ contains many simple Lie algebras as subalgebras. The list of such simple Lie algebras is deduced in [Mo], and we refer to Table 1 in [Ps] for the complete list. We believe that the finite-dimensional subalgebras of $\mathfrak{F}(n)$ are of great importance in geometry and that this will become manifest in view of **FP1** and the rest of this book.

Resuming our argument with (16.2), suppose that $(\mathfrak{g}_0, \mathfrak{g})$ is a Klein geometry with the filtration (16.2) and $o(\mathfrak{g}_0, \mathfrak{g}) = m$. Then $(\mathfrak{g}_k, \mathfrak{g})$ is another (effective) Klein geometry with the filtration

$$\cdots \subset \mathfrak{g}_{k+1} \subset \mathfrak{g}_k \subset \mathfrak{g} \tag{16.8}$$

as follows easily from (16.1). In particular, we obtain the following proposition:

Proposition 16.6 *Suppose $o(\mathfrak{g}_0, \mathfrak{g}) = m$. Then $o(\mathfrak{g}_k, \mathfrak{g}) = m - k, 0 \leq k \leq m$.*

According to our geometric intuition rearding the pair $(\mathfrak{g}_0, \mathfrak{g})$ explained above in terms of the vector fields, Proposition 16.6 states that if a transformation group acts on some space with order m, then, for $0 \leq k \leq m$, there is another space, a "prolonged jet space," on which it acts with order $m - k$. To find these spaces in group-theoretic terms, we now consider effective Klein geometries (H, G), where G is a connected Lie group and $H \subset G$ is a Lie subgroup. We set $G_0 = H$, let $\mathfrak{g}$ and $\mathfrak{g}_0$ denote the Lie algebras of G and G_0, and define inductively

$$G_{k+1} \stackrel{\text{def}}{=} \{g \in G_k \mid \mathrm{Ad}(g)x - x \in \widehat{\mathfrak{g}}_k,\ x \in \mathfrak{g}, 0 \leq k\} \tag{16.9}$$

where $\widehat{\mathfrak{g}}_k$ is the Lie algebra of G_k, so that $\widehat{\mathfrak{g}}_0 = \mathfrak{g}_0$. Thus, we obtain the decreasing filtration of Lie groups

$$\cdots \subset G_2 \subset G_1 \subset G_0 \subset G \tag{16.10}$$

and the decreasing filtration of their Lie algebras

$$\cdots \subset \widehat{\mathfrak{g}}_2 \subset \widehat{\mathfrak{g}}_1 \subset \mathfrak{g}_0 \subset \mathfrak{g} \tag{16.11}$$

Proposition 16.7 *(i) $G_k \subset G_0$ is normal, $k \geq 0$.*

(ii) $\widehat{\mathfrak{g}}_k = \mathfrak{g}_k$.

(iii) If G is connected, then $G_{m+1} = \{1\}$, where $m = o(\mathfrak{g}_0, \mathfrak{g})$.

Proof (i) As with (16.1), we first observe that G_{k+1} can be defined also by

$$G_{k+1} \stackrel{\text{def}}{=} \{g \in G_0 \mid \mathrm{Ad}(g)x - x \in \widehat{\mathfrak{g}}_k,\ x \in \mathfrak{g}, 0 \leq k\} \tag{16.12}$$

We define the representation

$$\begin{aligned} \mathrm{Ad}_k : G_0 &\longrightarrow GL(\mathfrak{g}/\widehat{\mathfrak{g}_k}) \\ \mathrm{Ad}_k(y)(x+\widehat{\mathfrak{g}_k}) &\stackrel{\mathrm{def}}{=} \mathrm{Ad}_y(x)+\widehat{\mathfrak{g}_k} \end{aligned} \tag{16.13}$$

and observe that $G_{k+1} = \mathrm{Ker}(\mathrm{Ad}_k)$, which proves (i).

(ii) If $\rho : H \to G$ is a homomorphism of Lie groups with the differential $d\rho : \mathfrak{h} \to \mathfrak{g}$, then the Lie algebra of the kernel $\mathrm{Ker}(\rho)$ is equal to $\mathrm{Ker}(d\rho)$. Therefore, the Lie algebra of G_1 is the kernel of ad_1, which is $\mathfrak{g}_1$. Iteration of this argument proves (ii).

(iii) Since $\mathfrak{g}_m = 0$, G_m is discrete, by (ii). Now $G_{m+1} \subset \mathrm{Ker}(\mathrm{Ad}) = Z(G) =$ the center of G, since G is connected. Thus, $G_{m+1} = \{1\}$, since (G_0, G) is effective. ■

Thus, we obtain the following interesting corollary:

Corollary 16.8 *Let H be a Lie group. Suppose there exists a connected Lie group $G \supset H$ such that (H, G) is effective. Then H has a faithful representation.*

It is desirable to find a simple sufficient condition that implies $G_m = \{1\}$ so that (16.2) and (16.10) stabilize at the same step. By successive liftings to the universal covers, we can make all the groups in (16.10) connected. In the examples that we know of, (16.2) and (16.10) stabilize at the same step even if G is not connected. To simplify our exposition, henceforth we will always assume that the groups in (16.10) are connected and write $o(G_0, G) = o(\mathfrak{g}_0, \mathfrak{g})$.

Now we have the tower of coset spaces and principal bundles

$$G \longrightarrow G/G_{m-1} \longrightarrow G/G_{m-2} \longrightarrow \dots \longrightarrow G/G_1 \longrightarrow G/G_0 \tag{16.14}$$

Therefore, we have

(i) The principal bundles $G \longrightarrow G/G_k$ with structure group G_k and $o(G_k, G) = m-k$, by Proposition 16.6.

(ii) The principal bundles $G/G_k \longrightarrow G/G_0$ with structure group G_0/G_k. In particular, we have the tower of structure groups

$$G_0 \longrightarrow G_0/G_{m-1} \longrightarrow G_0/G_{m-2} \longrightarrow \dots \longrightarrow G_0/G_1 \longrightarrow 1 \tag{16.15}$$

Our purpose is to understand the group-theoretic filtrations (16.2), (16.10) and the towers (16.14), (16.15) geometrically in terms of the k-arrows of the actions of G on the spaces in (16.14). After some further preparation in Chapter 17, this will achieved in Chapter 18. The picture turns out to be somewhat complicated, and it is useful at this point to introduce the abstract formalism of principal bundles and groupoids as bookkeeping devices and recall some basic facts.

Let $\pi : P \to M$ be a right principal bundle with structure group H. We define an equivalence relation on $P \times P$ as follows: $(a, b) \sim (c, d)$ if there exists an $h \in H$ satisfying $(ah, bh) = (c, d)$. We denote an equivalence class by $[a, b]$ and the set of equivalence classes by $[P \times P]$.

Definition 16.9 *$[a, b]$ is an arrow on M from $\pi(a)$ to $\pi(b)$ (induced by $\pi : P \to M$).*

So, $[P \times P]$ is the set of all arrows on M. We call $\pi(a)$ the source and $\pi(b)$ the target of the arrow $[a, b]$. We have the projection map

$$\begin{aligned} \pi &: [P \times P] \to M \times M \\ &: [a, b] \to (\pi(a), \pi(b)) \end{aligned} \tag{16.16}$$

and the source and target projections

$$\begin{aligned} \alpha &: [P \times P] \to M \\ &: [a, b] \to \pi(a) \end{aligned} \tag{16.17}$$

and

$$\begin{aligned} \beta &: [P \times P] \to M \\ &: [a, b] \to \pi(b) \end{aligned} \tag{16.18}$$

We now define

$$[b, c] \circ [a, b] \stackrel{\text{def}}{=} [a, c] \tag{16.19}$$

which is well-defined since H acts freely on P: $ah = a$ implies $h = e$. Thus, we can compose two compatible arrows. Let $[P \times P]^{\pi(a),\pi(b)}$ denote the set of all arrows from $\pi(a)$ to $\pi(b)$. Now (16.19) gives the composition map

$$[P \times P]^{\pi(b),\pi(c)} \times [P \times P]^{\pi(a),\pi(b)} \longrightarrow [P \times P]^{\pi(a),\pi(c)} \tag{16.20}$$

The set of arrows $[P \times P]^{\pi(a),\pi(a)}$ with the same source and target $\pi(a)$ is a group isomorphic to H, but not canonically. We call $[P \times P]^{\pi(a),\pi(a)}$ the vertex group of the transitive groupoid $[P \times P]$ at $\pi(a)$. If $H = \{1\}$, then $[P \times P]$ is the pair groupoid $M \times M$.

We also have the inversion map

$$\begin{aligned} \gamma &: [P \times P]^{\pi(a),\pi(b)} \longrightarrow [P \times P]^{\pi(b),\pi(a)} \\ &: [a, b] \longrightarrow [b, a] = [a, b]^{-1} \end{aligned} \tag{16.21}$$

and $[a, b]$ defines the conjugation map

$$\begin{aligned} &: [P \times P]^{\pi(a),\pi(a)} \longrightarrow [P \times P]^{\pi(b),\pi(b)} \\ &: [a, a] \longrightarrow [a, b] \circ [a, a] \circ [a, b]^{-1} \end{aligned} \tag{16.22}$$

We can recover $P \to M$ from $[P \times P]$ by fixing a base point $e \in M$ and considering the set of all arrows on M spreading out from e. We denote this principal bundle by $\beta : [P \times P]^{e,\bullet} \to M$, whose structure group is $[P \times P]^{e,e} \simeq H$. We have the obvious identification $(M, P, H, \pi) \simeq (M, [P \times P]^{e,\bullet}, [P \times P]^{e,e}, \beta)$.

We will denote the transitive groupoid $[P \times P]$ on M by $(M, [P \times P])$, emphasizing the base because shortly $[P \times P]$ will define groupoids over different bases.

There is a fundamental difference between a general principal bundle $P \to M$ and a homogeneous principal bundle $G \to G/H = M$: P does not act on M, whereas G does. Also, there is no canonical choice for the base point e for $[P \times P]^{e,\bullet} \to M$, whereas the coset of H is the canonical base point for $[G \times G]^{e,\bullet} \to M$ and $[G \times G]^{e,e} = H$. Since we will work with frames rather than coframes, we will assume that all the homogeneous principal bundles below are right principal bundles, and therefore the cosets are left cosets.

We now fix k, $0 \lneqq k \leq m$ and start with the principal bundle

$$\pi_{k,0} : G/G_k \longrightarrow G/G_0 \tag{16.23}$$

over the fixed base G/G_0 with structure group G_0/G_k. Note that $hG_k \in G_0/G_k$ acts on G/G_k on the right by $gG_k \to ghG_k$, which is well-defined since $G_k \subset G_0$ is normal. Thus, we have the groupoid $\big(G/G_0, [G/G_k \times G/G_k]\big)$.

Definition 16.10 *Arrows of $\big(G/G_0, [G/G_k \times G/G_k]\big)$ are k-arrows (of G on G/G_0).*

The following analogy will prepare us for Chapter 18: G acts transitively and effectively as a transformation group on G/G_0 as $xG_0 \to gxG, g \in G$. Thus, we can consider the k-arrow of the global diffeomorphism g from xG_0 to gxG_0. We will define a k-arrow in Chapter 17, but this is more or less clear from Chapters 1 and 15. Now we may consider all k-arrows on G/G_0 induced by the action of G, obtaining a groupoid of k-arrows on G/G_0. It will turn out in Chapter 18, that the group-theoretic counterpart of this purely jet-theoretic construction is $\big(G/G_0, [G/G_k \times G/G_k]\big)$, which keeps track of these k-arrows, justifying our terminology in Definition 16.10. Note that this correspondence will completely clarify the geometric content of the filtration (16.10). For $k = m$ we get $G \to G/G_0$ and the groupoid of m-arrows of G on G/G_0, where $o(G_0, G) = m$.

We denote an arrow of $\big(G/G_0, [G/G_k \times G/G_k]\big)$ by $[gG_k, g'G_k]_{(k,0)}$, incorporating the base G/G_0 into our notation: k in $(k, 0)$ denotes the order of the arrow and 0 denotes the level where this arrow is defined. Now $\pi_{k,0}[gG_k, g'G_k]_{(k,0)} = (gG_0, g'G_0)$ and $[gG_k, g'G_k]_{(k,0)}$ is a k-arrow from gG_0 to $g'G_0$ on G/G_0. Thus, we have the projection homomorphism of groupoids

$$\begin{aligned} \pi_{k+r,k} : \big(G/G_0, [G/G_{k+r} \times G/G_{k+r}]\big) &\longrightarrow \big(G/G_0, [G/G_k \times G/G_k]\big) \\ : [gG_{k+r}, g'G_{k+r}]_{(k+r,0)} &\longrightarrow [gG_k, g'G_k]_{(k,0)} \end{aligned} \tag{16.24}$$

which preserves the composition and inversion of arrows. In the same way as a 1-arrow projects to a 0-arrow in Part I, a $(k + r)$-arrow on G/G_0 projects to a k-arrow on G/G_0, and this jet-theoretic projection will correspond to the group-theoretic projection (16.24).

Note that the vertex groups of $\left(G/G_0, [G/G_k \times G/G_k]\right)$ are isomorphic to G_0/G_k, and (16.24) restricts to a Lie group homomorphism at the vertex groups. With the choice of the canonical basepoints as above, (16.24) restricts to

$$\pi_{k+r,k} : G_0/G_{k+r} \longrightarrow G_0/G_k \tag{16.25}$$

at the cosets of G_{k+r} and G_k.

Up to now, we have worked over the base G/G_0 and considered the arrows of G on G/G_0 of order k, $0 \le k \le m$. Unfortunately, the situation is more complicated, because G acts also on G/G_r, with $o(G_r, G) = m - r$. We call the action of G on G/G_r the rth prolongation of the action of G on G/G_0. These prolonged actions, which will be relevant in Chapters 17 and 18, are studied in the same way by replacing G/G_0 with G/G_r and repeating the above constructions with the new base G/G_r. However, there is a crucial relation between the prolonged action and the original action. Before turning to this issue, we will first construct certain vector bundles over G/G_0. So, let us briefly recall the concept of an associated vector bundle.

Let $P \to M$ be a right principal bundle with structure group H as above and let $\rho : H \to \mathrm{GL}(V)$ be a representation of H on a (finite-dimensional) vector space V. So, $\rho(ab) = \rho(a) \circ \rho(b)$ and $\rho(a^{-1}) = \rho(a)^{-1}$. We define an equivalence relation on $P \times V$ as follows: $(a, v) \sim (a', v')$ if there exists an $h \in H$ with $(a', v') = (ah, \rho(h)^{-1}v)$. We denote an equivalence class by $[a, v]$ and the set of equivalence classes by $(M, P \times_H V)$. We define $\pi : P \times_H V \to M$ by $\pi[a, v] = \pi(a)$. We fix some $a \in P$ with $\pi(a) = p$ and define $[a, v] + [a, v'] \overset{\text{def}}{=} [a, v + v']$, $c[a, v] \overset{\text{def}}{=} [a, cv]$ and easily check that these operations are independent of the choice of a. Thus, the fiber $\pi^{-1}(p)$ inherits the structure of a vector space, which can be identified with V by $v \longleftrightarrow [a, v]$.

Definition 16.11 *A vector in the fiber $\pi^{-1}(p)$ of $\pi : P \times_H V \to M$ is an infinitesimal V-arrow at p.*

The vector bundle $P \times_H V \to M$ is associated with the groupoid $\left(M, [P \times P]\right)$ in the following sense: An arrow from p to q induces a linear isomorphism $\pi^{-1}(p) \to \pi^{-1}(q)$ between the fibers. More precisely, for $[a, b]$ with $\pi(a) = p$ and $\pi(b) = q$, we define

$$\sharp[a, b]([a, v]) \overset{\text{def}}{=} [b, v] \tag{16.26}$$

If $\sharp[a, b] = \sharp[a', b']$, then $[a, b] = [a', b']$ if ρ is also faithful. Also,

$$\sharp\left([b, c] \circ [a, b]\right) = \sharp[b, c] \circ \sharp[a, b] \tag{16.27}$$

Therefore, we can "carry" infinitesimal V-arrows with arrows.

Now (16.13) gives the *faithful* representation (using the same notation)

$$\mathrm{Ad}_k : G_0/G_{k+1} \longrightarrow \mathrm{GL}(\mathfrak{g}/\mathfrak{g}_k), \qquad k \ge 0 \tag{16.28}$$

Since the principal bundle $G_{k+1} \to G/G_0$ has G_0/G_{k+1} as its structure group, we can construct its associated vector bundle $(G/G_0, G/G_{k+1} \times_{G_0/G_{k+1}} \mathfrak{g}/\mathfrak{g}_k)$ with respect to

(16.28). It is useful to recall here that $\mathfrak{g}/\mathfrak{g}_k$ can be canonically identified with the tangent space of G/G_k at the coset of G_k. We now have the association

$$(G/G_0, [G/G_{k+1} \times G/G_{k+1}]) \rightsquigarrow (G/G_0, G/G_{k+1} \times_{G_0/G_{k+1}} \mathfrak{g}/\mathfrak{g}_k) \tag{16.29}$$

and the associated vector bundle

$$\pi_{k,0} : (G/G_{k+1} \times_{G_0/G_{k+1}} \mathfrak{g}/\mathfrak{g}_k) \longrightarrow G/G_0 \tag{16.30}$$

For $k = 0$, (16.30) is the tangent bundle $T(G/G_0) \to G/G_0$, and the left-hand side of (16.29) is the groupoid of 1-arrows on G/G_0, so we can carry tangent vectors with 1-arrows as in Part I. In general, the left-hand side of (16.29) is the groupoid of $(k+1)$-arrows, and we make the following definition:

Definition 16.12 *A vector in the fiber* $\pi_{k,0}^{-1}(p)$ *of* (16.30) *is an infinitesimal k-arrow ("ik-a") at p.*

Note that a vector in the fiber $\pi_{k,0}^{-1}(p)$ is of the form $[gG_{k+1}, \xi + \mathfrak{g}_k]_{(k,0)}$ and $\pi_{k,0}(gG_{k+1}) = gG_0 = p$.

The terminology of Definition 16.12 is enforced by Definitions 16.10 and 16.11. However, it is better to call an *ik*-a a "*k*-jet of an infinitesimal generator (*k*-jig)" at *p*. We agree that a 0-jig at *p* is the value of the vector field at *p*, that is, a tangent vector at *p*. The analogy is this: Since G acts on G/G_0, some $\xi \in T_e(G)$ induces the global vector field $\overline{\xi} \in \mathfrak{X}(G/G_0) =$ the Lie algebra of all vector fields on G/G_0. The map $\xi \to \overline{\xi}$ is a homomorphism of Lie algebras, which is injective since G acts effectively. Further, $\overline{\xi}(p)$ ranges through $T_p(G/G_0)$ as ξ ranges through $T_e(G)$, since G acts transitively. Thus, we may define $j_k(\overline{\xi})_p$, and it turns out that $\pi_{k,0}^{-1}(p) = \left\{ j_k\left(\overline{\xi}\right)_p \mid \xi \in T_e(G) \right\}$; that is, (16.30) is the group-theoretic counterpart of another jet-theoretic construction. The term "*k*-jig," which is another formal name for the moment like a *k*-arrow, will be shown to be a true name in Chapter 18. Thus we can "carry" *k*-jigs with $(k+1)$-arrows.

We have the obvious projection maps

$$(G/G_0, G/G_{k+1+r} \times_{G_0/G_{k+1+r}} \mathfrak{g}/\mathfrak{g}_{k+r}) \xrightarrow{\pi_{k+r,r}} (G/G_0, G/G_{r+1} \times_{G_0/G_{r+1}} \mathfrak{g}/\mathfrak{g}_r) \tag{16.31}$$

defined by

$$\pi_{k+r,r}[gG_{k+1+r}, \xi + \mathfrak{g}_{k+r}]_{(k+r,0)} \stackrel{\text{def}}{=} [gG_{r+1}, \xi + \mathfrak{g}_r]_{(r,0)} \tag{16.32}$$

which are compatible with the projection and association maps.

Now, (16.30) has a very important subbundle, called the isotropy subbundle. We have

$$\mathrm{Ad}_k : G_0/G_k \longrightarrow \mathrm{GL}(\mathfrak{g}_0/\mathfrak{g}_k), \qquad k \geq 0 \tag{16.33}$$

which is simply the adjoint representation of the Lie group G_0/G_k on its Lie algebra $\mathfrak{g}_0/\mathfrak{g}_k$. This gives the association

$$\big(G/G_0, [G/G_k \times G/G_k]\big) \leftrightsquigarrow (G/G_0, G/G_k \times_{G_0/G_k} \mathfrak{g}_0/\mathfrak{g}_k) \tag{16.34}$$

and the vector bundle

$$\pi_{k,0} : G/G_k \times_{G_0/G_k} \mathfrak{g}_0/\mathfrak{g}_k \longrightarrow G/G_0 \tag{16.35}$$

Since $\pi_{k,0}[gG_{k+1}, \xi + \mathfrak{g}_k]_{(k,0)} = [gG_0, \xi + \mathfrak{g}_0]_{(0,0)}$, a vector in the fiber $\pi_{k,0}^{-1}(p)$ of (16.35) projects to the zero vector on the tangent space $T_p(G/G_0)$ if and only if $\xi \in \mathfrak{g}_0$. Therefore, any k-jig in the fiber of (16.30) over p that projects to zero at $T_p(G/G_0)$ defines the k-jig $[gG_k, \xi + \mathfrak{g}_k]$ in the fiber of (16.35). The converse also holds, and (16.35) is a subbundle of (16.30). We observe that the fibers of (16.35) are the Lie algebras of the vertex groups of the groupoid on the left-hand side of (16.34). Therefore, the sections of (16.35) are endowed with a bracket defined pointwise; that is, this bracket does not involve differentiation of the sections. However, there is also an important *differential* bracket called the Spencer bracket on the sections of (16.30) that restricts to this pointwise bracket on the sections of (16.35). These facts become more transparent if one works directly with jets. In fact, as we will see, the corresponding jet constructions in Chapters 17 and 18 will almost trivialize the above group-theoretic constructions. This is quite ironic, since the formalism of principal bundles and their associated vector bundles is much better known than that of jets.

We now come to the prolonged actions and their associated vector bundles. We lift the base G/G_0 to G/G_r and consider the principal bundle

$$G/G_{r+k} \longrightarrow G/G_r \tag{16.36}$$

with structure group G_r/G_{r+k}. Thus, we construct the corresponding groupoids and their projections as above.

The following definition is now a logical consequence of Definition 16.10:

Definition 16.13 *The arrow* $[gG_{r+k}, g'G_{r+k}]_{(k,r)}$ *of* $\big(G/G_r, [G/G_{r+k} \times G/G_{r+k}]\big)$ *is a k-arrow (of G) on G/G_r from gG_r to $g'G_r$.*

The analogy is that G now acts on G/G_r, and $\big(G/G_r, [G/G_{r+k} \times G/G_{r+k}]\big)$ keeps track of the k-arrows of G.

We now observe the crucial fact that the k-arrow in Definition 16.13 defines the $(k+1)$-arrow $[gG_{r+k}, g'G_{r+k}]_{(k+1,r-1)}$ on G/G_{r-1} from gG_{r-1} to $g'G_{r-1}$, defines the $(k+2)$-arrow $[gG_{r+k}, g'G_{r+k}]_{(k+2,r-2)}$ on G/G_{r-2} from gG_{r-2} to $g'G_{r-2}$, ... and finally defines the $(k+r)$-arrow $[gG_{r+k}, g'G_{r+k}]_{(k+r,0)}$ on G/G_0 from gG_0 to $g'G_0$. For $r = m$, (16.36) defines the pair groupoid $[G, [G \times G]] = G \times G$, since $G_{m+k} = G_m = \{1\}$. The 0-arrow $[g, g']_{(0,m)}$ on G from g to g' now defines a 1-arrow $[g, g']_{(1,m-1)}$ on G/G_{m-1} from gG_{m-1} to $g'G_{m-1}, \ldots$ and finally the m-arrow $[g, g']_{(m,0)}$ on G/G_0 from gG_0 to $g'G_0$.

Replacing k with $r + k$ in (16.13) and modding out the kernel, we obtain the faithful representation

$$\mathrm{Ad}_{r+k} : G_r/G_{r+k+1} \longrightarrow \mathrm{GL}(\mathfrak{g}/\mathfrak{g}_{r+k}), \qquad k \geq 0 \tag{16.37}$$

We construct the associated vector bundle

$$\pi_{r+k,r} : G/G_{r+k+1} \times_{G_r/G_{r+k+1}} \mathfrak{g}/\mathfrak{g}_{r+k} \longrightarrow G/G_r \tag{16.38}$$

and have the association

$$\big(G/G_r, [G/G_{r+k+1} \times G/G_{r+k+1}]\big) \leftrightsquigarrow \big(G/G_r, G/G_{r+k+1} \times_{G_r/G_{r+k+1}} \mathfrak{g}/\mathfrak{g}_{r+k}\big) \tag{16.39}$$

which gives (16.29) for $r = 0$.

Definition 16.14 *A vector $[gG_{r+k+1}, x + \mathfrak{g}_{r+k}]_{(k,r)}$ in the fiber $\pi^{-1}_{r+k,r}(p)$ is an ik-a (or k-jig) at p.*

An element of $\pi^{-1}_{r-k,r}(p)$ is of the form $[gG_{r+k+1}, x + \mathfrak{g}_{r+k}]_{(k,r)}$, where $gG_r = p$. Everything done above over the base G/G_0 can be done over G/G_r in a way compatible with projections and associations.

We now come to a delicate point: The filtrations (16.2) and (16.10) are determined by (G_0, G). Further, all the above arrows are determined by the 0-arrows of G on G. Therefore, the primary object is (G_0, G), and all the rest is a consequence. This raises the question of what these filtrations and the above talk about arrows are really good for. The answer is that the presence of G makes all the "curvatures" vanish in the above picture.

We will conclude with two remarks:

1. There is considerable structure hidden in the association (16.29). For instance, G acts on the sections of (16.29), which is an infinite-dimensional vector space, in the obvious way. The vector bundle (16.30) is called a homogeneous vector bundle, and the action of G on its sections is called the induced representation of G. For the moment, assume that m-jig is a true name, so that any infinitesimal generator of G on G/G_0 is globally determined by its m-jig at, say, the coset of G_0. This gives a finite-dimensional subspace of the space of sections that is stable under the action of G. This representation is easily seen to be the adjoint representation of G on its Lie algebra $\mathfrak{g}$. We believe that this phenomenon of "localization" underlies the fact that some globally defined cohomology groups turn out to be isomorphic to some algebraically defined cohomology groups as in [Bo2].
2. In the representation theory of Lie algebras, there are two fundamental concepts: irreducibility and faithfulness. Let $\rho : \mathfrak{h} \rightarrow V$ be an irreducible representation and $\mathfrak{k}$ be an arbitrary Lie algebra. We define the Lie algebra $\mathfrak{h} \oplus \mathfrak{k}$ and the representation $\overline{\rho} : \mathfrak{h} \oplus \mathfrak{k} \rightarrow V$ by acting with ρ on $\mathfrak{h}$ and acting trivially on $\mathfrak{k}$. Clearly, $\overline{\rho}$ is another irreducible representation. Equally clear is the fact that the irreducible representation $\overline{\rho}$ contains a redundancy that can be arbitrarily large. Therefore,

it seems that to concentrate solely on the concept of irreducibility is not very much in accordance with the spirit of representation theory, whose motivation is to represent. At the other extreme, we may want to represent without any loss of information by requiring faithfulness. However, if a Lie algebra $\mathfrak{h}$ (always finite-dimensional) admits a completely reducible and faithful representation on some (finite-dimensional) V, then it is easy to show that $\mathfrak{h} = \mathfrak{s} \oplus \mathfrak{z}$ where $\mathfrak{s}$ is semisimple and $\mathfrak{z}$ is the center of $\mathfrak{h}$. This decomposition is taken as the definition of a reductive Lie algebra. Therefore, if we insist on both these concepts, then we must leave out the very large class of nonabelian nilpotent Lie algebras. Naturally, the question arises whether there is a geometric approach to representation theory that keeps a reasonable balance between these two fundamental concepts.

Now let $(\mathfrak{g}_0, \mathfrak{g})$ and $(\mathfrak{g}_0', \mathfrak{g}')$ be two (effective) Klein geometries of order m.

Definition 16.15 $(\mathfrak{g}_0, \mathfrak{g})$ *is isomorphic to* $(\mathfrak{g}_0', \mathfrak{g}')$ *if there exists an isomorphism* $\gamma : \mathfrak{g} \to \mathfrak{g}'$ *such that* $\gamma(\mathfrak{g}_i) = \mathfrak{g}_i'$ *for* $0 \leq i \leq m-1$.

Now suppose $\mathfrak{h}$ is given and $\mathfrak{h} \subset \mathfrak{g}$ for some Lie algebra $\mathfrak{g}$. We have the representation

$$\begin{aligned} \rho : \mathfrak{h} &\longrightarrow \mathrm{gl}(\mathfrak{g}/\mathfrak{h}) \\ \rho(h)(x+\mathfrak{h}) &\overset{\text{def}}{=} [h,x] + \mathfrak{h} \end{aligned} \tag{16.40}$$

We call (16.40) the homogeneous representation of $\mathfrak{h}$ determined by $\mathfrak{g}$. Note that $(\mathfrak{h}, \mathfrak{g})$ need not be effective in (16.40). However, if $(\mathfrak{g}_0, \mathfrak{g})$ is effective, then we have the faithful representations

$$\mathrm{ad}_k : \mathfrak{g}_0/\mathfrak{g}_{k+1} \longrightarrow \mathrm{gl}(\mathfrak{g}/\mathfrak{g}_k), \qquad k \geq 0 \tag{16.41}$$

which are the linearizations of (16.28). We call (16.41) the homogeneous representations of the Klein geometry $(\mathfrak{g}_0, \mathfrak{g})$.

We can easily show the following proposition:

Proposition 16.16 *If* $(\mathfrak{g}_0, \mathfrak{g}) \simeq (\mathfrak{g}_0', \mathfrak{g}')$*, then the homogeneous representations of* $(\mathfrak{g}_0, \mathfrak{g})$ *and* $(\mathfrak{g}_0', \mathfrak{g}')$ *are equivalent for all k.*

Do we lose anything if we restrict our attention to homogeneous representations? Surprisingly the answer is negative:

Proposition 16.17 *Let* $\rho : \mathfrak{h} \to V$ *be any representation. Then there exists a Lie algebra* $\mathfrak{g} \supset \mathfrak{h}$ *such that* $\rho : \mathfrak{h} \to V$ *is equivalent to the homogeneous representation of* $\mathfrak{h}$ *determined by* $\mathfrak{g}$*. Further,* ρ *is faithful if and only if* $(\mathfrak{h}, \mathfrak{g})$ *is effective and* $o(\mathfrak{h}, \mathfrak{g}) = 1$.

Proof We define $\mathfrak{g} \overset{\text{def}}{=} \mathfrak{h} \oplus V$ with the bracket $[(h_1, v_1), (h_2, v_2)] \overset{\text{def}}{=} ([h_1, h_2], \rho(h_1)v_2 - \rho(h_2)v_1)$ and check that $\mathfrak{g}$ is a Lie algebra. Now $(\mathfrak{h},0)$ is a subalgebra, which we identify with $\mathfrak{h}$. It is easy to check that the obvious isomorphism $V \simeq \mathfrak{g}/\mathfrak{h}$ establishes the required equivalence. We observe that the kernel of the homogeneous representation is $\mathrm{Ker}(\rho)$, which implies the second statement. ■

The converse of Proposition 16.17 is false. For instance, consider the Lie groups $G(1) = O(n+1)$, $G(0) = O(n) \ltimes \mathbb{R}^n$, and $G(-1) = O(1,n)$, so that we have

$$G(1)/O(n) = S^n, \qquad G(0)/O(n) = \mathbb{R}^n, \qquad G(-1)/O(n) = \mathbb{H}^n \tag{16.42}$$

Now the Klein geometries $(\mathfrak{o}(n), \mathfrak{g}(1))$, $(\mathfrak{o}(n), \mathfrak{g}(0))$, and $(\mathfrak{o}(n), \mathfrak{g}(-1))$ are not isomorphic according to Definition 16.15, whereas the three homogeneous representations of $\mathfrak{o}(n)$ determined by $\mathfrak{g}(1)$, $\mathfrak{g}(0)$, and $\mathfrak{g}(-1)$ are isomorphic to the standard representation of $\mathfrak{o}(n)$. So, the concept of isomorphism of Klein geometries is stronger than the standard concept of equivalence of representations: The first concept respects the order of jets, whereas the second does not.

For a Lie algebra $\mathfrak{h}$, consider a filtration

$$0 \subset \mathfrak{h}_k \subset \cdots \subset \mathfrak{h}_2 \subset \mathfrak{h}_1 \subset \mathrm{Nil}(\mathfrak{h}) \subset \mathfrak{h} \tag{16.43}$$

satisfying

$$[\mathfrak{h}_i, \mathfrak{h}_j] \subset \mathfrak{h}_{i+j}, \qquad i,j \geq 0 \tag{16.44}$$

where $\mathfrak{h} = \mathfrak{h}^0$. We call (16.44) a nilfiltration of $\mathfrak{h}$ and denote it by $\mathfrak{N}(\mathfrak{h})$. For instance, we may choose $\dim(\mathfrak{h}_k/\mathfrak{h}_{k+1}) = 1$, in which case the length of $\mathfrak{N}(\mathfrak{h})$ is maximal. At the other extreme, we may choose $\mathfrak{h}_1 = 0$. We have seen that any Lie algebra $\mathfrak{g} \supset \mathfrak{h}$ such that $(\mathfrak{h}, \mathfrak{g})$ is effective induces the nilfiltration $\mathfrak{N}(\mathfrak{h}, \mathfrak{g})$ of $\mathfrak{h}$ according to (16.2) and Proposition 16.4. As a modification of **C1**, we now make the following conjecture:

C2: For any nilfiltration $\mathfrak{N}(\mathfrak{h})$ of $\mathfrak{h}$, there exists a Lie algebra $\mathfrak{g} \supset \mathfrak{h}$ such that $(\mathfrak{h}, \mathfrak{g})$ is effective and induces the given nilfiltration; that is, $\mathfrak{N}(\mathfrak{h}) = \mathfrak{N}(\mathfrak{h}, \mathfrak{g})$.

Note that **C2** is a far reaching generalization of Ado's Theorem, because any such nilfiltration defines a faithful representation of $\mathfrak{h}$.

Now fix some $\mathfrak{h}$ and a nilfiltration $\mathfrak{N}(\mathfrak{h})$. We call k admissible for $\mathfrak{N}(\mathfrak{h})$ if there exists a Lie algebra $\mathfrak{g} \supset \mathfrak{h}$ with $\dim \mathfrak{g}/\mathfrak{h} = k$ that induces $\mathfrak{N}(\mathfrak{h})$.

Definition 16.18 *For an admissible k for $\mathfrak{N}(\mathfrak{h})$, the uniformization number $\mu(\mathfrak{h}, \mathfrak{N}(\mathfrak{h}), k)$ (in brief, μ) is the number of isomorphism classes of $(\mathfrak{h}, \mathfrak{g}')$ satisfying* $\dim \mathfrak{g}'/\mathfrak{h} = k$ *and* $\mathfrak{N}(\mathfrak{h}, \mathfrak{g}') = \mathfrak{N}(\mathfrak{h})$.

The well-known examples in (16.42) show that the nonisomorphic Klein geometries $(\mathfrak{h}, \mathfrak{g})$ and $(\mathfrak{h}, \mathfrak{g}')$ may define the same nilfiltrations and $\mu(\mathfrak{o}(n), 0, n) = 3$. Clearly, $\mu(0, \mathfrak{N}(0), k)$ is the number of the isomorphism classes of all Lie algebras of dimension k and is therefore infinite for $k \geq 3$. Note that Definition 16.18 gives an "infinitesimal count" of the possible homogeneous spaces that can arise from a geometric structure $\mathcal{P}$ when $\mathfrak{R} = 0$, where $\mathfrak{h}$ is the Lie algebra of the stabilizer H and $k = \dim M$, both fixed beforehand. If $\mu(\mathfrak{h}, \mathfrak{N}(\mathfrak{h}), k) = 1$, then there exists a unique model with these choices, and we do not lose any generality by fixing this model beforehand as in the classical approach via Cartan connections. Therefore, to justify that our model-free approach advertised in the Introduction is a genuine generalization of the classical approach, we must affirmatively answer the following question:

FP2: Is $\mu \gneqq 1$ in "sufficiently many" cases (with large $o(\mathfrak{h}, \mathfrak{g})$)?

Also note that μ is the infinitesimal and geometric analog of the classification problem of principal bundles with structure group H over a fixed base of dimension k, which has a well-known homotopy-theoretic solution.

We hope to have given the reader some optimism that such an approach may indeed exist.

17

The Universal Jet Groupoids

We constructed in Chapter 16 the groupoids $(G/G_r, [G/G_{r+k}, G/G_{r+k}])$ and gave their arrows a formal name: the k-arrows on G/G_r induced by the action of G on G/G_r. That this formal name is a true name will be proved in Chapter 18. Assuming this fact for the moment, we fix some r, k, set $G/G_r = N$, and consider *all possible* k-arrows on N induced by *all* local diffeomorphisms of N. We denote this set by $\mathcal{U}_k(N)$, which is easily seen to be a groupoid. In this way, we get the inclusion of groupoids $(G/G_r, [G/G_{r+k}, G/G_{r+k}]) \subset \mathcal{U}_k(N)$. This realization of $(G/G_r, [G/G_{r+k}, G/G_{r+k}])$ as a subgroupoid suggests that the group-theoretic constructions in Chapter 15 are the restrictions of certain jet-theoretic constructions on $\mathcal{U}_k(N)$. A parallel argument is valid on the infinitesimal level. This is indeed true, and the purpose of this chapter is to recall these jet-theoretic constructions. The study of this *universal calculus* on $\mathcal{U}_k(N)$ and its algebroid $J_{k-1}(N)$ was initiated by D. C. Spencer and his co-workers around 1960–70 (see [Sp], [GS1], [G], [KS], and [P1], and references therein). Our elementary exposition in this chapter is local and arranged according to our needs. Unfortunately, this highly developed technology still remains widely unknown among differential geometers, to the extent that the universal groupoids $\mathcal{U}_k(N)$ and their subgroupoids, also called jet groupoids, are not given even as examples in many modern works on groupoids, even though they are surely the most important examples of groupoids since they correspond to geometric PDEs on manifolds and belong to the core of Lie theory.

We start with the definition of the jet groups $G_k(n)$, where k is the order of jets and n is the number of base variables. We observed in Chapter 1that $G_1(n) = \mathrm{GL}(n, \mathbb{R})$, and $G_2(n)$ is defined by (15.3) with $x = y = z$. Let o denote the origin of $\mathbb{R}^n$. We consider the set of all local diffeomorphisms of $\mathbb{R}^n$ that fix o. For two such local diffeomorphisms f and g, we define $f \sim_k g$ if

$$\frac{\partial^m f(o)}{\partial x^{j_s} \cdots \partial x^{j_1}} = \frac{\partial^m g(o)}{\partial x^{j_s} \cdots \partial x^{j_1}}, \qquad 1 \leq s \leq k \tag{17.1}$$

where (x^i) are the coordinates of $\mathbb{R}^n$. Clearly, $\sim_k$ is an equivalence relation. We denote the set of equivalence classes by $G_k(n)$ and an element of $G_k(n)$ by $j_k(f)^{o,o}$, where f is a representative. Omitting the source and target from our notation (which are both o), $j_k(f)$ has the coordinate representation

$$j_k(f) = \left(f^i_j, f^i_{j_2 j_1}, \dots, f^i_{j_k \dots j_1}\right) \tag{17.2}$$

Using multi-index notation, we sometimes write (17.2) as $\left(f^i_\mu\right)$, $1 \leq |\mu| \leq k$. We define

$$j_k(g) \circ j_k(f) \stackrel{\text{def}}{=} j_k(g \circ f) \tag{17.3}$$

Now, (17.1) together with the chain rule shows that (17.23) is well-defined. Indeed, if $j_k(f) = \left(f^i_j, f^i_{j_2 j_1}, \dots, f^i_{j_k \dots j_1}\right)$ and $j_k(g) = \left(g^i_j, g^i_{j_2 j_1}, \dots, g^i_{j_k \dots j_1}\right)$, then (17.3) is of the form

$$\left(g^i_j, g^i_{j_2 j_1}, \dots, g^i_{j_k \dots j_1}\right) \circ \left(f^i_j, f^i_{j_2 j_1}, \dots, f^i_{j_k \dots j_1}\right) = \left(h^i_j, h^i_{j_2 j_1}, \dots, h^i_{j_k \dots j_1}\right) \tag{17.4}$$

where the right-hand side of (17.4) is computed using the chain rule as follows: We start with the formula $\partial z^i / \partial x^j = (\partial z^i / \partial y^a)(\partial y^a / \partial x^j)$, differentiate it k times with respect to the x's using the chain rule, evaluate the resulting formulas at $x = y = z = 0$, and make the substitutions

$$\frac{\partial^s z^i}{\partial x^{j_s} \dots \partial x^{j_1}}(0) = h^i_{j_s \dots j_1}, \qquad \frac{\partial^s z^i}{\partial y^{j_s} \dots \partial y^{j_1}}(0) = g^i_{j_s \dots j_1}$$
$$\frac{\partial^s y^i}{\partial x^{j_s} \dots \partial x^{j_1}}(0) = f^i_{j_s \dots j_1}$$

for $1 \leq s \leq k$. For instance,

$$\begin{aligned}
h^i_j &= g^i_a f^a_j \\
h^i_{mj} &= g^i_{ab} f^a_m f^b_j + g^i_a f^a_{mj} \\
h^i_{lmj} &= g^i_{abc} f^a_l f^b_m f^c_j + g^i_{ab} f^a_{lm} f^b_j + g^i_{ab} f^a_m f^b_{lj} + g^i_{ba} f^b_l f^a_{mj} + g^i_a f^a_{lmj} \\
&\vdots \\
h^i_{j_k \dots j_1} &= g^i_{a_1 \dots a_k} f^{a_1}_{j_1} \cdots f^{a_k}_{j_k} + \dots + g^i_a f^a_{j_k \dots j_1}
\end{aligned} \tag{17.5}$$

The first formula in (17.5) gives the operation of $G_1(n)$, which is the usual matrix multiplication, the first two formulas together give the operation of $G_2(n)$, and so on . . . finally, all the formulas in (17.5) together give the operation of $G_k(n)$. We call an element of $G_k(n)$ a k-arrow in $\mathbb{R}^n$ with source and target at 0.

Now, $G_k(n)$ is a Lie group—in fact, an algebraic group, as can be seen from (17.5). The identity is $j_k(Id) = \left(\delta^i_j, 0, 0, \dots, 0\right)$ and $j_k\left(f\right)^{-1} = j_k\left(f^{-1}\right)$. Note that it is not at all easy to invert using (17.5) for large k. We define $G_0(n) \stackrel{\text{def}}{=} \{0\}$. We observe that the first j formulas in (17.5) involve only derivatives of order $\leq j$. This simple fact has important consequences. As a first observation, we have the projection homomorphism

$$\pi_{k+r,r} : G_{k+r}(n) \longrightarrow G_k(n) \tag{17.6}$$

and the exact sequence of groups

$$1 \longrightarrow K_{k+r,k}(n) \longrightarrow G_{k+r}(n) \longrightarrow G_k(n) \longrightarrow 1 \tag{17.7}$$

In coordinates, (17.7) is given by

$$\left(\delta^i_j, 0, \ldots, 0, f^i_{jk+1\cdots j1}, \ldots, f^i_{jk+r\cdots j1}\right) \to \left(f^i_j, \ldots, f^i_{jk+r\cdots j1}\right) \to \left(f^i_j, \ldots, f^i_{jk\cdots j1}\right) \tag{17.8}$$

Clearly, $G_{k+r}(n)/K_{k+r,k}(n) \simeq G_k(n)$, where $K_{r+s,0}(n) = G_{r+s}(n)$.

Next, we observe that the right-hand side of the jth formula in (17.5) contains exactly two terms involving jth-order derivatives as a sum. This fact implies that $K_{k+1,k}(n)$ is a vector group for $k \geq 1$. We have the tower of projection homomorphisms

$$\ldots \longrightarrow G_k(n) \longrightarrow G_{k-1}(n) \longrightarrow \ldots \longrightarrow G_2(n) \longrightarrow G_1(n) \longrightarrow 1 \tag{17.9}$$

and the decreasing filtration of ideals

$$1 = K_{k,k}(n) \subset K_{k,k-1}(n) \subset \cdots \subset K_{k,2}(n) \subset K_{k,1}(n) \subset K_{k,0}(n) = G_k(n) \tag{17.10}$$

Now, in the exact sequence (17.7), we have $G_k(n) \simeq G_{k+r+s}(n)/K_{k+r+s,k}(n)$ and $G_{k+r}(n) \simeq G_{k+r+s}(n)/K_{k+r+s,k+r}(n)$. Thus, we conclude

$$K_{k+s,k}(n) \simeq K_{k+s+r,k}(n)/K_{k+s+r,k+s}(n) \tag{17.11}$$

For $s = 1$, we deduce from (17.11) that the successive quotients in (17.10) starting from $K_{k,1}(n)$ are abelian, and therefore $K_{k,1}(n)$ in the exact sequence

$$1 \longrightarrow K_{k,1}(n) \longrightarrow G_k(n) \longrightarrow G_1(n) \longrightarrow \{o\} \tag{17.12}$$

is solvable. More is true: An inspection of (17.5) shows that we have the splitting

$$\begin{aligned} \varepsilon : G_1(n) &\longrightarrow G_k(n) \\ : \left(f^i_j\right) &\longrightarrow \left(f^i_j, 0, \ldots, 0\right) \end{aligned} \tag{17.13}$$

and $G_k(n)$ is the semidirect product

$$G_k(n) = \varepsilon G_1(n) \ltimes K_{k,1}(n) \tag{17.14}$$

An algebraic group is known to be the semidirect product of its nilradical and a maximal reductive subgroup, and (17.14) is a special case of this decomposition. Therefore, $K_{k,1}(n)$ is in fact the nilradical of $G_k(n)$, which can be shown directly using (17.5).

Definition 17.1 (17.10) *is the canonical nilfiltration of* $G_k(n)$.

We observe that the canonical nilfiltration of $G_k(n)$ is induced by jets and not by some Klein pair $(G_k(n), G)$ as in Chapter 16.

Now, certain group-theoretic constructions have jet-theoretic counterparts and conversely. For instance, $G_{k+r}(n)$ acts on $G_k(n)$ by $x \rightarrow \pi_{k+r,k}(g) \circ x \circ \pi_{k+r,k}(g)^{-1}$, and this adjoint action can be expressed in the terms of the left and right actions of $G_{k+r}(n)$ on the left coset space $G_{k+r}(n)/K_{k+r,k}(n) \simeq G_k(n)$ for $k \geq 1$.

There are remarkable combinatorics hidden in the formulas (17.5). For instance, we will derive the formula for the Schwarzian derivative from (17.5). An element of $G_3(1)$ is of the form (a, b, c), where $a, b, c \in \mathbb{R}$ and $a \neq 0$. Now (17.5) gives

$$(a_1, a_2, a_3) \circ (b_1, b_2, b_3) = (c_1, c_2, c_3) \tag{17.15}$$

where

$$\begin{aligned} c_1 &= a_1 b_1 \\ c_2 &= a_2 (b_1)^2 + a_1 b_2 \\ c_3 &= a_3 (b_1)^3 + 3a_2 b_1 b_2 + a_1 b_3 \end{aligned} \tag{17.16}$$

We define the map (using the same notation ε)

$$\begin{aligned} \varepsilon : G_2(1) &\longrightarrow G_3(1) \\ : (a, b) &\longrightarrow \left(a, b, \frac{3b^2}{2a}\right) \end{aligned} \tag{17.17}$$

Using (17.16), a straightforward but surprising computation shows that ε is a homomorphism. Therefore, we have the semidirect product

$$G_3(1) = \varepsilon G_2(1) \ltimes K_{3,2}(1) \tag{17.18}$$

Consider (say) the right coset space $G_3(1)/\varepsilon G_2(1)$ and let $[(a, b, c)]$ denote the coset $\varepsilon G_2(1)(a, b, c)$. Now

$$\begin{aligned} [(a, b, c)] &= [\varepsilon(a, b)^{-1} \circ (a, b, c)] \\ &= \left[\left(a, b, \frac{3b^2}{2a}\right)^{-1} \circ (a, b, c)\right] \\ &= \left[\left(1, 0, \frac{c}{a} - \frac{3bc}{2a^2}\right)\right] \end{aligned} \tag{17.19}$$

Recalling that $(a, b, c) = j_3(f)^o = \left(f'(o), f''(o), f'''(o)\right)$, we observe that the right-hand side of (17.19) is

$$\left[\left(1, 0, \frac{f'''}{f'} - \frac{3f''f'''}{2(f')^2}\right)\right] = [(1, 0, S(f)]$$

where S is the Schwarzian derivative! Therefore, any right coset has a unique representative of the form $(1, 0, S(f))$, and the scalars $S(f)$ parametrize $G_3(1)/\varepsilon G_2(1)$. Now, $G_3(1)$ acts on $G_3(1)/\varepsilon G_2(1)$, and therefore

$$[(1, 0, S(f)]g = [(1, 0, S(f \circ g)] \tag{17.20}$$

which expresses $S(f \circ g)$ in terms of $g \in G_3(1)$ and $S(f)$. So, (17.20) is the transformation rule of $S(f)$, which can be found in the classical textbooks and can be computed explicitly using (17.16). This computation will be carried out for arbitrary n in Chapter 19. The question is, of course: Where do we get (17.17) from? There is a fundamental principle here: Any splitting "inside $G_{k+1}(n) \to G_k(n)$" determines a "structure object for a geometry of order k." For instance, the splitting (17.13) defines the structure object for affine geometry: We will leave it to the reader to check that (see Chapter 19) $G_2(n)/\varepsilon G_1(n)$ is parametrized by $\Gamma = \left(\Gamma^i_{jk}\right)$, and the action of $G_2(n)$ on $G_2(n)/\varepsilon G_1(n)$ gives (15.6), but now with *symmetric* Γ^i_{jk}! The trivial splitting

$$\begin{aligned} \varepsilon : \{o\} &\longrightarrow G_1(n) \\ : \{o\} &\longrightarrow \{e\} \end{aligned} \tag{17.21}$$

defines the structure object for parallelizable manifolds in Part I. Note that $G_1(n)/\varepsilon\{o\}$ is parametrized by invertible matrices (w^i_j), and $G_1(n)$ acts on $G_1(n)/\varepsilon\{o\}$ by

$$\left[w^i_j\right]g = \left[w^i_a g^a_j\right] \tag{17.22}$$

which is the transformation rule (7.2).

The splittings (17.13) and (17.17) may give the impression that splitting of the jet groups is a common phenomenon, but this is not so: $G_k(n)$ does not sit inside $G_{k+1}(n)$ for $k, n \geq 2$ (see [KMS], 13.8). However, there is an abundance of splittings "inside jet groups" coming from the Klein pairs (H, G) in Chapter 16. Leaving the details to Chapter 18, we now take a look at the Lie algebras of the jet groups $G_k(n)$.

Let $\xi = (\xi^i(x))$ be a vector field defined around o and $j_k(\xi)^o$ be the k-jet of ξ at o. In the same way in which we defined a k-arrow as an equivalence class using coordinates, we can define the object $j_k(\xi)^o$ using coordinates. The idea is the same: Two vector fields defined around o define the same k-jet at o if their derivatives up to order k agree at o in some (hence all) coordinates. The independence of coordinates can be shown using (17.34) below. We denote $j_k(\xi)^o$ in brief by $j_k(\xi)$. We have the representation

$$j_k(\xi) = \left(\xi^i, \xi^i_j, \xi^i_{j_2 j_1}, \ldots, \xi^i_{j_k \ldots j_1}\right) \tag{17.23}$$

where ξ is a representative and

$$\xi^i_{j_s \ldots j_1} = \frac{\partial^s \xi(o)}{\partial x^{j_s} \cdots \partial x^{j_1}}, \qquad 0 \leq s \leq k \tag{17.24}$$

Let $J_k(\mathbb{R}^n)^{o} = J_k(\mathbb{R}^n)$ denote the set of all k-jets of vector fields at o. Clearly, $J_k(\mathbb{R}^n)$ is a vector space. We have the projection map

$$\pi_{k+s,k} : J_{k+s}(\mathbb{R}^n) \longrightarrow J_k(\mathbb{R}^n) \tag{17.25}$$

and the exact sequence of vector spaces

$$0 \longrightarrow J_{k+s,k}(\mathbb{R}^n) \longrightarrow J_{k+s}(\mathbb{R}^n) \longrightarrow J_k(\mathbb{R}^n) \longrightarrow 0 \tag{17.26}$$

In coordinates, (17.26) is given by

$$\left(0, \ldots, 0, \xi^i_{j_{k+1}\cdots j_1}, \ldots, \xi^i_{j_k\cdots j_1}\right) \to \left(\xi^i, \ldots, \xi^i_{j_k\cdots j_1}, \ldots, \xi^i_{j_{k+s}\cdots j_1}\right) \to \left(\xi^i, \ldots, \xi^i_{j_k\cdots j_1}\right) \tag{17.27}$$

We now have the point bracket $\{\,,\,\}_o = \{\,,\,\}$

$$\{\,,\,\} : J_{k+1}(\mathbb{R}^n) \times J_{k+1}(\mathbb{R}^n) \longrightarrow J_k(\mathbb{R}), \qquad k \geq 0 \tag{17.28}$$

defined as follows: Suppose $\alpha = j_{k+1}(\xi) = \left(\xi^i, \ldots, \xi^i_{j_{k+1}\cdots j_1}\right)$, $\beta = j_{k+1}(\eta) = \left(\eta^i, \ldots, \eta^i_{j_{k+1}\cdots j_1}\right)$ belong to $J_{k+1}(\mathbb{R}^n)$. We define

$$\{\alpha, \beta\} \stackrel{\text{def}}{=} j_k[\xi, \eta] \tag{17.29}$$

where $[\xi, \eta]$ is the usual bracket of vector fields defined by (5.16). To express (17.29) in coordinates, we differentiate this bracket formula k-times and evaluate at $x = o$, which gives

$$\begin{aligned}
\{\alpha, \beta\}^i &= \xi^a \eta^i_a - \eta^a \xi^i_a \\
\{\alpha, \beta\}^i_j &= \xi^a \eta^i_{ja} + \xi^a_j \eta^i_a - \eta^a \xi^i_{ja} - \eta^a_j \xi^i_a \\
\{\alpha, \beta\}^i_{lj} &= \xi^a \eta^i_{lja} + \xi^a_l \eta^i_{ja} + \xi^a_{lj} \eta^i_a + \xi^a_j \eta^i_{la} \\
&\quad - \eta^a \xi^i_{lja} - \eta^a_l \xi^i_{ja} - \eta^a_{lj} \xi^i_a - \eta^a_j \xi^i_{la} \\
&\;\;\vdots \\
\{\alpha, \beta\}^i_{j_k\cdots j_1} &= \xi^a \eta^i_{j_k\cdots j_1 a} + \cdots - \eta^a \xi^i_{j_k\cdots j_1 a}
\end{aligned} \tag{17.30}$$

Clearly, $\{\alpha, \beta\} = -\{\beta, \alpha\}$. Note that $\{\,,\,\}$ does not define a Lie algebra structure on $J_{k+1}(\mathbb{R}^n)$, since it reduces the order of jets by one. As with (17.5), we observe that (17.30) commutes with projections. We define $J_{k,0}(\mathbb{R}^n) \stackrel{\text{def}}{=} \{x \in J_k(\mathbb{R}^n) \mid \pi_{k,0}(x) = 0\}$. So, $J_{k,0}(\mathbb{R}^n)$ consists of the elements of the form $\left(0, \xi^i_j, \ldots, \xi^i_{j_k\cdots j_1}\right)$, that is, the k-jets of vector fields that project to zero on the tangent space $T_0(\mathbb{R}^n) = J_0(\mathbb{R}^n)$ as in (16.35). Setting $\xi^i = \eta^i = 0$ in (17.30), we see that $\{\,,\,\}$ restricts to the bracket

$$\{\,,\,\} : J_{k,0}(\mathbb{R}^n) \times J_{k,0}(\mathbb{R}^n) \longrightarrow J_{k,0}(\mathbb{R}), \qquad k \geq 0 \tag{17.31}$$

and, using (17.30), we can easily show that $(J_{k,0}(\mathbb{R}^n), \{\ ,\ \})$ is a Lie algebra endowed with the bracket (17.31). Note that the second formula in (17.30) gives minus the usual bracket of matrices on $\mathfrak{gl}(n, \mathbb{R}) = \mathfrak{g}_1(n)$. The next proposition is not difficult to show using (17.5) and (17.31):

Proposition 17.2 *$(J_{k,0}(\mathbb{R}^n), \{\ ,\ \})$ is naturally isomorphic to the Lie algebra of $G_k(n)$.*

We now have the decreasing filtration of ideals

$$0 = J_{k,k}(\mathbb{R}^n) \subset J_{k,k-1}(\mathbb{R}^n) \subset \cdots \subset J_{k,2}(\mathbb{R}^n) \subset J_{k,1}(\mathbb{R}^n) \subset J_{k,0}(\mathbb{R}^n) \tag{17.32}$$

We set $J_{k,s}(\mathbb{R}^n) = 0$ for $s \gneqq k$. Careful inspection of (17.30) shows that

$$\{J_{k,r}(\mathbb{R}^n), J_{k,s}(\mathbb{R}^n)\} \subset J_{k,r+s}(\mathbb{R}^n) \tag{17.33}$$

and it is easy to see that $J_{k,s}(\mathbb{R}^n)$ is the Lie algebra of $K_{k,s}(n)$. Clearly, everything done above with the nilfiltration (17.10) using the group operation (17.5) of $G_k(n)$ has a counterpart for the nilfiltration (17.32) using the bracket $\{\ ,\ \}$ defined by (17.31). Therefore, (17.32) and (17.10) are the universal jet analogs of (16.2) and (16.10) for $\mathfrak{g}_0$ and G_0; that is, we have the analogy $G_0 \leftrightsquigarrow G_k(n)$ and $\mathfrak{g}_0 \leftrightsquigarrow J_{k,0}(\mathbb{R}^n)$, where k is arbitrary.

We now want to construct the adjoint representation of $G_k(n)$ on its Lie algebra $J_{k,0}(\mathbb{R}^n)$ in terms of jets. Let $j_{k+1}(f) = \left(f^i_j, \ldots, f^i_{j_{k+1}\cdots j_1}\right) \in G_{k+1}(n)$ and $j_k(\xi) \in \left(\xi^i, \ldots, \xi^i_{j_k\cdots j_1}\right) \in J_k(\mathbb{R}^n)$. We differentiate the well-known formula $\eta^i(y) = (\partial f^i/\partial x^a)\xi^a(x)$, $y = f(x)$, k times with respect to the x's and evaluate at $x = o$. Keeping in mind that f and ξ are representatives, we get

$$\begin{aligned}
\eta^i &= f^i_a \xi^a \\
\eta^i_a f^a_j &= f^i_{ja}\xi^a + f^i_a \xi^a_j \\
\eta^i_{la} f^a_j + \eta^i_a f^a_l &= f^i_{lja}\xi^a + f^i_{ja}\xi^a_l + f^i_{la}\xi^a_j + f^i_a \xi^a_{lj} \\
&\ \ \vdots \\
\eta^i_{j_{k-1}\cdots j_2 a} f^a_{j_1} + \cdots &= f^i_{j_k\cdots j_1 a}\xi^a + \cdots + f^i_a \xi^a_{j_k\cdots j_1}
\end{aligned} \tag{17.34}$$

Going from the top to the bottom in (17.34), we can uniquely solve for the η's in terms of the f's and ξ's. Therefore, $j_k(\eta) = \left(\eta^i, \ldots, \eta^i_{j_k\cdots j_1}\right)$ is uniquely determined by $j_{k+1}(f) = \left(f^i_j, \ldots, f^i_{j_{k+1}\cdots j_1}\right)$ and $j_k(\xi) = \left(\xi^i, \ldots, \xi^i_{j_k\cdots j_1}\right)$. We define

$$\mathrm{Adj}_{k+1}(f)(j_k(\xi)) \overset{\text{def}}{=} j_k(\eta) \tag{17.35}$$

Thus, we can act on k-jets of vector fields with $(k + 1)$-arrows. Further, if two $(k + 1)$-arrows have the same action, then they must be equal. Therefore, (17.35) gives the faithful representation

$$\mathrm{Ad} : G_{k+1}(n) \longrightarrow \mathrm{GL}(J_k(\mathbb{R}^n)) \tag{17.36}$$

There is an elegant coordinate-free definition of (17.36) that will be useful in Chapter 18. Consider the diffeomorphism $g(t,x) \stackrel{\text{def}}{=} \left(f \circ \exp(t\xi) \circ f^{-1}\right)(x)$ defined for all x in some neighborhood U of o and $t \in (-\epsilon, \epsilon)$, where f and ξ are as in (17.35). Since $g(0,x) = x$, we can define the vector field η on $f(U)$ by $\eta(x) \stackrel{\text{def}}{=} [\partial g(t,x)/\partial t]_{t=0}$. We can now show that the right-hand side of

$$\mathrm{Adj}_{k+1}(f)(j_k(\xi)) = j_k\left(\frac{\partial}{\partial t}\left[f \circ e^{t\xi} \circ f^{-1}\right]_{t=0}\right) \tag{17.37}$$

depends only on $j_{k+1}(f)$ and $j_k(\xi)$, and (17.37) is equivalent to (17.35). Note that (17.37) gives the formula

$$j_k(\xi) = j_k\left(\left[\frac{\partial e^{t\xi}}{\partial t}\right]_{t=0}\right) = \frac{\partial}{\partial t}[j_k(e^{t\xi(x)})]_{t=0} \tag{17.38}$$

which allows us to define $j_k(\xi)^o$ in a coordinate-free way once we define $j_k(h)^{p,q}$ in a coordinate-free way for a local diffeomorphism h with $h(p) = q$. This coordinate-free approach to jets is very elegant and useful, but it sometimes hides essential points related to coordinates: For instance, knowledge of the existence of the splitting (17.17) does not tell us that it is the Schwarzian derivative.

Using (17.37) (or using (17.30) and (17.34) directly), we can easily deduce

$$\mathrm{Adj}_k(f)\{j_k(\xi), j_k(\phi)\} = \{\mathrm{Adj}_{k+1}(f)(j_k(\xi)), \mathrm{Adj}_{k+1}(f)(j_k(\phi)\} \tag{17.39}$$

for $k \geq 1$, where $j_k(f) = \pi_{k+1,k} j_{k+1}(f)$. Note that both sides of (17.39) live in $J_{k-1}(\mathbb{R}^n)$.

Now something crucial: If we set $\xi^i = 0$ in (17.34), we get not only $\eta^i = 0$, but also $f^i_{j_k \cdots j_1 a}\xi^a = 0$. Therefore, $\left(0, \eta^i_j, \ldots, \eta^i_{j_k \cdots j_1}\right)$ is determined by $\left(0, \xi^i_j, \ldots, \xi^i_{j_k \cdots j_1}\right)$ and by $j_k(f)$, and we do not need $j_{k+1}(f)$ as in (17.35). This fact shows that we have the representation

$$\mathrm{Ad}\, G_k(n) : J_{k,0}(\mathbb{R}^n) \longrightarrow J_{k,0}(\mathbb{R}^n) \tag{17.40}$$

Note that for $k = 1$, the second formula of (17.34) gives $\xi \to f \circ \xi \circ f^{-1} = \eta$, where $\circ$ is multiplication of matrices. This is no coincidence, because we have the following proposition:

Proposition 17.3 (17.40) *is the adjoint representation of* $G_k(n)$ *on its Lie algebra* $J_{k,0}(\mathbb{R}^n)$.

It remains to construct the analogs of G and its Lie algebra $\mathfrak{g}$ in Chapter 16. This is done by choosing a smooth manifold M, repeating the above constructions at all points of M, and then gluing together in a smooth way. We first need to define the universal groupoid $\mathcal{U}_k(M) = \mathcal{U}_k$ of order k on M, which, as a set, is the set of all k-arrows on M. We consider the set of all local diffeomorphisms of M that map p to q. Choosing coordinates (x), (y) around p, q, we define $f \sim_k g$ if f and g have the same derivatives at p of all orders up to and including k. This is an equivalence relation, and each class has a unique representation of the form

$$j_k(f)^{p,q} = \left(x^i, y^i, f^i_j, \ldots, f^i_{j_k \ldots j_1}\right) \tag{17.41}$$

We write (17.41) also as $\left(x^i, f^i_\mu\right)$, $0 \leq |\mu| \leq k$, where $f^i_\mu = f^i$ for $|\mu| = 0$. Using the group operation (17.5) of $G_k(n)$, we can easily show that $\sim_k$ is independent of coordinates. An equivalence class is called a k-arrow from p to q, which we denote by $j_k(f)^p$ and also by $j_k(f)^{p,q}$, $q = f(p)$. We denote the set of k-arrows from p to q by $\mathcal{U}^{p,q}_k$ and define $\mathcal{U}_k \overset{\text{def}}{=} \cup_{p,q} \mathcal{U}^{p,q}_k$. The representation (17.41) shows that $\mathcal{U}_k$ is a smooth manifold. Finally, the composition map $\mathcal{U}^{q,r}_k \times \mathcal{U}^{p,q}_k \longrightarrow \mathcal{U}^{p,r}_k$ and the inversion map $\mathcal{U}^{p,q}_k \longrightarrow \mathcal{U}^{q,p}_k$ endow $\mathcal{U}_k$ with the structure of a differentiable groupoid.

Definition 17.4 $\mathcal{U}_k = \mathcal{U}_k(M)$ *is the universal jet groupoid on M of order k.*

We defined $\mathcal{U}_k$ using k-arrows of *local* diffeomorphisms. However, for any k-arrow from p to q, it can be shown that there exists a *global* diffeomorphism f of M that maps p to q and defines this k-arrow. Therefore, $\mathcal{U}_k$ is the groupoid of k-arrows on M induced by the action of Diff(M). In short, we can choose the representative f in $j_k(f)^{p,q}$ in Diff(M). We may view $\mathcal{U}_k$ as a very weak (even if we choose $k = \infty$ in the smooth category) approximation to the action of Diff(M) on M. On the other hand, we will show in Chapter 18 that the action of $G \subset \text{Diff}(M)$ on $G/H = M$ is uniquely determined by $m = o(H, G) = (\mathfrak{h}, \mathfrak{g})$.

We call $\mathcal{U}^{p,p}_k$ the vertex group of $\mathcal{U}_k$ at p. A choice of coordinates around p identifies $\mathcal{U}^{p,p}_k$ with $G_k(n)$, $n = \dim M$. Therefore, everything stated above for $G_k(n)$ is valid for $\mathcal{U}^{p,p}_k$ for all $p \in M$. Note that any $j_k(f)^{p,q}$ induces an isomorphism $\text{Adj}_k(f)^{p,q} : \mathcal{U}^{p,p}_k \to \mathcal{U}^{q,q}_k$ by conjugation $\text{Ad}\, g : x \to g \circ x \circ g^{-1}$, and all vertex groups are isomorphic (but not canonically). Clearly, $\text{Adj}_k(f)^{p,q}$ maps the nilfiltration of $\mathcal{U}^{p,p}_k$ bijectively onto the nilfiltration of $\mathcal{U}^{q,q}_k$. In particular, $\text{Adj}_k(f)^{p,p}$ is an inner automorphism of $\mathcal{U}^{p,p}_k$ that preserves the nilfiltration of $\mathcal{U}^{p,p}_k$.

We have the tower of groupoids and projection homomorphisms

$$\ldots \longrightarrow \mathcal{U}_k \longrightarrow \ldots \longrightarrow \mathcal{U}_2 \longrightarrow \mathcal{U}_1 \longrightarrow \mathcal{U}_0 = M \times M \tag{17.42}$$

which restricts to the vertex groups and gives (17.9) in coordinates. As before, we denote the projection $\mathcal{U}_{k+r} \to \mathcal{U}_r$ by $\pi_{k+r,r}$, which preserves composition and inversion of arrows.

We fix a basepoint $e \in M$ and define $\mathcal{U}^{e,\bullet}_k \overset{\text{def}}{=} \cup_{p \in M} \mathcal{U}^{e,p}_k$, which is the set of all k-arrows spreading out from e. We define the projection $\pi_{k,0} : \mathcal{U}^{e,\bullet}_k \to M$ with fiber $\pi^{-1}_{k,0}(p) = \mathcal{U}^{e,p}_k$. The group $\mathcal{U}^{e,e}_k \simeq G_k(n)$ acts on $\mathcal{U}^{e,\bullet}_k$ on the right by composition at the source, in view of the composition map $\mathcal{U}^{e,p}_k \times \mathcal{U}^{e,e}_k \to \mathcal{U}^{e,p}_k$, and $\pi_{k,0} : \mathcal{U}^{e,\bullet}_k \to M$ is a right principal bundle with structure group $\mathcal{U}^{e,e}_k$. Clearly, $\mathcal{U}^{e,\bullet}_0 \approx M$. Note that $\mathcal{U}^{p,p}_k$ acts on the fiber $\mathcal{U}^{e,p}_k$ on the *left* by the composition $\mathcal{U}^{p,p}_k \times \mathcal{U}^{e,p}_k \to \mathcal{U}^{e,p}_k$ at the target. For $x \in \mathcal{U}^{e,p}_k$ and $a \in \mathcal{U}^{e,e}_k$, $x \circ a = b \circ x$ for a unique $b \in \mathcal{U}^{p,p}_k$ that is $b = x \circ a \circ x^{-1}$. Also note that if we act on $x, y \in \mathcal{U}^{e,\bullet}_k$ by some $a \in \mathcal{U}^{e,e}_k$, then $x \circ y^{-1} = x \circ a \circ (y \circ a)^{-1}$, which explains the equivalence relation $(x, y) \sim (xa, ya)$ on $P \times P$ in the definition of $[P \times P]$ in Chapter 16.

We recall the formal names in Chapter 16: G defines m-arrows on G/G_0 and, more generally, $(m - r)$-arrows on G/G_r. To be consistent with our notation $(G/G_r, [G/G_{r+k}, G/G_{r+k}])$, henceforth we will denote $\mathcal{U}_k = \mathcal{U}_k(M)$ by $(M, \mathcal{U}_k(M)) = (\mathcal{U}_0^{e,\bullet}(M), \mathcal{U}_k(M))$, because $\mathcal{U}_{r+k}(M)$ defines also k-arrows on $\mathcal{U}_r^{e,\bullet}(M)$ by prolonging the action of $\mathrm{Diff}(M)$ on M to an obvious action on $\mathcal{U}_r^{e,\bullet}(M)$ as follows: Some $j_r(h)^{p,q}$ defines a bijection $\mathcal{U}_r^{e,p}(M) \to \mathcal{U}_r^{e,q}(M)$ by composing at the target as $j_r(f)^{e,p} \to j_r(h)^{p,q} \circ j_r(f)^{e,p}$. Let $a, b \in \mathcal{U}_r^{e,\bullet}(M)$, with $\pi(a) = p$ and $\pi(b) = q$, and let $j_r(h)^{p,q} \in \mathcal{U}_r^{p,q}$ be the unique k-arrow satisfying $j_r(h)^{p,q}(a) = b, h \in \mathrm{Diff}(M)$. Thus, h "prolongs" to some diffeomorphism $\rho_r(h) : \mathcal{U}_r^{e,\bullet}(M) \to \mathcal{U}_r^{e,\bullet}(M)$ by composition at the target that satisfies $\rho_r(h)(a) = b$. In particular, note that this prolonged action of $\mathrm{Diff}(M)$ on $\mathcal{U}_r^{e,\bullet}(M)$ is transitive for all r. Clearly, a Lie group G cannot do this, since $\dim \mathcal{U}_r^{e,\bullet}(M)$ increases without bound. Now, $j_k(\rho_r(h))^{a,b}$ makes perfect sense, and it turns out that $j_k(\rho_r(h))^{a,b}$ depends only on $j_{r+k}(h)^{p,q}$. To summarize, any $(r + k)$-arrow from p to q on M lifts to k-arrows on $\mathcal{U}_r^{e,\bullet}(M)$ whose sources and targets are in the fibers $\pi^{-1}(p)$ and $\pi^{-1}(q)$. Such lifts are closed under composition and inversion, and we get the groupoid $(\mathcal{U}_r^{e,\bullet}(M), \mathcal{U}_{r+k}(M))$. It is not difficult to check that the vertex groups of $(\mathcal{U}_r^{e,\bullet}(M), \mathcal{U}_{r+k}(M))$ can be identified with $K_{r+k,r}$ defined by (17.7). Clearly, $(\mathcal{U}_r^{e,\bullet}, \mathcal{U}_{r+k}(M)) \subset (\mathcal{U}_r^{e,\bullet}, \mathcal{U}_k(\mathcal{U}_r^{e,\bullet}))$ is a subgroupoid.

We now want to "linearize" $(\mathcal{U}_r^{e,\bullet}, \mathcal{U}_{r+k}(M))$. We do this first for $r = 0$. The idea is to replace the "point transformations" by the "infinitesimal point transformations," as the old masters would say. In modern language, this amounts to passing from the groupoid $(M, \mathcal{U}_k(M))$ to its algebroid.

Let $p \in M, \xi = (\xi^i(x))$ be a vector field defined near p and let $j_k(\xi)^p$ be the k-jet of ξ at p. We let $(J_kT)^p$ denote the set of all k-jets of vector fields at p and define the bundle of vectors $J_kT \stackrel{\text{def}}{=} \cup_{p\in M} (J_kT)^p \to M$ whose fiber over p is $(J_kT)^p$. Now $J_kT \to M$ is easily seen to be a locally trivial vector bundle. A local section of $J_kT \to M$ over (U, x) is of the form $(x^i, \xi^i_\mu(x)) = (x^i, \xi^i(x), \xi^i_j(x), \xi^i_{j_2j_1}(x), \ldots, \xi^i_{j_k\cdots j_1}(x))$. For any $\overline{x} \in U$, there exists a vector field $\eta = (\eta^i(x))$ defined near $\overline{x}$ satisfying $\xi^i_\mu(\overline{x}) = \partial^{|\mu|}\eta^i(\overline{x})/\partial x^\mu$, by the definition of $j_k(\xi)^{\overline{x}}$. However, η depends on $\overline{x}$, and we may not be able to find a vector field η satisfying this equality for all $x \in U$; that is, we may not have $\partial^{|\mu|}\xi^i(x)/\partial x^{|\mu|} = \xi^i_\mu(x)$ on U for a local section. This fact suggest the following definition: For a vector field $\xi = (\xi^i(x))$ on (U, x), we define

$$j_k(\xi) = \left(x^i, \xi^i, \xi^i_j, \xi^i_{j_2j_1}, \ldots, \xi^i_{j_k\cdots j_1}\right) \tag{17.43}$$

$j_k(\xi)^x \stackrel{\text{def}}{=} \left(x^i, \partial^{|\mu|}\xi^i(x)/\partial x^\mu\right)$, and $j_k(\xi)$ is a section of $J_kT \to M$ over (U, x) called the kth prolongation of ξ. The definition of $j_k(\xi)$ is independent of coordinates, and therefore $j_k(\xi)$ is a globally defined section of $J_kT \to M$ if ξ is a globally defined vector field on M.

Now a choice of coordinates around $p \in M$ identifies $(J_kT)^p$ with $J_k(\mathbb{R}^n)^o = J_k(\mathbb{R}^n)$ studied above. Everything stated for $J_k(\mathbb{R}^n)$ is valid for the fiber $(J_kT)^p$ of the vector bundle $J_kT \stackrel{\text{def}}{=} \cup_p (J_kT)^p \to M$ for any $p \in M$. In particular, we have the nilfiltrations at all points, and the point bracket

$$\{\, ,\, \}_p : (J_{k+1}T)^p \times (J_{k+1}T)^p \longrightarrow (J_kT)^p \tag{17.44}$$

is given in coordinates by (17.30) and restricts to a bracket of Lie algebras

$$\{\, ,\, \}_p : (J_{k,0}T)^p \times (J_{k,0}T)^p \longrightarrow (J_{k,0}T)^p \tag{17.45}$$

Some $j_{k+1}(f)^{p,q}$ induces the isomorphism

$$\mathrm{Adj}_{k+1}(f)^{p,q} : (J_kT)^p \longrightarrow (J_kT)^q \tag{17.46}$$

defined by (17.34); that is, we can carry k-jets of vector fields by $(k+1)$-arrows as in Chapter 16. We express this fact by saying that $(J_kT)^p \to M$ is associated with $(M, \mathcal{U}_{k+1}(M))$. This association respects the nilfiltrations. Note that we do not need the concepts of principal bundles and associated vector bundles to formulate this association.

Now (17.44) and (17.45) define the brackets $J_{k+1}T \times J_{k+1}T \to J_kT$ and $J_{k,0}T \times J_{k,0}T \to J_{k,0}T$ on the sections of $J_{k+1}T \to M$ and $J_{k,0}T \to M$. Note that $J_{k,0}T \to M$ is a bundle of Lie algebras whose fiber $(J_{k,0}T)^p$ over p is the Lie algebra of $\mathcal{U}_k^{p,p}$, by Proposition 17.2. These brackets are defined pointwise and do not involve differentiation. We will denote these vector bundles also by (M, J_kT) and $(M, J_{k,0}T)$.

The relation between $(M, \mathcal{U}_{k+1}(M))$ and (M, J_kT) is not confined to the association (17.46) and is more involved. First, we have a differential bracket on the sections of (M, J_kT), called the Spencer bracket, defined as follows. We have the Spencer operator

$$D : J_{k+1}T \longrightarrow T^* \otimes J_kT \tag{17.47}$$

defined by

$$\left(\xi^i, \xi^i_j, \ldots, \xi^i_{j_{k+1}\ldots j_1}\right) \longrightarrow \left(\frac{\partial \xi^i}{\partial x^r} - \xi^i_r, \ldots, \frac{\partial \xi^i_{j_k\ldots j_1}}{\partial x^r} - \xi^i_{rj_k\ldots j_1}\right) \tag{17.48}$$

Now let ξ_k and η_k (using the notation of [P1]) be two sections of (M, J_kT). We lift them to the sections ξ_{k+1} and η_{k+1} of $(M, J_{k+1}T)$ and define

$$[\xi_k, \eta_k] \stackrel{\text{def}}{=} \{\xi_{k+1}, \eta_{k+1}\} + i_{\xi_1} D(\eta_{k+1}) - i_{\eta_1} D(\xi_{k+1}) \tag{17.49}$$

where i denotes the interior bracket and $\xi_1 = \pi_{k,1}(\xi_k)$. Using (17.30) and (17.48), we can easily show that the right-hand side of (17.49) does not depend on the lifts ξ_{k+1} and η_{k+1}. The Spencer bracket satisfies the Jacobi identity, is compatible with the projection of jets, and satisfies $[j_k(\xi), j_k(\eta)] = j_k[\xi, \eta]$ for the vector fields ξ and η on M. It reduces to $\{\, ,\, \}$ on the sections of $(M, J_{k,0}T)$. How do the point bracket and the Spencer bracket behave with respect to the arrows of $(M, \mathcal{U}_k(M))$? We refer to [P1] for more details.

To understand the meaning of the Spencer bracket, we will first define the vector bundle $(\mathcal{U}_r^{e,\bullet}, J_{r+k}T)$. The key fact is that an $(r+k)$-jet of a vector field at $p \in M$ defines a k-jet of a vector field at any $a \in \mathcal{U}_r^{e,\bullet}$ that projects to p. This is done easily by first integrating a vector field to its 1-parameter group of local diffeomorphisms, prolonging these local diffeomorphisms to $\mathcal{U}_r^{e,\bullet}$ by the above construction, and differentiating with respect to t at $t = 0$.

We first set $k = 0$ in $(\mathcal{U}_r^{e,\bullet}, J_{r+k}T)$. Now $\mathcal{U}_r^{e,e}$ acts on $\mathcal{U}_r^{e,\bullet}$ and therefore induces an action on the tangent vectors at the points in the fiber over p. This action commutes with the above prolonged actions, because the first is defined by composition at the source whereas the second is defined by composition at the target. We easily deduce the important fact that the fiber $(M, J_rT)^p$ over $p \in M$ is in 1–1 correspondence with a set of $\mathcal{U}_r^{e,e}$-invariant vectors in the fiber $\mathcal{U}_r^{e,p}$ over p. This gives the following proposition:

Proposition 17.5 *There is a 1–1 correspondence between*

(i) global sections of $J_rT \to M$

and

(ii) $\mathcal{U}_r^{e,e}$-invariant vector fields on $\mathcal{U}_r^{e,\bullet} \to M$.

Let $\mathfrak{X}_{\text{inv}}(\mathcal{U}_r^{e,e}) \subset \mathfrak{X}(\mathcal{U}_r^{e,e})$ denote the set of $\mathcal{U}_r^{e,e}$-invariant vector fields on the principal bundle $\mathcal{U}_r^{e,\bullet} \to M$. Clearly, $\mathfrak{X}_{\text{inv}}(\mathcal{U}_r^{e,\bullet})$ is closed under the usual bracket of vector fields and is a Lie algebra. According to Proposition 17.5, this bracket induces a bracket on the sections of (M, J_rT), which turns out to be (17.49)!

For the general case, we replace $M = \mathcal{U}_0^{e,\bullet}$ by $\mathcal{U}_r^{e,\bullet}$ and define the fiber $(\mathcal{U}_r^{e,\bullet}, J_{r+k}T)^a$ over a by $(\mathcal{U}_r^{e,\bullet}, J_{r+k}T)^a \stackrel{\text{def}}{=} \{j_k(\rho_r\xi)^a \mid \pi_{r,0}(a) = p\}$, where $\rho_r(\xi)$ denotes the prolongation of the vector field ξ on M to $\mathcal{U}_r^{e,\bullet}$. This gives the vector bundle $(\mathcal{U}_r^{e,\bullet}, J_{r+k}T) \stackrel{\text{def}}{=} \cup_{a\in\mathcal{U}_r^{e,\bullet}} (\mathcal{U}_r^{e,\bullet}, \mathcal{U}_{r+k}(T))^a$, which is clearly a subbundle of $(\mathcal{U}_r^{e,\bullet}, J_k(\mathcal{U}_r^{e,\bullet}))$ and is easily seen to be associated with $(\mathcal{U}_r^{e,\bullet}, \mathcal{U}_{r+k})$. Now Proposition 17.5 generalizes and gives a 1–1 correspondence between the global sections of $(M, J_{r+k}T)$ and global sections of $(\mathcal{U}_r^{e,\bullet}, J_{r+k}T)$ that are $\mathcal{U}_r^{e,e}$-invariant. For $k = 0$, we get Proposition 17.5. Clearly, $(\mathcal{U}_r^{e,\bullet}, J_{r+k}T)$ is the "linearization" of $(\mathcal{U}_r^{e,\bullet}, \mathcal{U}_{r+k})$ in the sense that (M, J_kT) is the linearization of $(M, \mathcal{U}_r(M))$. We will leave further details to the interested reader.

18

Embeddings of Klein Geometries into Universal Jet Groupoids

The purpose of this chapter is to show that the formal names "k-arrow" and "k-jet of an infinitesimal generator (k-jig)" given to the elements of the groupoids and vector bundles in Chapter 16 are true names. This will be shown by embedding the group theory in Chapter 16 into the jet theory in Chapter 17.

Let G be a transitive Lie transformation group of a smooth manifold M. Let $\mathfrak{X}(M)$ denote the Lie algebra of all vector fields on M and $\overline{\mathfrak{g}} \subset \mathfrak{X}(M)$ the Lie algebra of the infinitesimal generators of the action of G. In more detail, the action of a smooth curve in G with tangent vector $\xi \in T_eG = \mathfrak{g}$ defines a smooth curve passing through each point $x \in M$, and the tangent vectors of these curves define a vector field $\overline{\xi}$ on M. This gives the map

$$\phi : \mathfrak{g} = T_eG \to \mathfrak{X}(M) \tag{18.1}$$

defined by $\phi(\xi) \overset{\text{def}}{=} \overline{\xi}$. Now, ϕ is a homomorphism of Lie algebras that is injective since G acts effectively. We let $\overline{\mathfrak{g}} \overset{\text{def}}{=} \operatorname{Im}\phi \subset \mathfrak{X}(M)$, and therefore $\mathfrak{g} \simeq \overline{\mathfrak{g}}$ as Lie algebras. Since G acts transitively, for any point in M and any tangent vector at this point, there exists a vector field in $\overline{\mathfrak{g}}$ whose value at this point is that given tangent vector. Thus, the values of the elements of $\overline{\mathfrak{g}}$ at p span the tangent space T_pM for all $p \in M$.

We now choose a base point $p \in M$. Let G_0 be the stabilizer at p with Lie algebra $\mathfrak{g}_0$. Thus, we get the nilfiltration (2.2) of $\mathfrak{g}_0$ determined by $\mathfrak{g}$. We recall the vector space $(J_kT)^p$ defined in Chapter 17, which is the set of all k-jets of vector fields at p. We define the map

$$\begin{aligned} j_k(\cdot)^p : \overline{\mathfrak{g}} &\longrightarrow (J_kT)^p \\ : \overline{\xi} &\longrightarrow j_k(\overline{\xi})^p, \qquad k \geq 0 \end{aligned} \tag{18.2}$$

We denote the image of (18.2) by $J_k(\overline{\mathfrak{g}})^p \subset (J_kT)^p$, so that

$$j_k(\cdot)^{\overline{e}} : \overline{\mathfrak{g}} \longrightarrow J_k(\overline{\mathfrak{g}})^p \subset (J_kT)^p \tag{18.3}$$

An Alternative Approach to Lie Groups and Geometric Structures, Ercüment H. Ortaçgil (2018).

The following important fact clarifies the relation between the group theory in Chapter 16 and the jet theory in Chapter 17:

Proposition 18.1 $j_k\left(\overline{\xi}\right)^p = 0 \Longleftrightarrow \xi \in \mathfrak{g}_k, x \in \mathfrak{g}, k \geq 0.$

Proof We proceed by induction. Recalling the identification $G/G_0 = M$ and $p = o =$ the coset of G_0, for $k = 0$ the claim is that $\overline{\xi}(o) = 0$ if and only if $\xi \in \mathfrak{g}_0$. Since $T_o(G/G_0) = \mathfrak{g}/\mathfrak{g}_0$, this is clear. Now,

$$\xi \in \mathfrak{g}_{k+1} \Longleftrightarrow \xi \in \mathfrak{g}_k \quad \text{and} \quad [\xi, \eta] \in \mathfrak{g}_k \qquad \text{for all } \eta \in \mathfrak{g} \tag{18.4}$$

by the definition (16.1) of $\mathfrak{g}_{k+1}$. By the induction hypothesis, the right-hand side of (18.4) is equivalent to

$$j_k\left(\overline{\xi}\right)^o = 0 \quad \text{and} \quad j_k\overline{[\xi, \eta]} = 0 \qquad \text{for all } \eta \in \mathfrak{g} \tag{18.5}$$

Since (18.1) is a homomorphism, $\overline{[\xi, \eta]} = [\overline{\xi}, \overline{\eta}]$. Now, the definition (17.29) of $\{\ ,\ \}_o$ shows that $j_k\left([\overline{\xi}, \overline{\eta}]\right)^o = \{j_{k+1}(\overline{\xi})^o, j_{k+1}(\overline{\eta})^o\}_o$. Therefore, (18.5) is equivalent to

$$j_k\left(\overline{\xi}\right)^o = 0 \quad \text{and} \quad \left\{j_{k+1}(\overline{\xi})^o, j_{k+1}(\overline{\eta})^o\right\}_o = 0 \qquad \text{for all } \eta \in \mathfrak{g} \tag{18.6}$$

We choose a coordinate system (x) around o. Now, $\left\{j_{k+1}(\overline{\xi})^o, j_{k+1}(\overline{\eta})^o\right\}_o$ is given by (17.30). Since $j_k\left(\overline{\xi}\right)^o = 0$, all terms in (17.30) vanish except $\overline{\xi}^a\overline{\eta}^i_{ajk\cdots j_1}$. Therefore, (18.6) is equivalent to $\overline{\eta}^a\overline{\xi}^i_{ajk\cdots j_1} = 0$ for all $\eta \in \mathfrak{g}$. Since $J_0(\overline{\mathfrak{g}})^o = T_o(G/G_0)$, $(\overline{\eta}^i)$ takes all values in $T_o(G/G_0)$. Therefore, the last statement is equivalent to $\overline{\xi}^i_{j_{k+1}j_k\cdots j_1} = 0$, that is, $j_{k+1}(\overline{\xi})^o = 0$, which completes the inductive step. ■

Proposition 18.1 gives the isomorphism of vector spaces

$$\theta_k : \mathfrak{g}/\mathfrak{g}_k \longrightarrow J_k(\overline{\mathfrak{g}})^p \tag{18.7}$$

where $\theta_k = j_k(\cdot)^p \circ \phi$. Since the infinitesimal generator $\overline{\xi}$ is globally determined on M by $j_m(\overline{\xi})^p$ by Proposition 18.1, where $m = o(\mathfrak{g}_0, \mathfrak{g})$, we have the isomorphism of Lie algebras

$$\begin{aligned} \varepsilon : J_m(\overline{\mathfrak{g}})^p &\longrightarrow J_{m+1}(\overline{\mathfrak{g}})^p \\ : j_m(\overline{\xi})^p &\longrightarrow j_{m+1}(\overline{\xi})^p \end{aligned} \tag{18.8}$$

Thus, we obtain the following diagram:

$$\begin{array}{ccccccccccc} & & \mathfrak{g} & \longrightarrow & \mathfrak{g}/\mathfrak{g}_{m-1} & \longrightarrow & \mathfrak{g}/\mathfrak{g}_{m-2} & \longrightarrow & \cdots & \longrightarrow & \mathfrak{g}/\mathfrak{g}_0 \\ & & \downarrow & & \downarrow & & \downarrow & & \downarrow & & \downarrow \\ \varepsilon\left(J_m(\overline{\mathfrak{g}})^p\right) & \simeq & J_m(\overline{\mathfrak{g}})^p & \longrightarrow & J_{m-1}(\overline{\mathfrak{g}})^p & \longrightarrow & J_{m-1}(\overline{\mathfrak{g}})^p & \longrightarrow & \cdots & \longrightarrow & J_0(\overline{\mathfrak{g}})^p \\ \cap & & \cap & & \cap & & \cap & & \cap & & \cap \\ \left(J_{m+1}T\right)^p & \longrightarrow & J_m(T)^p & \longrightarrow & J_{m-1}(T)^p & \longrightarrow & J_{m-2}(T)^p & \longrightarrow & \cdots & \longrightarrow & J_0(T)^p \end{array} \tag{18.9}$$

where the top vertical maps are isomorphisms. The projections $\mathfrak{g}/\mathfrak{g}_{r+s} \to \mathfrak{g}/\mathfrak{g}_r$ in the top sequence are defined algebraically by $x + \mathfrak{g}_{r+s} \to x + \mathfrak{g}_r$ and in the bottom sequence by the projection of jets $\pi_{r+s,r} : J_{r+s}(T)^e \to J_r(T)^e$. The proof of Proposition 18.1 shows that $\pi_{r+s,r}$ restricts to the middle sequence and coincides with the projection induced from the top sequence. Therefore,(18.9) commutes. Fixing a column of (18.9) gives the diagram

$$\begin{array}{ccccc}
\mathfrak{g}/\mathfrak{g}_{k+1} & \times & \mathfrak{g}/\mathfrak{g}_{k+1} & \xrightarrow{[\,,\,]} & \mathfrak{g}/\mathfrak{g}_k \\
\| & & \| & & \| \\
J_{k+1}(\overline{\mathfrak{g}})^p & \times & J_{k+1}(\overline{\mathfrak{g}})^p & \xrightarrow{\{\,,\,\}_p} & J_k(\overline{\mathfrak{g}})^p \\
\cap & & \cap & & \cap \\
(J_{k+1}T)^p & \times & (J_{k+1}T)^p & \xrightarrow{\{\,,\,\}_p} & (J_kT)^p
\end{array} \tag{18.10}$$

The top bracket is defined by $[x + \mathfrak{g}_{k+1}, y + \mathfrak{g}_{k+1}] \overset{\text{def}}{=} [x, y] + \mathfrak{g}_k$, which is well-defined since $[\mathfrak{g}_{k+1}, \mathfrak{g}] \subset \mathfrak{g}_k$. The bottom bracket is the point bracket defined by (17.29). It is now easy to see from the proof of Proposition 18.1 that both brackets agree in the middle row of (18.10), so that (18.10) commutes. For $k = m$, the top bracket in (18.10) gives the Lie algebra $(\mathfrak{g}, [\,,\,])$. To recover $(\mathfrak{g}, [\,,\,])$ from the bottom bracket in (18.10), we define the torsion bracket

$$T(\cdot,\cdot)_{\overline{e}} : J_m(\overline{\mathfrak{g}})^p \times J_m(\overline{\mathfrak{g}})^p \longrightarrow J_m(\overline{\mathfrak{g}})^p \tag{18.11}$$

by the formula

$$T(j_m(\overline{\xi})^p, j_m(\overline{\eta})^p)_p \overset{\text{def}}{=} \{\varepsilon\,(j_m(\overline{\xi})^p), \varepsilon\,(j_m(\overline{\eta})^p)\}_p \tag{18.12}$$

and obtain the isomorphism

$$(\mathfrak{g}, [\,,\,]) \simeq (J_m(\overline{\mathfrak{g}})^p, T(\cdot,\cdot)_p) \tag{18.13}$$

as in Part I. We also deduce the commutative diagram

$$\begin{array}{ccccccccc}
0 & \longrightarrow & \mathfrak{g}_k/\mathfrak{g}_{k+s} & \longrightarrow & \mathfrak{g}/\mathfrak{g}_{k+s} & \longrightarrow & \mathfrak{g}/\mathfrak{g}_k & \longrightarrow & 0 \\
 & & \| & & \| & & \| & & \\
0 & \longrightarrow & J_{k+s,k}(\overline{\mathfrak{g}})^p & \longrightarrow & J_{k+s}(\overline{\mathfrak{g}})^p & \longrightarrow & J_k(\overline{\mathfrak{g}})^p & \longrightarrow & 0 \\
 & & \cap & & \cap & & \cap & & \\
0 & \longrightarrow & (J_{k+s,k}T)^p & \longrightarrow & (J_{k+s}T)^p & \longrightarrow & (J_sT)^p & \longrightarrow & 0
\end{array} \tag{18.14}$$

By restricting (18.9) to $J_{m,0}(\overline{\mathfrak{g}})^p$, we get the commutative diagram of the Lie algebras:

$$\begin{array}{ccccccccccc} & & \mathfrak{g}_0 & \longrightarrow & \mathfrak{g}_0/\mathfrak{g}_{m-1} & \longrightarrow & \cdots & \longrightarrow & \mathfrak{g}_0/\mathfrak{g}_1 & \longrightarrow & 0 \\ & & \| & & \| & & \| & & \| & & \\ \varepsilon\left(J_{m,0}(\overline{\mathfrak{g}})^p\right) & \simeq & J_{m,0}(\overline{\mathfrak{g}})^p & \longrightarrow & J_{m-1,0}(\overline{\mathfrak{g}})^p & \longrightarrow & \cdots & \longrightarrow & J_{1,0}(\overline{\mathfrak{g}})^p & \longrightarrow & 0 \\ \cap & & \cap & & \cap & & \cap & & \cap & & \\ \left(J_{m+1,0}T\right)^p & \longrightarrow & J_{m,0}(T)^p & \longrightarrow & J_{m-1,0}(T)^p & \longrightarrow & \cdots & \longrightarrow & J_{1,0}(T)^p & \longrightarrow & 0 \end{array} \tag{18.15}$$

Similarly, (18.10) restricts from $\mathfrak{g}/\mathfrak{g}_k$ to $\mathfrak{g}_0/\mathfrak{g}_k$, and we obtain the isomorphism of Lie algebras

$$(\mathfrak{g}_0/\mathfrak{g}_k, [\,,\,]) \simeq (J_{k,0}(\overline{\mathfrak{g}})^p, \{\,,\,\}_p\} \subset (J_{k,0}(T)^p, \{\,,\,\}_p) \tag{18.16}$$

which does not use ε and $T(\,,\,)$, since $\{\,,\,\}_p$ already restricts to the Lie algebra $J_{k,0}(T)^p$ by (17.31), which is the Lie algebra of $\mathcal{U}_k^{p,p}$ according to Proposition 17.2.

To consider the vector bundles, we first need to settle the issue of arrows. Recalling that we identify M with G/G_0 and p with the left coset o of G_0, we have the nilfiltration (16.14) of G_0 determined by G. Now, G acts on M by $xG_0 \to gxG_0$, and G_0 stabilizes $p = o$. Thus, we have the homomorphism $\varphi : G \to \mathrm{Diff}(M)$, which is injective since G acts effectively. We denote $\varphi(G)$ by $\overline{G}(M)$. Thus, $\overline{g}(p) = p$ if and only if $g \in G_0$ and $j_k(\overline{g})^{p,p} = j_k(\overline{g})^p$ is defined as a k-arrow with the source and target at p. Clearly, $j_k(\overline{g})^p \in \mathcal{U}_k^{p,p} \simeq G_k(n)$.

So, we obtain the group homomorphism (using the same notation)

$$\begin{aligned} j_k(\cdot)^p : \overline{G}_0 &\longrightarrow \mathcal{U}_k^{p,p}(M), \qquad k \geq 1 \\ : \overline{g} &\longrightarrow j_k(\overline{g})^p \end{aligned} \tag{18.17}$$

We denote the image of (18.17) by $\overline{G}_k^{p,p}(M)$, so that

$$\theta_k = j_k(\cdot)^p \circ \varphi : G_0 \longrightarrow \overline{G}_k^{p,p}(M) \subset \mathcal{U}_k^{p,p}(M) \tag{18.18}$$

It is not difficult to show that (18.17) is a continuous map, so it is a Lie group homomorphism. However, $\overline{G}_k^{p,p}(M) \subset \mathcal{U}_k^{p,p}(M)$ need not be a closed subgroup. This point will be relevant below.

The next proposition is the group analog of Proposition 18.1.

Proposition 18.2 $\theta_k(g) = 1 = \theta_k(Id)^p \Longleftrightarrow g \in G_k$.

Proof We recall the faithful representation (17.35):

$$\mathrm{Ad} : \mathcal{U}_{k+1}^{p,p} \longrightarrow \mathrm{GL}((J_k T)^p) \tag{18.19}$$

Composing j_{k+1} in (18.17) with (18.19), we obtain the faithful representation

$$G_0 \longrightarrow \mathrm{GL}((J_k T)^p) \tag{18.20}$$

We claim that (18.20) restricts to

$$G_0 \longrightarrow \mathrm{GL}(J_k(\overline{\mathfrak{g}})^p) \tag{18.21}$$

Indeed, from (17.37), we deduce

$$\begin{aligned}
\mathrm{Adj}_{k+1}\left(\overline{g}\right)^p\left(j_k\left(\overline{\xi}\right)^p\right) &= j_k\left(\left(\frac{\partial}{\partial t}\left(\overline{g}\circ\exp\left(t\overline{\xi}\right)\circ\overline{g}^{-1}\right)\right)_{t=0}\right)^p \\
&= j_k\left(\frac{\partial}{\partial t}\left(\overline{g\circ\exp\left(t\xi\right)\circ g^{-1}}\right)_{t=0}\right)^p \\
&= j_k\left(\overline{\mathrm{Ad}\left(g\right)\xi}\right)^p
\end{aligned} \tag{18.22}$$

which implies (18.21). Now

$$\begin{aligned}
j_k(\overline{g})^p = j_k(\mathrm{Id})^p & \\
&\Longleftrightarrow j_k(\overline{g})^p \text{ acts as identity on } J_k(\overline{\mathfrak{g}})^p \\
&\Longleftrightarrow j_k(\overline{\mathrm{Ad}(g)\xi})^p = j_k(\overline{\xi})^p && \text{by (18.22)} \\
&\Longleftrightarrow j_k(\overline{\mathrm{Ad}(g)\xi - \xi})^p = 0 \\
&\Longleftrightarrow \mathrm{Ad}(g)\xi - \xi \in \mathfrak{g}_k && \text{by Proposition 18.1} \\
&\Longleftrightarrow g \in G_{k+1} && \text{by the definition of } G_{k+1}
\end{aligned} \tag{18.23}$$

which establishes the claim. ■

Since $\overline{g} \in \overline{G}_0$ is determined by $j_m(\overline{g})^p$ by Proposition 18.2, we obtain the isomorphism

$$\begin{aligned}
\varepsilon : \overline{G}_m^{p,p}(M) &\longrightarrow \overline{G}_{m+1}^{p,p}(M) \\
: j_m(\overline{g})^p &\longrightarrow j_{m+1}(\overline{g})^p
\end{aligned} \tag{18.24}$$

Thus, we obtain the following commutative diagram (omitting M from our notation):

$$\begin{array}{ccccccccccc}
 & & G_0 & \longrightarrow & G_0/G_{m-1} & \longrightarrow & \cdots & \longrightarrow & G_0/G_1 & \longrightarrow & 1 \\
 & & \downarrow & & \downarrow & & \downarrow & & \downarrow & & \\
\varepsilon(\overline{G}_m^{p,p}) & \simeq & \overline{G}_m^{p,p} & \longrightarrow & \overline{G}_{m-1}^{p,p} & \longrightarrow & \cdots & \longrightarrow & \overline{G}_1^{p,p} & \longrightarrow & 1 \\
\cap & & \cap & & \cap & & \cap & & \cap & & \\
\mathcal{U}_{m+1}^{p,p} & \longrightarrow & \mathcal{U}_m^{p,p} & \longrightarrow & \mathcal{U}_{m-1}^{p,p} & \longrightarrow & \cdots & \longrightarrow & \mathcal{U}_1^{p,p} & \longrightarrow & 1
\end{array} \tag{18.25}$$

where the top vertical maps are isomorphisms of Lie groups. Thus, (18.25) injects the structure groups of the principal bundles in Chapter 16 into the vertex groups of the universal jet groupoids. Clearly, (18.15) is the linearization of (18.25).

We now want to derive the analog of (18.9). We define the map

$$\begin{aligned} [G/G_0, [G/G_k \times G/G_k]] &\longrightarrow \mathcal{U}_k(M) \\ [g_1 G_k, g_2 G_k] &\longrightarrow j_k \left(\overline{g}_2 \circ \overline{g}_1^{-1}\right)^{p,q} \end{aligned} \tag{18.26}$$

where $g_1 G_0 = p$ and $g_2 G_0 = q$. Using the definitions, we easily check that (18.26) preserves composition and inversion of arrows and is a homomorphism of groupoids. Moreover, it is injective, by Proposition 18.2. We denote the image of (18.26) by $(M, \overline{G}_k(M)) \subset (M, \mathcal{U}_k(M))$. Clearly, the vertex group of $(M, \overline{G}_k(M))$ at p is $\overline{G}_k^{p,p}(M)$ defined by (18.18). Now, (18.24) gives a splitting for all m-arrows of $\overline{G}_m(M)$. Thus, we get the isomorphism of groupoids

$$\varepsilon : (M, \overline{G}_m(M)) \longrightarrow (M, \overline{G}_{m+1}(M)) \tag{18.27}$$

and obtain the commutative diagram of groupoids

$$\begin{array}{ccccc} & & (M, \varepsilon\overline{G}_m(M) & \subset & (M, \mathcal{U}_{m+1}(M)) \\ & & \| & & \downarrow \\ (G/G_0, [G \times G]) & \longrightarrow & (M, \overline{G}_m(M)) & \subset & (M, \mathcal{U}_m(M)) \\ \downarrow & & \downarrow & & \downarrow \\ (G/G_0, [G/G_{m-1} \times G/G_{m-1}]) & \longrightarrow & (M, \overline{G}_{m-1}(M)) & \subset & (M, \mathcal{U}_{m-1}(M)) \\ \downarrow & & \downarrow & & \downarrow \\ (G/G_0, [G/G_{m-2} \times G/G_{m-2}]) & \longrightarrow & (M, \overline{G}_{m-2}(M)) & \subset & (M, \mathcal{U}_{m-2}(M)) \\ \downarrow & & \downarrow & & \downarrow \\ \vdots & \longrightarrow & \vdots & \subset & \vdots \\ \downarrow & & \downarrow & & \downarrow \\ (G/G_0, [G/G_2 \times G/G_2]) & \longrightarrow & (M, \overline{G}_2(M)) & \subset & (M, \mathcal{U}_2(M)) \\ \downarrow & & \downarrow & & \downarrow \\ (G/G_0, [G/G_1 \times G/G_1]) & \longrightarrow & (M, \overline{G}_1(M)) & \subset & (M, \mathcal{U}_1(M)) \\ \downarrow & & \downarrow & & \downarrow \\ G/G_0 \times G/G_0 & = & M \times M & = & M \times M \end{array} \tag{18.28}$$

where the horizontal arrows are isomorphisms of groupoids. Thus, (18.28) injects the group-theoretic constructions of Chapter 16 into the jet-theoretic construction of Chapter 17, and the projections in the middle column are compatible with those induced from the first and third columns. Note the crucial fact that it is not possible to recover the splitting at the top of the middle column by working solely within the first column: We must introduce a "torsion-free connection" to do this (see the Appendix). Another crucial fact is the following: The horizontal arrows in (18.28) are immersions, and therefore the middle column is an immersed submanifold of the third column, which need not be closed as a subset.

It is now time to be more precise with our terminology: A subset $\mathcal{P}_k \subset \mathcal{U}_k$ that is closed with respect to the composition and inversion of arrows induced from $\mathcal{U}_k$ is a subgroupoid. If $\mathcal{P}_k$ projects onto $M \times M = \mathcal{U}_0$, then $\mathcal{P}_k$ is transitive. If $\mathcal{P}_k \subset \mathcal{U}_k$ is an embedded submanifold with smooth arrow operations, then $\mathcal{P}_k \subset \mathcal{U}_k$ is a differentiable (or Lie) subgroupoid. Henceforth, we will make the following assumption:

A: The image of the first column in (18.28) is an embedded submanifold of the third column. Thus, the second column is a Lie subgroupoid of the third column. In particular, the image of j_k in (18.17) is a closed subgroup of $\mathcal{U}_k^{p,p}$ for all k.

This regularity assumption is made in all studies of pseudogroups via their jets (see, e.g., [GS1] and [OP]). However, the situation is rather special in our case: Whether **A** holds depends only on the pair (G_0, G)! We believe that **A** holds if G is algebraic and G_0 is an algebraic subgroup, in which case the image of (18.17) is also an algebraic subgroup. An example is given in [Wg], where **A** fails.

The assumption **A** entails interesting results for the embedding of certain Lie groups into jet groups.

Proposition 18.3 *Let (G_0, G) be a (effective) Klein geometry with G connected. If* **A** *holds, then G_0 is a Lie subgroup of $G_{m+1}(n)$, where $n = \dim G/G_0$ and $m = o(\mathfrak{g}_0, \mathfrak{g})$.*

Proposition 18.3 is a restatement of (18.17) in view of Proposition 16.7(iii).

There is one case when the image of (18.17) is clearly a closed subgroup, namely, if G_0 is compact.

Proposition 18.4 *If G_0 is compact and G is connected, then $o(G_0, G) \leq 1$.*

Proof By Proposition 16.7(iii), (18.17) embeds G_0 into $G_{m+1}(n)$, where $n = \dim G/G_0$ Identifying G_0 with its image, we restrict the projection homomorphism $\pi_{m+1,m} : G_{m+1}(n) \to G_m(n)$ to G_0. Since $K_{m+1,m}(n)$ is a vector group and G_0 is compact, $K_{m+1,m}(n) \cap G_0$ is trivial. Therefore, the restriction of $\pi_{m+1,m}$ to G_0 is an isomorphism. Iterating this argument, we see that $\pi_{m+1,1}$ projects G_0 isomorphically into $G_1(n)$. Therefore, (18.17) is an embedding already for $k = 1$. ■

We now come to the prolonged actions by replacing the base G/G_0 with G/G_r for some r with $0 \leq r \leq m$. Keeping in mind that $o(G_r, G) = m - r$ by Proposition 16.6, the repetition of the above arguments gives the following commutative diagram of groupoids (omitting $M = G/G_r$ from our notation in the second and third columns):

$$
\begin{array}{ccccc}
 & & (\overline{G}_r^{o,\bullet}, \varepsilon\overline{G}_m) & \subset & (\mathcal{U}_r^{o,\bullet}, \mathcal{U}_{m+1}) \\
 & & \| & & \downarrow \\
(G/G_r, [G \times G]) & \longrightarrow & (\overline{G}_r^{o,\bullet}, \overline{G}_m) & \subset & (\mathcal{U}_r^{o,\bullet}, \mathcal{U}_m) \\
\downarrow & & \downarrow & & \downarrow \\
(G/G_r, [G/G_{m-1} \times G/G_{m-1}]) & \longrightarrow & (\overline{G}_r^{o,\bullet}, \overline{G}_{m-1}) & \subset & (\mathcal{U}_r^{o,\bullet}, \mathcal{U}_{m-1}) \\
\downarrow & & \downarrow & & \downarrow \\
(G/G_r, [G/G_{m-2} \times G/G_{m-2}]) & \longrightarrow & (\overline{G}_r^{o,\bullet}, \overline{G}_{m-2}) & \subset & (\mathcal{U}_r^{o,\bullet}, \mathcal{U}_{m-2}) \\
\downarrow & & \downarrow & & \downarrow \\
\vdots & \longrightarrow & \vdots & \subset & \vdots \\
\downarrow & & \downarrow & & \downarrow \\
(G/G_r, [G/G_{r+2} \times G/G_{r+2}]) & \longrightarrow & (\overline{G}_r^{o,\bullet}, \overline{G}_{r+2}) & \subset & (\mathcal{U}_r^{o,\bullet}, \mathcal{U}_{r+2}) \\
\downarrow & & \downarrow & & \downarrow \\
(G/G_r, [G/G_{r+1} \times G/G_{r+1}]) & \longrightarrow & (\overline{G}_r^{o,\bullet}, \overline{G}_{r+1}) & \subset & (\mathcal{U}_r^{o,\bullet}, \mathcal{U}_{r+1}) \\
\downarrow & & \downarrow & & \downarrow \\
G/G_r \times G/G_r & = & \overline{G}_r^{o,\bullet} \times \overline{G}_r^{e,\bullet} & \subset & \mathcal{U}_r^{o,\bullet} \times \mathcal{U}_r^{e,\bullet}
\end{array}
\tag{18.29}
$$

where $o \in M = G/G_r$ is the coset of G_r, and (18.28) is recovered for $r = 0$. To explain the middle column, we recall that $(\mathcal{U}_r^{e,\bullet}(M), \mathcal{U}_{r+k}(M))$ is defined in Chapter 17 by prolonging the action of $\mathrm{Diff}(M)$ to $\mathcal{U}_r^{e,\bullet}(M)$ and is the groupoid of k-arrows on $\mathcal{U}_r^{e,\bullet}(M)$ defined by $\mathcal{U}_{r+k}(M) = (M, \mathcal{U}_{r+k})$. In the same way, $\left(\overline{G}_r^{e,\bullet}(M), \overline{G}_{r+k}(M)\right)$ is defined by prolonging the action of $\overline{G}(M) \subset \mathrm{Diff}(M)$ to $\overline{G}_r^{e,\bullet}(M)$. We should be careful with the meaning of $\subset$ in (18.29), which is not an inclusion of sets but an inclusion of pairs with the obvious definition.

Now the associated vector bundles in Chapter 16 become almost trivial in view of (18.28) and (18.29). We define

$$J_k(\overline{\mathfrak{g}}) \stackrel{\text{def}}{=} \cup_{x \in M} J_k(\overline{\mathfrak{g}})^x, \qquad k \geq 0 \tag{18.30}$$

where $J_k(\overline{\mathfrak{g}})^x \stackrel{\text{def}}{=} \left\{ j_k\left(\overline{\xi}\right)^x \mid \xi \in \mathfrak{g} \right\}$, and obtain the vector bundle

$$J_k(\overline{\mathfrak{g}}) \longrightarrow M \tag{18.31}$$

which is the vector bundle $(G/G_0, G/G_{k+1} \times_{G_0/G_{k+1}} \mathfrak{g}_0/\mathfrak{g}_k)$ constructed in Chapter 16. Thus, we obtain the diagram

$$
\begin{array}{ccccc}
 & & \varepsilon J_m(\overline{\mathfrak{g}}) & \subset & J_{m+1}T \\
 & & \| & & \downarrow \\
(G/G_0, G \times_{G_0} \mathfrak{g}) & \longrightarrow & J_m(\overline{\mathfrak{g}}) & \subset & J_m T \\
\downarrow & & \downarrow & & \downarrow \\
(G/G_0, G \times_{G_0} \mathfrak{g}/\mathfrak{g}_{m-1}) & \longrightarrow & J_{m-1}(\overline{\mathfrak{g}}) & \subset & J_{m-1}T \\
\downarrow & & \downarrow & & \downarrow \\
(G/G_0, G/G_{m-1} \times_{G_0/G_{m-1}} \mathfrak{g}/\mathfrak{g}_{m-2}) & \longrightarrow & J_{m-2}(\overline{\mathfrak{g}}) & \subset & J_{m-2}T \\
\downarrow & & \downarrow & & \downarrow \\
\vdots & \longrightarrow & \vdots & \subset & \vdots \\
\downarrow & & \downarrow & & \downarrow \\
(G/G_0, G/G_2 \times_{G_0/G_2} \mathfrak{g}/\mathfrak{g}_1) & \longrightarrow & J_1(\overline{\mathfrak{g}}) & \subset & J_1 T \\
\downarrow & & \downarrow & & \downarrow \\
(G/G_0, G/G_1 \times_{G_0/G_1} \mathfrak{g}/\mathfrak{g}_0) & \longrightarrow & J_0(\overline{\mathfrak{g}}) & \subset & J_0 T = T \\
\downarrow & & \downarrow & & \downarrow \\
G/G_0 & = & M & = & M
\end{array}
\tag{18.32}
$$

which is the linearization of (18.28). The horizontal arrows are isomorphisms of vector bundles over $M = G/G_0$. The first column of (18.32) is associated with the first column of (18.28) as explained in Chapter 16, and the third column of (18.32) is associated with the third column of (18.28) as explained in Chapter 17. Notice how the association of the first columns of (18.28) and (18.32) simplifies if it is viewed as the restriction of the association of the third columns to the second columns. For instance, to explain the association of the groupoid $\overline{\mathcal{G}}_m(M)$ with the vector bundle $J_m(\overline{\mathfrak{g}}) \to M$, let $p, q \in M$ and choose $j_m(\overline{g})^{p,q} \in \overline{\mathcal{G}}_m^{p,q}(M)$. The splitting (18.27) together with (17.46) shows that we have the map $\mathrm{Ad}\,\varepsilon j_m(\overline{g})^{p,q} : (J_m T)^p \to (J_m T)^q$. Now, (18.21) shows that this map restricts to $\mathrm{Ad}\,\varepsilon j_m(\overline{g})^{p,q} : J_k(\overline{\mathfrak{g}})^p \to J_k(\overline{\mathfrak{g}})^q$, which is the asserted association. If $\mathrm{Ad}\,\varepsilon j_m(\overline{g})^{p,q} = \mathrm{Ad}\,\varepsilon j_m(\overline{h})^{p,q}$, then $\varepsilon j_m(\overline{g})^{p,q} = \varepsilon j_m(\overline{h})^{p,q}$, and therefore $\overline{g} = \overline{h}$ or $g = h$. To see the role of Ad of G in this picture, we note that $\mathrm{Ad}\,\varepsilon j_m(\overline{g})^{p,q}(j_k(\overline{\xi})^p) = j_k(\overline{\eta})^q$ is equivalent to $\mathrm{Ad}\,\xi = \eta$, as can be seen from (18.22).

More generally, linearization of (18.29) gives

$$
\begin{array}{ccccc}
 & & \varepsilon J_m(\overline{\mathfrak{g}}(r)) & \subset & (\mathcal{U}_r^{o,\bullet}, J_{m+1}T)) \\
 & & \| & & \downarrow \\
(G/G_r, G\times_{G_r}\mathfrak{g}) & \longrightarrow & J_m(\overline{\mathfrak{g}}(r)) & \subset & (\mathcal{U}_r^{o,\bullet}, J_m T) \\
\downarrow & & \downarrow & & \downarrow \\
(G/G_r, G\times_{G_r}\mathfrak{g}/\mathfrak{g}_{m-1}) & \longrightarrow & J_{m-1}(\overline{\mathfrak{g}}(r)) & \subset & (\mathcal{U}_r^{o,\bullet}, J_{m-1}T) \\
\downarrow & & \downarrow & & \downarrow \\
(G/G_r, G/G_{m-1}\times_{G_r/G_{m-1}}\mathfrak{g}/\mathfrak{g}_{m-2}) & \longrightarrow & J_{m-2}(\overline{\mathfrak{g}}(r)) & \subset & (\mathcal{U}_r^{o,\bullet}, J_{m-2}T) \\
\downarrow & & \downarrow & & \downarrow \\
\vdots & \longrightarrow & \vdots & \subset & \vdots \\
\downarrow & & \downarrow & & \downarrow \\
(G/G_r, G/G_{r+2}\times_{G_r/G_{r+2}}\mathfrak{g}/\mathfrak{g}_{r+1}) & \longrightarrow & J_1(\overline{\mathfrak{g}}(r)) & \subset & (\mathcal{U}_r^{o,\bullet}, J_{r+1}T) \\
\downarrow & & \downarrow & & \downarrow \\
(G/G_r, G/G_{r+1}\times_{G_r/G_{r+1}}\mathfrak{g}/\mathfrak{g}_r) & \longrightarrow & J_0(\overline{\mathfrak{g}}(r)) & \subset & (\mathcal{U}_r^{o,\bullet}, J_r T) = T(\mathcal{U}_r^{o,\bullet}) \\
\downarrow & & \downarrow & & \downarrow \\
G/G_r & = & \overline{G}_r^{o,\bullet} & \subset & \mathcal{U}_r^{o,\bullet}
\end{array}
\tag{18.33}
$$

where $\overline{\mathfrak{g}}(r)$ denotes the Lie algebra of the infinitesimal generators of $\overline{G}(M)$ on $\overline{G}_r^{o,\bullet}$ obtained by prolonging the action of $\overline{G}(M)$ on $M = \overline{G}_0^{o,\bullet}$ to $\overline{G}_r^{o,\bullet}$. For $r = 0$, (18.33) reduces to (18.32). The right column of (18.33) is defined in Chapter 17 and is associated with the right column of (18.29), and the left column of (18.33) is associated with the left column of (18.29) as in Chapter 16. The horizontal arrows are isomorphisms of vector bundles over $\overline{G}_r^{o,\bullet}(M) \simeq G/G_r$. The association of the middle columns of (18.29) and (18.33) again has a simple interpretation.

So much now for the filtrations (6.2) and (6.14) and for the first columns of (18.28) and (18.29), as they have served their purpose. In the next chapter, our main concern will be to define the middle column of (18.28) as an object independent of the first column, satisfying certain axioms.

19

The Definition of a Prehomogeneous Geometry (PHG)

The diagram (18.28) suggests a natural definition of a geometric structure: Axiomatize the second column of (18.28) as a subset of the third column, forgetting that this subset arises from the first column, which has no curvature. Therefore, the idea is to concentrate on the properties of the second column that can be formulated independently of the first column. We imagine the first column as a *flat* and *globalizable* model for this subset, keeping in mind the possibility that we may have *other* models for the *same* subset.

For this purpose, we denote $\varepsilon\overline{G}_m(M)$ in (18.28) by $\mathcal{P}_{m+1}(M)$ (or, in brief, $\mathcal{P}_{m+1}$) and $\overline{G}_k(M)$ by $\mathcal{P}_k$ for $0 \leq k \leq m$. Therefore, the subset $\mathcal{P}_k \subset \mathcal{U}_k$ is a subgroupoid for $0 \leq k \leq m+1$, $\mathcal{P}_0 = M \times M$, and the restriction of the projection map $\pi_{j+1,j} : \mathcal{U}_{j+1} \to \mathcal{U}_j$ gives a surjection $\mathcal{P}_{j+1} \to \mathcal{P}_j$ for $0 \leq j \leq m-1$ and a bijection for $j = m$; that is, $\mathcal{P}_{m+1} \approx \mathcal{P}_m$. Further, m is the smallest integer j satisfying $\mathcal{P}_{j+1} \approx \mathcal{P}_j$. All these properties of $\mathcal{P}_{m+1} \subset \mathcal{U}_{m+1}$ are formulated independently of the first column of (18.28).

Now let M be a smooth manifold with $\dim M = \dim G/G_0$. The above fact suggests that a prehomogeneous geometry (PHG) on M should be, first of all, a subgroupoid $\mathcal{P}_{m+1} \subset \mathcal{U}_{m+1}$ satisfying the following two axioms:

A1: $\mathcal{P}_j \overset{\text{def}}{=} \pi_{m+1,j}(\mathcal{P}_{m+1}) \subset \mathcal{U}_j$ is a closed and embedded submanifold with smooth arrow operations of composition and inversion for $0 \leq j \leq m+1$; that is, it is a *differentiable* (or *Lie*) subgroupoid.

A2: $\mathcal{P}_0 = M \times M$, $\mathcal{P}_{m+1} \approx \mathcal{P}_m$, and m is the smallest integer j satisfying $\mathcal{P}_{j+1} \approx \mathcal{P}_j$, or, equivalently, the projection map $\mathcal{P}_{j+1} \to \mathcal{P}_j$ is also injective for $j = m$ but not injective for $0 \leq j \leq m-1$.

Note that the last condition of **A2** is well-defined since $\mathcal{P}_{m+1}$ is transitive. We will also write $\mathcal{P}_{m+1} = \varepsilon\,\mathcal{P}_m$, where ε is defined by **A2**.

We also want to incorporate the stabilizer G_0 of the coset $o \in G/G_0$ in the first column of (18.28) into the definition of a PHG. For this purpose, let $\mathcal{H}_r \subset \mathcal{U}_r$ be any transitive subgroupoid on M. We choose some $p \in M$ and a coordinate neighborhood (U, x) around p that

An Alternative Approach to Lie Groups and Geometric Structures, Ercüment H. Ortaçgil (2018).

identifies, as we recall from Chapter 17, the vertex group $\mathcal{U}_r^{p,p}$ with the rth-order jet group $G_r(n)$, $n = \dim M$. In this way, $\mathcal{H}_r^{p,p} \subset \mathcal{U}_r^{p,p}$ is identified with a subgroup $\mathcal{H}_{r,x}^{p,p} \subset G_r(n)$, which depends both on p and on the coordinates (x). This subgroup need not be closed, since we assume only transitivity. Let $\{\mathcal{H}_{r,x}^{p,p}\}$ denote the conjugacy class of $\mathcal{H}_{r,x}^{p,p}$ inside $G_r(n)$, that is, the set of all subgroups of $G_r(n)$ that are conjugate to $\mathcal{H}_{r,x}^{p,p}$. We easily see that $\{\mathcal{H}_{r,x}^{p,p}\}$ is independent of both p and (x). Indeed, fixing p and applying a coordinate change $(x) \to (y)$ conjugates $\mathcal{H}_{r,x}^{p,p}$ to $\mathcal{H}_{r,y}^{p,p}$ by the r-arrow of $(x) \to (y)$. Similarly, if we change p to q, any r-arrow of $\mathcal{H}_r$ from p to q pulls back $\mathcal{H}_r^{p,p}$ to $\mathcal{H}_r^{q,q}$ by conjugation and acts as a coordinate change. Thus, we denote $\{\mathcal{H}_{r,x}^{p,p}\}$ by $\{\mathcal{H}_r\}$.

Definition 19.1 *For a transitive subgroupoid $\mathcal{H}_r \subset \mathcal{U}_r$ on M, the conjugacy class $\{\mathcal{H}_r\}$ inside $G_r(n)$, $n = \dim M$, defined above is the vertex class of $\mathcal{H}_r$.*

Therefore, $\{\mathcal{P}_{m+1}\}$ is defined for $\mathcal{P}_{m+1} \subset \mathcal{U}_{m+1}$. In this case, $\mathcal{P}_{m+1,x}^{p,p} \subset G_{m+1}(n)$ is a closed subgroup for all p and (x), by **A1**. For an effective Klein geometry (G_0, G) of order m, we define its vertex class $\{(G_0, G)\}$ using the injection

$$\begin{aligned} \theta_{m+1} : G_0 &\longrightarrow \mathcal{U}_{m+1}^{o,o}(G/G_0) \approx G_{m+1}(n) \\ : g &\longrightarrow j_{m+1}(\overline{g})^o \end{aligned} \tag{19.1}$$

defined by (18.17) for $k = m + 1$ as the conjugacy class $\{\theta_{m+1}(G_0)\}$ of the image $\theta_{m+1}(G_0) = \varepsilon\theta_m(G_0)$. Thus, we have

$$\{(G_0, G)\} = \{\theta_{m+1}(G_0)\} \tag{19.2}$$

Note that $\{(G_0, G)\}$ is independent of the point $o \in M$, where G_0 is the stabilizer of o. We may be more familiar with one group in the set $\{(G_0, G)\}$ than its conjugates. In this case, we call this subgroup a distinguished element of $\{(G_0, G)\}$.

We are now ready for the following definition:

Definition 19.2 *A prehomogeneous geometry (PHG) of order m on M is a subgroupoid $\mathcal{P}_{m+1} \subset \mathcal{U}_{m+1}$ satisfying* **A1**, **A2**, *and*

A3: *There exists an (effective) Klein geometry (G_0, G) with $\dim G/G_0 = M$ and $\{\mathcal{P}_{m+1}\} = \{(G_0, G)\}$. The vertex class $\{(G_0, G)\}$ is the vertex model for $\mathcal{P}_{m+1}$.*

This definition raises an important question: Do all splittings inside jet groups arise from Klein geometries? More precisely, does any $\mathcal{P}_{m+1} \subset \mathcal{U}_{m+1}$ satisfying **A1** and **A2** have at least one vertex model? That is, is **A3** a consequence of **A1** and **A2**? We believe that the answer to this nontrivial question is affirmative and can be shown using Proposition 9.7 in [P1]. Even though this question is of considerable theoretical importance, we will bypass it by taking **A3** as an axiom.

We now pose the following question:

Q: Let (G_0, G) have order m and let M be any smooth manifold with $\dim M = \dim G/G_0$. Can we give a method for constructing some PHG $\mathcal{P}_{m+1} \subset \mathcal{U}_{m+1}$ on M satisfying $\{\mathcal{P}_{m+1}\} = \{(G_0, G)\}$ that is suitable for local computations?

Consider the principal frame bundle $\mathcal{U}_{m+1}^{p,\bullet} \to M$ of order $m+1$ with structure group $\mathcal{U}_{m+1}^{p,p} \approx G_{m+1}(n)$ for some $p \in M$. According to (19.1), $\theta_{m+1}(G_0) \subset \mathcal{U}_{m+1}^{o,o}(G/G_0) \approx G_{m+1}(n)$. Therefore, $\theta_{m+1}(G_0) \subset \mathcal{U}_{m+1}^{p,p}$ modulo the coordinates around p and o. We may arrange these coordinates so that $\theta_{m+1}(G_0)$ is a distinguished subgroup if necessary. Now suppose we can reduce the structure group $G_{m+1}(n)$ of $\mathcal{U}_{m+1}^{p,\bullet} \to M$ to $\theta_{m+1}(G_0)$. Passing from this reduced principal bundle to the associated groupoid $\mathcal{P}_{m+1} \subset \mathcal{U}_{m+1}$ with $\mathcal{P}_{m+1}^{p,p} = \theta_{m+1}(G_0)$, we see that $\mathcal{P}_{m+1}$ satisfies **A1–A3** with $\{\mathcal{P}_{m+1}\} = \{(G_0, G)\}$. All $\mathcal{P}_{m+1} \subset \mathcal{U}_{m+1}$ satisfying **A1–A3** with $\{\mathcal{P}_{m+1}\} = \{(G_0, G)\}$ are obtained in this way by reducing the structure group (if certain topological obstructions do not exist). However, note that this important and well-known global construction does not satisfy the second requirement of **Q**, since it does not provide a local computational method.

To answer the above question, we first recall the main hypothesis of the formal theory of PDEs: A PDE is a submanifold of some jet space. In our case, we fix a row in (18.28) (or, more generally, in (18.29)). Now $\mathcal{U}_r(M)$ is the jet space, the submanifold $(M, [G/G_r, G/G_r]) \simeq \overline{G}_r(M) \subset \mathcal{U}_r(M)$, is the PDE of order r, and the transformation group G of M is the *global* solution space of this PDE. To make this point more transparent, let $p, q \in M$. Choosing coordinates, a k-arrow from p to q has the representation (17.41) or (x^i, f^i_μ), $0 \le |\mu| \le k$. Since $\mathcal{P}_r \subset \mathcal{U}_r$ is a closed submanifold for $0 \le r \le m+1$, *locally* it is given as the zero set of some independent functions Φ^j_r defined on some coordinate neighborhood $W \subset \mathcal{U}_r$ by

$$\mathcal{P}_r : \Phi^j_r(x^i, f^i_\mu) = 0 \tag{19.3}$$

where $0 \le |\mu| \le r$, $1 \le j \le \dim \mathcal{U}_r - \dim \mathcal{P}_r$. The equations (19.3) impose no restriction on (x^i, y^i), since $\mathcal{P}_r$ is transitive. So, for any (x^i, y^i), (19.3) has solutions that project to $(x^i, y^i) \in \mathcal{P}_0 = M \times M$. For any solution (x^i, f^i_μ) of (19.3) for $r = m$, there is a unique solution (x^i, g^i_μ) of (19.3) for $r = m+1$ with $g^i_\mu = f^i_\mu$, $0 \le |\mu| \le m$, since $\mathcal{P}_{m+1} \approx \mathcal{P}_m$. For $r = 0$, the set of functions $\{\Phi^j_r\}_{1 \le j \le \dim \mathcal{U}_r - \dim \mathcal{P}_r}$ is empty, since $\mathcal{P}_0 = \mathcal{U}_0 = M \times M$. For some particular example, this set may also be empty for some $r \ge 1$, as we will see in the examples below. The functions $\{\Phi^j_r\}$ in (19.3) are clearly not unique but very special. Using a shorthand notation, if $\Phi(x, y, p) = 0$ and $\Phi(y, z, q) = 0$, then $\Phi(x, z, r) = 0$, where $(y, z, q) \circ (x, y, p) = (x, y, r)$ and $\circ$ denotes composition of arrows: They are the so-called "differential invariants" and have a subtle local theory, although we will barely touch on it here. Therefore, a PHG is, by definition, a particular type of nonlinear PDE, and **Q** asks for some set of defining functions as in (19.3) A solution of (19.3) is a local diffeomorphism $y = y(x)$ of M that satisfies (19.3) identically for all x upon the substitution $f^i_\mu = \partial y^i / \partial y^\mu$. More precisely, we have the following definition:

Definition 19.3 $\mathcal{P}_r \subset \mathcal{U}_r$ *is locally solvable (or completely integrable) if, for any initial condition* $(\widetilde{x}^i, \widetilde{f}^i_\mu) \in \mathcal{P}_r$, $\mathcal{P}_r$ *admits local solutions satisfying this initial condition: There exists a local diffeomorphism* $y = y(x)$ *satisfying* $(\partial^{|\mu|} y^i / \partial x^\mu)(\widetilde{x}) = \widetilde{f}^i_\mu$ *and* $\Phi^j_r(x^i, \partial^{|\mu|} y^i / \partial x^\mu) = 0$ *identically in* x.

Clearly, $\mathcal{U}_r$ is locally (in fact globally) solvable for all r. If $\mathcal{P}_r(M) = \overline{G}_r(M)$ as in (18.28), then the transformations coming from the first column of (18.28) give *global* solutions of (19.3), and (19.3) is globally solvable with arbitrary initial conditions. Note that in this case $\mathcal{P}_r$ may have more solutions than those coming from the first column. For instance, $\mathcal{P}_0 = M \times M$, and *any* local diffeomorphism is a solution of $\mathcal{P}_0$. However, $\mathcal{P}_1 = \varepsilon(\mathcal{P}_0)$ in Part I, and $\mathcal{P}_1$ has unique local solutions with arbitrary initial conditions if and only if $\mathcal{R} = 0$. We want to define the nonlinear curvature $\mathcal{R}$ such that $\mathcal{R} = 0$ if and only if $\mathcal{P}_{m+1}$ is *uniquely* locally solvable. The uniqueness will imply the existence of "analytic continuation" of local solutions, and if the pseudogroup $\mathcal{G}$ of local solutions globalizes to a transformation group $\mathcal{G}^{\text{ex}}$ ($= \mathcal{G}$ with the same abuse of notation as in Part I) with stabilizer $\mathcal{G}_p$ at some $p \in M$, then the vertex class $\{(\mathcal{G}, \mathcal{G}_p)\} = \{\mathcal{P}_{m+1}\}$ will be the vertex class $\{(G, G_0)\} = \{\mathcal{P}_{m+1}\}$ given by **A3**, but we allow the possibility that G and $\mathcal{G}$ may be nonisomorphic.

Since $\mathcal{P}_{j+1} \subset \mathcal{U}_{j+1}$ surjects onto $\mathcal{P}_j \subset \mathcal{U}_j$, the functions Φ^j_r in (19.3) consist of r sets of independent functions as follows:

$$
\begin{aligned}
&\Phi^j_r\left(x^i, f^i_\mu\right) = 0, && 1 \le j \le j_r, && 0 \le |\mu| \le r \\
&\Phi^j_{r-1}\left(x^i, f^i_\mu\right) = 0, && 1 \le j \le j_{r-1}, && 0 \le |\mu| \le r-1 \\
&\quad\vdots \\
&\Phi^j_2\left(x^i, f^i_\mu\right) = 0 && 1 \le j \le j_2, && 0 \le |\mu| \le 2 \\
&\Phi^j_1\left(x^i, f^i_\mu\right) = 0, && 1 \le j \le j_1, && 0 \le |\mu| \le 1
\end{aligned}
\tag{19.4}
$$

The last equation in (19.4) defines $\mathcal{P}_1$, the last two equations together define $\mathcal{P}_2$, …, and together they all define $\mathcal{P}_r$. We have $j_1 + \cdots + j_r = \dim \mathcal{U}_r - \dim \mathcal{P}_r$.

We now come to some examples of PHGs.

Example 1

We know from Part I that M is parallelizable if and only if it admits a splitting ε. Now $\mathcal{P}_0 = \mathcal{U}_0 = M \times M$ and $\mathcal{P}_1 = \varepsilon(\mathcal{P}_0) \subset \mathcal{U}_1$ satisfies **A1** and **A2**. Using the structure object $w = (w^i_j)$, (7.5) gives

$$
\mathcal{P}_1 : f^a_j w^i_a(y) - w^i_j(x) = 0 \tag{19.5}
$$

for the n^2 functions $(\Phi_1)^i_j$ in (19.3). Note that these functions are globally defined on $\mathcal{U}_1$, since w is a globally defined geometric object on M. The vertex class can be identified with $G_1(n)$, and any pair $(\{1\}, G)$ is a vertex model for $\mathcal{P}_1$, where G is an arbitrary Lie group. Therefore, $\mathcal{P}_1$ also satisfies also **A3**, having infinitely many vertex models.

There is an important lesson in (19.5): If we can construct a geometric object on M and express $\mathcal{P}_{m+1} \subset \mathcal{U}_{m+1}$ as the subset preserving this geometric object, then we automatically get the functions in (19.3) that are *globally* defined on $\mathcal{U}_r$. We will always follow this strategy. In fact, the following steps constitute a general algorithm for studying a PHG giving a complete answer to **Q**:

Step 1: Start with some effective Klein geometry (G_0, G) of order m and explicitly determine the image of (19.1).

Step 2: Find a set of functions that separate the points of $G_{m+1}(n)/\theta_{m+1}(G_0)$ and therefore serve as coordinates on $G_{m+1}(n)/\theta_{m+1}(G_0)$.

Step 3: Using the action of $G_{m+1}(n)$ on $G_{m+1}(n)/\theta_{m+1}(G_0)$, find the transformation rule of these coordinates.

Step 4: Using this transformation rule, define a geometric object on M and express $\mathcal{P}_{m+1} \subset \mathcal{U}_{m+1}$ as the subset preserving this geometric object.

Example 2

We consider the transformations of $\mathbb{R}^n$ of the form $x \to Ax + a$, where $A \in \mathrm{GL}(n, \mathbb{R}) = G_1(n)$ and $a \in \mathbb{R}^n$. We denote this transformation by (A, a). We check that $(A, a)^{-1} = (A^{-1}, -A^{-1}a)$ and $(A, a) \circ (B, b) = (AB, Ab + a)$, and therefore $\mathrm{Aff}(\mathbb{R}^n) \overset{\mathrm{def}}{=} \{(A, a) \mid A \in G_1(n),\ a \in \mathbb{R}^n\}$ is a transitive transformation group of $\mathbb{R}^n$. The stabilizer of $\mathrm{Aff}(\mathbb{R}^n)$ at, say, the origin $o \in \mathbb{R}^n$ is the subgroup $\{(A, a) \mid A \in G_1(n),\ a = 0\}$, which we write as $(G_1(n), 0)$ and identify with $G_1(n)$. We can realize $\mathrm{Aff}(\mathbb{R}^n)$ as a matrix group. We define G to be the subgroup of $\mathrm{GL}(n + 1, \mathbb{R})$ consisting of the matrices of the form $\begin{bmatrix} A & a \\ 0 & 1 \end{bmatrix}$, where we interpret $a \in \mathbb{R}^n$ as a column vector of length n. We have the isomorphism $(A, a) \leftrightarrow \begin{bmatrix} A & a \\ 0 & 1 \end{bmatrix}$, which identifies $\mathrm{Aff}(\mathbb{R}^n)$ with G and the stabilizer $G_1(n)$ with the subgroup $\begin{bmatrix} A & 0 \\ 0 & 1 \end{bmatrix}$. Using the notation of Chapter 16, we denote this stabilizer by G_0. Thus, we obtain the Klein geometry (G_0, G), which we may also write as $(\mathbb{R}^n, \mathrm{Aff}(n))$, with the obvious identification $(\mathbb{R}^n, \mathrm{Aff}(\mathbb{R}^n)) \approx (G/G_0, G)$.

We now want to determine the filtration (16.14) and the order $o(G_0, G)$ if it is defined (note that G_0 and G are not connected). Once this is done, the next step is to inject G_0 into some jet group by (19.1) and explicitly determine the image as in Step 1. In the present simple case, neither task poses any difficulty: In coordinates, (A, a) is given by $(x^i) \to (y^i)$, $y^i = A^i_j x^j + a^i$. Therefore, $\partial y^i/\partial x^j = A^i_j$ and $\partial^2 y^i/\partial x^j \partial x^k = 0$. From $\partial y^i/\partial x^j = A^i_j$, we deduce that if (A, a) maps p to q, then (A, a) is determined by its 1-arrow from p to q, and therefore this 1-arrow determines the 2-arrow of (A, a). Therefore, above any 1-arrow, there is a unique 2-arrow, and $o(G_0, G) = 1$. Choosing $p = q = o$, we see that G_0 injects into $G_2(n)$ as $f = (f^i_j) \to \left(f^i_j, \varepsilon^i_{jk}\left(f^s_r\right)\right) = (f, \varepsilon(f))$ for some splitting ε. Now $\partial^2 y^i/\partial x^j \partial x^k = 0$ shows that $\varepsilon = 0$, and this injection is given by (16.14) for $k = 2$. Note that the image

$\theta_2(G_0) = (G_1(n), 0) = \varepsilon\theta_1(G_0)$, which we write in brief as $\varepsilon G_1(n)$, is our distinguished subgroup and is a closed and algebraic subgroup of $G_2(n)$.

To recapitulate, we have

$$\left(f^i_j, f^i_{jk}\right) \in \varepsilon G_1(n) \iff f^i_{jk} = 0 \tag{19.6}$$

as the defining equations of the distinguished subgroup $\varepsilon G_1(n) \subset G_2(n)$. Thus, (19.6) completes Step 1.

We now consider the left coset space $G_2(n)/\varepsilon G_1(n)$. Let $f = \left(f^i_j, f^i_{jk}\right)$ and $g = \left(g^i_j, g^i_{jk}\right) \in G_2(n)$. Using the group operation (17.5) of $G_2(n)$, we find $\left(f^i_j, f^i_{jk}\right)^{-1} = \left(\overline{f}^i_j, -\overline{f}^i_a f^a_{bc} \overline{f}^b_j \overline{f}^c_k\right)$, where $f^i_a \overline{f}^a_j = \delta^i_j$. Therefore, $\left(f^i_j, f^i_{jk}\right)^{-1} \left(g^i_j, g^i_{jk}\right) = \left(\overline{f}^i_a g^a_j, \overline{f}^i_a g^a_{jk} - \overline{f}^i_a f^a_{bc} \overline{f}^b_d \overline{f}^c_e g^d_j g^e_k\right)$ and

$$\begin{aligned} f^{-1}g = \left(f^i_j, f^i_{jk}\right)^{-1} \left(g^i_j, g^i_{jk}\right) &\in \varepsilon G_1(n) \\ \iff \overline{f}^i_a g^a_{jk} - \overline{f}^i_a f^a_{bc} \overline{f}^b_d \overline{f}^c_e g^d_j g^e_k &= 0 \\ \iff f^i_{ab} \overline{f}^a_j \overline{f}^b_k &= g^i_{ab} \overline{g}^a_j \overline{g}^b_k \end{aligned} \tag{19.7}$$

We define the rational function

$$\begin{aligned} \Gamma : G_2(n) &\longrightarrow \mathbb{R}^{n^2(n+1)/2} \\ : f = \left(f^i_j, f^i_{jk}\right) &\longrightarrow \Gamma^i_{jk}(f) \stackrel{\text{def}}{=} f^i_{ab} \overline{f}^a_j \overline{f}^b_k \end{aligned} \tag{19.8}$$

From (19.7) and (19.8), we deduce

$$f^{-1}g \in \varepsilon G_1(n) \iff \Gamma(f) = \Gamma(g) \tag{19.9}$$

Therefore, Γ restricts to the well-defined injective map

$$\begin{aligned} \Gamma : G_2(n)/\varepsilon G_1(n) &\longrightarrow \mathbb{R}^{n^2(n+1)/2} \\ : f\varepsilon G_1(n) &\longrightarrow \Gamma(f) \end{aligned} \tag{19.10}$$

which is also surjective. Now, the components Γ^i_{jk} of Γ are symmetric and define global coordinates on $G_2(n)/\varepsilon G_1(n)$. Note that Γ vanishes only at the coset of $\varepsilon G_1(n)$. Therefore, Γ separates the points of $G_2(n)/\varepsilon G_1(n)$, as required by Step 2. In more detail, the algebraic subgroup $\varepsilon G_1(n) \subset G_2(n)$ acts freely on $G_2(n)$ by left multiplication, and the rational function Γ separates the orbits, which are the left cosets. Since $G_2(n)/\varepsilon G_1(n) \approx \mathbb{R}^k$ is homotopically trivial, we can always reduce the structure group $G_2(n)$ to $\varepsilon G_1(n)$ and define some $\mathcal{P}_2 \subset \mathcal{U}_2$.

Now, $G_2(n)$ acts on $G_2(n)/\varepsilon G_1(n)$ on the left. To find this action in terms of coordinates, we note that $f = \left(f^i_j, f^i_{jk}\right) \in G_2(n)$ maps $x = \left(x^i_j, x^i_{jk}\right) \varepsilon G_1(n)$ with coordinates Γ^i_{jk} to fx, which is

$$
\begin{aligned}
\left(f^i_j, f^i_{jk}\right)\left(x^i_j, x^i_{jk}\right)\varepsilon G_1(n) &= \left(f^i_j, f^i_{jk}\right)\left(x^i_j, x^i_{jk}\right)\left[\left(f^i_j, 0\right)\left(x^i_j, 0\right)\right]^{-1}\varepsilon G_1(n) \\
&= \left(f^i_j, f^i_{jk}\right)\left(x^i_j, x^i_{jk}\right)\left(x^i_j, 0\right)^{-1}\left(f^i_j, 0\right)^{-1}\varepsilon G_1(n) \\
&= \left(f^i_j, f^i_{jk}\right)\left(\delta^i_j, \Gamma^i_{jk}\right)\left(f^i_j, 0\right)^{-1}\varepsilon G_1(n) \\
&= \left(\delta^i_j, \overline{\Gamma}^i_{jk}\right)\varepsilon G_1(n)
\end{aligned}
\tag{19.11}
$$

where $\overline{\Gamma}^i_{jk}$ are the coordinates of fx. Using the group operation of $G_2(n)$, we deduce from (19.11)

$$
\overline{\Gamma}^i_{jk} = f^i_a \Gamma^a_{bc} \overline{f}^b_j \overline{f}^c_k + f^i_{ab} \overline{f}^a_j \overline{f}^b_k
\tag{19.12}
$$

which completes Step 3. We observe that (5.2) and (19.12) are the *same* transformation rules, but their components have *quite different meanings*.

Using (19.12), we define a family of geometric objects Γ on M as follows: On a coordinate neighborhood (U, x), Γ has symmetric components $\Gamma^i_{jk}(x)$. A coordinate change $(x) \to (y)$ transforms these components according to

$$
\Gamma^i_{jk}(y) = \frac{\partial y^i}{\partial x^a}\Gamma^a_{bc}(x)\frac{\partial x^b}{\partial y^j}\frac{\partial x^c}{\partial y^k} + \frac{\partial^2 y^i}{\partial x^a \partial x^b}\frac{\partial x^a}{\partial y^j}\frac{\partial x^b}{\partial y^k}
\tag{19.13}
$$

Transforming $\Gamma^i_{jk}(x)$ to $\Gamma^i_{jk}(y)$ and then to $\Gamma^i_{jk}(z)$ using (19.13) gives the same result by transforming $\Gamma^i_{jk}(x)$ to $\Gamma^i_{jk}(z)$, because acting on $G_2(n)/\varepsilon G_1(n)$ first by f and then by g is the same as acting by $g \circ f$. Thus, Γ is a geometric object. It is possible to cover an arbitrary smooth M by coordinate neighborhoods (U_α, x_α) together with some components $\Gamma^i_{jk}(x_\alpha)$ on U_α subject to the transformation rule (19.13) on all possible overlaps: If $\Gamma^i_{jk}(x)$ and $\overline{\Gamma}^i_{jk}(x)$ both satisfy (19.13) and $\lambda \in \mathbb{R}$, then (19.13) shows that $\lambda\Gamma^i_{jk}(x) + (1 - \lambda)\overline{\Gamma}^i_{jk}(x)$ also satisfies (19.13). Therefore, we can piece together some arbitrarily chosen components by a standard partition-of-unity argument, showing that any M admits some Γ.

We call Γ an affine structure object on M. It now follows easily from the definitions that any affine structure object Γ on M defines some PHG $\mathcal{P}_2$ with vertex class $\{\mathcal{P}_2\} = \{(G_1(n), 0), G_2(n)\}$ whose 2-arrows preserve Γ, and conversely any such $\mathcal{P}_2 \subset \mathcal{U}_2$ defines some Γ by defining its value $\Gamma(p)$ at p as a parametrization of the coset space $\mathcal{U}_2^{p,p} / \mathcal{P}_2^{p,p}$. Note that any $p \in M$ has some coordinates around it, called regular coordinates at p, where all components of Γ vanish and the vertex inclusion $\mathcal{P}_2^{p,p} \subset \mathcal{U}_2^{p,p}$ becomes $\varepsilon G_1(n) \subset G_2(n)$. We call $\mathcal{P}_2$ an affine PHG on M. We now have

$$
\mathcal{P}_2 : \Phi^i_{2,jk} \stackrel{\text{def}}{=} f^i_a \Gamma^a_{bc}(x) \overline{f}^b_j \overline{f}^c_k + f^i_{ab} \overline{f}^a_j \overline{f}^b_k - \Gamma^i_{jk}(y) = 0
\tag{19.14}
$$

concluding Step 4. Note that $\mathcal{P}_1 = \mathcal{U}_1$, and therefore the equations $\Phi_1 = 0$ in (19.4) are vacuous. Clearly, $\mathcal{P}_2 \simeq \mathcal{P}_1$ or $\mathcal{P}_2 = \varepsilon(\mathcal{P}_2)$. Equivalently, above any 1-arrow $\left(x^i, y^i, f^i_j\right)$, there is a unique 2-arrow $\left(x^i, y^i, f^i_j, f^i_{jk}\right)$ that is found by solving $\left(f^i_{jk}\right)$ uniquely from (19.14).

The definition of $\mathcal{P}_2$ by (19.14) has some computational advantages over the above global bundle reduction argument. For instance, we can define the *tensor* R by the right-hand side of (5.11) using the structure object Γ. We observe that the right-hand side of (5.11) is now equal to the right-hand side of (6.9), because Γ^i_{jk} is symmetric in j, k.

Example 3

Some $(A, a) \in \mathrm{Aff}(\mathbb{R}^n)$ belongs to the isometry group $\mathrm{Iso}(\mathbb{R}^n)$ if and only if $A \in \mathrm{O}(n)$; that is, $A^T A = I$ or, equivalently, $(A^T)^i_a A^a_j = A^a_i A^a_j = \delta_{ij}$. Arguing exactly as in Example 2, we inject $\mathrm{O}(n)$ into $G_2(n)$ with the image $\varepsilon\mathrm{O}(n) \subset G_2(n)$ given by

$$f = \left(f^i_j, f^i_{jk}\right) \in \varepsilon\mathrm{O}(n) \iff f^a_i f^a_j = \delta^{ij},\ f^i_{jk} = 0 \tag{19.15}$$

Now,

$$\begin{aligned} f^{-1}g = \left(f^i_j, f^i_{jk}\right)^{-1}\left(g^i_j, g^i_{jk}\right) &\in \varepsilon\mathrm{O}(n) \\ \iff \bar{f}^i_a g^a_{jk} - \bar{f}^i_a f^a_{bc}\bar{f}^b_d \bar{f}^c_e g^d_j g^e_k = 0, &\qquad \bar{f}^c_a g^a_i \bar{f}^c_b g^b_j = \delta_{ij} \\ \iff f^i_{ab}\bar{f}^a_j \bar{f}^b_k = g^i_{ab}\bar{g}^a_j \bar{g}^b_k, &\qquad f^a_i f^a_j = g^a_i g^a_j \end{aligned} \tag{19.16}$$

Note that $\mathbf{g}_{ij} \overset{\text{def}}{=} f^a_i f^a_j$ is symmetric and positive-definite, since f^i_j is invertible. Also, $\mathbf{g}^{ia}\mathbf{g}_{aj} = \delta^i_j$, where $\mathbf{g}^{ij} \overset{\text{def}}{=} f^i_a f^j_a$. Using (19.16), we define the rational map

$$\begin{aligned} \mathbf{g} &: G_2(n) \longrightarrow \mathbb{R}^k \\ &: \left(f^i_j, f^i_{jk}\right) \longrightarrow \left(\mathbf{g}_{ij}(f), \mathbf{g}^i_{jk}(f)\right) \overset{\text{def}}{=} \left(f^a_i f^a_j, f^i_{ab}\bar{f}^a_j \bar{f}^b_k\right) \end{aligned} \tag{19.17}$$

where $k = \frac{1}{2}n(n+1)^2$. By the definition of $\mathbf{g}$, we have

$$f^{-1}g \in \varepsilon\mathrm{O}(n) \iff \mathbf{g}(f) = \mathbf{g}(g) \tag{19.18}$$

Thus, $\mathbf{g}$ restricts to the well-defined injective (not surjective this time) map

$$\begin{aligned} \mathbf{g} &: G_2(n)/\varepsilon\mathrm{O}(n) \longrightarrow \mathbb{R}^k \\ &: \left(f^i_j, f^i_{jk}\right)\varepsilon\mathrm{O}(n) \longrightarrow \left(\mathbf{g}_{ij}(f), \mathbf{g}^i_{jk}(f)\right) \end{aligned} \tag{19.19}$$

whose image is an open subset of $\mathbb{R}^k$. Therefore, the components $\left(\mathbf{g}_{ij}, \mathbf{g}^i_{jk}\right)$ of $\mathbf{g}$ serve as global coordinates on $G_2(n)/\varepsilon\mathrm{O}(n)$.

Using the action of $G_2(n)$ on $G_2(n)/\varepsilon\mathrm{O}(n)$, we easily get

$$\bar{\mathbf{g}}_{ij} = \bar{f}^a_i \bar{f}^b_j \mathbf{g}_{ab} \tag{19.20}$$

$$\bar{\mathbf{g}}^i_{jk} = f^i_a \mathbf{g}^a_{bc}\bar{f}^b_j \bar{f}^c_k + f^i_{ab}\bar{f}^a_j \bar{f}^b_k \tag{19.21}$$

Using (19.20) and (19.21), we define a geometric object $\mathbf{g} = \left(\mathbf{g}_{ij}, \mathbf{g}^i_{jk}\right)$. A partition-of-unity argument shows that any M admits such a geometric object. Now let $\mathcal{P}_2 \subset \mathcal{U}_2$ satisfy **A1–A3** with $\{\mathcal{P}_2\} = \{(G_0, G)\}$; that is, the vertex inclusions $\mathcal{P}_2^{p,p} \subset \mathcal{U}_2^{p,p}$ become $\varepsilon O(n) \subset G_2(n)$ in regular coordinates around p. We call $\mathcal{P}_2$ a Riemannian PHG. We define the geometric object $\mathbf{g}(p)$ by setting $\left(\mathbf{g}_{ij}(p), \mathbf{g}^i_{jk}(p)\right) \stackrel{\text{def}}{=} (\delta_{ij}, 0)$ in some regular coordinates around p. As in Example 2, $\mathbf{g}$ is defined on M, and $\mathcal{P}_2$ consists of those 2-arrows of $\mathcal{U}_2$ that preserve $\mathbf{g}$. The fact that (19.21) becomes (19.12) if we set $\mathbf{g}^i_{jk} = \Gamma^i_{jk}$ may give the impression that $\mathbf{g}$ is composed of two separate geometric objects, but this is not so, since $\left(\mathbf{g}_{ij}(p), \mathbf{g}^i_{jk}(p)\right) = (\delta_{ij}, 0)$ in regular coordinates. Therefore, $\mathbf{g}^i_{jk}(p)$ has pointwise the property of the Christoffel symbols but locally $\mathbf{g}^i_{jk}(x)$ need not be determined in terms of the derivatives of $\mathbf{g}_{ij}(x)$.

We now have

$$\Phi^i_{2,jk} = f^i_a \mathbf{g}^a_{bc}(x) \bar{f}^b_j \bar{f}^c_k + f^i_{ab} \bar{f}^a_j \bar{f}^b_k - \mathbf{g}^i_{jk}(y) = 0 \tag{19.22}$$

$$\Phi_{1,ij} = \bar{f}^a_i \bar{f}^b_j \mathbf{g}_{ab}(x) - \mathbf{g}_{ij}(y) = 0 \tag{19.23}$$

where $\mathcal{P}_1$ is defined by (19.23) and $\mathcal{P}_2$ is defined by (19.22) and (19.23). As in Example 2, if the 2-arrow $(x, y^i, f^i_j, f^i_{jk})$ solves (19.22) and (19.23), then f^i_{jk} is uniquely determined by (19.22), so that above any 1-arrow there is a unique 2-arrow.

Example 4

Let P^n be the real projective space of dimension n: P^n is the set of lines in $\mathbb{R}^{n+1}$ passing through the origin $\mathbf{o} \in \mathbb{R}^{n+1}$. Some $x \in P^n$ is represented by the nonzero $(n+1)$-tuple $(x^1, \ldots, x^{n+1})$. Now, $A \in GL(n+1, \mathbb{R}) = G_1(n+1)$ acts on P^n by $x \to A\bar{x}$, where $\bar{x}$ is any representative for x, which we interpret as a column vector. Since A and λA define the same transformation for $\lambda \neq 0$, we may assume $\det A = 1$; that is, $A \in SL(n+1, \mathbb{R})$. There is still an ambiguity if $n+1$ is even, because A and $-A$ define the same transformation. To make the action effective, we define

$$G \stackrel{\text{def}}{=} SL(n+1, \mathbb{R})/Z \tag{19.24}$$

where $Z = \{-1, 1\}$ if $n+1$ is even and $Z = \{1\}$ if $n+1$ is odd. We can easily check that the stabilizer G_0 of the point represented by $(0, \ldots, 0, 1)$ is

$$G_0 = \left\{ \begin{bmatrix} A & 0 \\ \xi & a \end{bmatrix}, A \in G_1(n), \ \xi \in \mathbb{R}^n, \ a \det A = 1 \right\} \Big/ Z \tag{19.25}$$

Clearly, $G/G_0 \approx P^n$ with the obvious identification $\approx$ that is compatible with the respective actions of G.

We are again lucky in this well-known example: The action of G on $G/G_0 = P^n$ can be identified with the action of the well-known linear fractional transformations on P^n in affine coordinates, and we accept the gift, expressing $g \in G$ as a linear fractional transformation near the point $o \in P^n$ represented by $(0, \ldots, 0, 1)$ in the affine coordinates containing o and differentiating this action formula three times at o. This straightforward but tedious computation gives the injective homomorphism

$$
\begin{gathered}
G_0 \xrightarrow{\theta_3} G_3(n) \\
\begin{bmatrix} A & 0 \\ \xi & a \end{bmatrix} \longrightarrow \left(A^i_j,\ a\left(A^i_j\xi_k + A^i_k\xi_j\right),\ 2a^2\left(A^i_j\xi_k\xi_m\right)_{(jkm)}\right)
\end{gathered}
\tag{19.26}
$$

where (jkm) denotes symmetrization. For $n+1$ even, we define (19.26) using the unique representative with $\det A \gneqq 0$. Using the group operation of $G_3(n)$ given by (17.5), the reader can check by another tedious (but amusing!) computation that $\theta_3(G_0) \subset G_3(n)$ is indeed a subgroup. For $n = 1$, (19.26) becomes $(A, 2aA\xi, 6a^2A\xi^2)$. Setting $B = 2aA\xi$, we get $\left(A, B, \frac{3}{2}B^2/A\right)$, which is the splitting (17.23) giving the Schwarzian derivative. Note that $\theta_3(G_0)$ surjects onto $G_2(1) = \mathbb{R} \times \mathbb{R}$ for $n = 1$, but $A^i_{jk} \stackrel{\text{def}}{=} a\left(A^i_j\xi_k + A^i_k\xi_j\right)$ does not give the whole $G_2(n)$ for $n \gneqq 1$.

Now (19.26) shows that the third-order derivatives are determined uniquely in terms of the lower-order derivatives, not only for $n = 1$ but for all n. To see this, we multiply

$$
A^i_{jk} = a\left(A^i_j\xi_k + A^i_k\xi_j\right) \tag{19.27}
$$

by $B = A^{-1}$ and get

$$
B^i_a A^a_{jk} = a\left(\delta^i_j\xi_k + \delta^i_k\xi_j\right) \tag{19.28}
$$

Summing over i, j gives

$$
\frac{1}{a(n+1)} B^i_a A^a_{jk} = \xi_k \tag{19.29}
$$

Substituting (19.29) into (19.26), we get the required result.

Turning back to our standard notation for arrows, we conclude from (19.29), (19.28), and (19.26) that

$$
\left(f^i_j, f^i_{jk}, f^i_{jkm}\right) \in \theta_3(G_0) \iff
$$

$$
f^i_{jk} - \frac{1}{n+1} f^i_j \overline{f}^b_a f^a_{bk} - \frac{1}{n+1} f^i_k \overline{f}^b_a f^a_{bj} = 0 \tag{19.30}
$$

$$
f^i_{jkm} - \frac{2}{(n+1)^2}\left(f^i_j \overline{f}^b_a f^a_{bk} \overline{f}^d_c f^c_{dm}\right)_{(jkm)} = 0 \tag{19.31}
$$

Therefore, (19.30) and (19.31) give the defining equations of $\theta_3(G_0) \subset G_3(n)$. Note that $\theta_1(G_0) = G_1(n)$ and $\theta_3(G_0) \subset G_3(n)$ is again a closed algebraic subgroup. Clearly, $o(G_0, G) = 2$ in this example and G_1 in the filtration (16.14) can easily be computed explicitly. Also note that the formulas in (19.31) generalize the Schwarzian derivative to n dimensions: For $n = 1$, (19.30) is vacuous and (19.31) gives the Schwarzian derivative.

For Step 2, we recall an important theoretical fact: If an algebraic group acts on an algebraic manifold freely, then we can always separate the orbits by rational functions. This is a special case of Rosenlicht's Theorem. Up to now, we have found these rational functions by an ad hoc trick, and we try it again: We compute $\left(f^i_j, f^i_{jk}\right)^{-1}\left(g^i_j, g^i_{jk}\right)$ and impose the condition (19.30). Denoting the subgroup of $G_2(n)$ obtained in this way by H, a straightforward computation gives

$$\left(f^i_j, f^i_{jk}\right)^{-1}\left(g^i_j, g^i_{jk}\right) \in H \iff P^i_{jk}\left(f_1, f_2\right) = P^i_{jk}\left(g_1, g_2\right) \tag{19.32}$$

where $\left(f_1, f_2\right) = (f^i_j, f^i_{jk})$ and

$$P^i_{jk}(f_1, f_2) \overset{\text{def}}{=} f^i_{bc}\overline{f}^b_j\overline{f}^c_k - \frac{1}{n+1}\delta^i_j f^r_{bc}\overline{f}^b_r\overline{f}^c_k - \frac{1}{n+1}\delta^i_k f^r_{bc}\overline{f}^b_r\overline{f}^c_j \tag{19.33}$$

Therefore, P^i_{jk} define coordinates on $G_2(n)/H$. However, if we compute $\left(f^i_j, f^i_{jk}, f^i_{jkl}\right)^{-1}\left(g^i_j, g^i_{jk}, g^i_{jkl}\right)$ and impose (19.30) and (19.31), our ad hoc method fails and it is not possible to separate the variables as before and get the rational functions $P^i_{jkl}(f_1, f_2, f_3)$ with the property that $\left(P^i_{jk}(f_1, f_2), P^i_{jkl}(f_1, f_2, f_3)\right)$ will separate the the orbits. Knowing that such rational functions do exist, henceforth we will pretend that we know $P^i_{jkl}(f_1, f_2, f_3)$ and therefore $\mathcal{P}_3 \approx \mathcal{P}_2$. We will comment more on this important point below.

Another straightforward computation shows that the action of $G_2(n)$ on $G_2(n)/H$ is given by

$$\overline{P}^i_{jk} = f^i_a P^a_{bc}\overline{f}^b_j\overline{f}^c_k + f^i_{ab}\overline{f}^a_j\overline{f}^b_k - \frac{1}{n+1}\delta^i_j f^a_{bc}\overline{f}^b_a\overline{f}^c_k - \frac{1}{n+1}\delta^i_k f^a_{bc}\overline{f}^b_a\overline{f}^c_j \tag{19.34}$$

Now, (19.34) defines a geometric object $P = \left(P^i_{jk}\right)$ whose components transform as

$$\begin{aligned} P^i_{jk}(y) = {} & \frac{\partial y^i}{\partial x^a}P^a_{bc}(x)\frac{\partial x^b}{\partial y^j}\frac{\partial x^c}{\partial y^k} + \frac{\partial^2 y}{\partial x^a \partial x^b}\frac{\partial x^a}{\partial y^j}\frac{\partial x^b}{\partial y^k} \\ & - \left(\frac{1}{n+1}\delta^i_j\frac{\partial y^a}{\partial x^b \partial x^c}\frac{\partial x^b}{\partial y^a}\frac{\partial x^c}{\partial y^k}\right)_{(jk)} \end{aligned} \tag{19.35}$$

where (jk) denotes symmetrization. We call $P = (P^i_{jk})$ a projective geometric object (or, rather, half of it, since it has also the components P^i_{jkl} that define $\mathcal{P}_3 \simeq \mathcal{P}_2$). Clearly, a partition-of-unity argument applies to (19.35), and any M admits a projective geometric object. There is a 1–1 correspondence between projective geometric objects on M and subgroupoids $\mathcal{P}_2 \subset \mathcal{U}_2$ satisfying **A1**, **A2**, and the condition that the vertex inclusions $\mathcal{P}_2^{p,p} \subset \mathcal{U}_2^{p,p}$ become $H \subset G_2(n)$ in regular coordinates around p. The submanifold $\mathcal{P}_2 \subset \mathcal{U}_2$ consists of those 2-arrows that preserve P, that is, those $\left(x^i y^i, f_j^i, f_{jk}^i\right) \in \mathcal{U}_2$ satisfying (19.34), where $\overline{P}_{jk}^i = P_{jk}^i(y)$ and $P_{jk}^i = P_{jk}^i(x)$. Note that $\mathcal{P}_1 = \mathcal{U}_1$.

If we ignore the last term in (19.35), then it becomes identical with (19.12) and (19.21). In fact, we declare two affine structure objects Γ and $\overline{\Gamma}$ on M to be related if there exists a 1-form ξ on M satisfying

$$\overline{\Gamma}_{jk}^i = \Gamma_{jk}^i + \delta_j^i \xi_k + \delta_k^i \xi_j \tag{19.36}$$

Now, (19.36) has a coordinate-free meaning and defines an equivalence relation on the affine structure objects on M. We fix an equivalence class, choose a representative Γ, and define

$$P_{jk}^i \overset{\text{def}}{=} \Gamma_{jk}^i - \frac{1}{n+1}\left(\delta_j^i \Gamma_{ak}^a + \delta_k^i \Gamma_{aj}^a\right) \tag{19.37}$$

We can easily check that all representatives define the same projective geometric object P. Therefore, any equivalence class defines uniquely a projective geometric object and conversely. Historically, a projective structure is defined in this way. This definition is quite useful, because it makes it possible to apply our notions developed in tensor calculus, like affine connection, covariant differentiation, and curvature, to the present situation. However, to check that a concept is independent of the representative sometimes becomes cumbersome. More seriously, this definition is clearly not very satisfactory theoretically, because it is peculiar to only this and some similar examples and is not the consequence of a general approach to geometric structures. We would like to remark here that most of the known local formulas in the theory of projective structures are contained in [Ei], written in 1927.

We define and study a conformal PHG along the same lines using the Klein geometry in Example 4.2 of [Kb].

A word is now appropriate about moving frames: Suppose that the transformation group G acts transitively on M with order m. Then G acts *freely* on the principal bundle $\mathcal{U}_m^{e,\bullet}(M)$ as follows: $g \in G$ maps the m-arrow $j_m(f)^{e,x}$ to the m-arrow $j_m(g \circ f)^{e,g(x)} = j_m(g)^{x,g(x)} \circ j_m(f)^{e,x}$. This is the extension of the prolonged action from $\overline{G}_r(M)$ in (18.29) to $\mathcal{U}_r^{e,\bullet}(M)$ for $r = m$. Therefore, any transitive action of order m on M prolongs to a free action on the jet space $\mathcal{U}_m^{e,\bullet}(M)$. This sets the stage for the application of a very powerful algorithm recently developed in [FO1] and [FO2] and applied successfully to a plethora of examples. We refer to the expository article [O3] to give the reader an idea about the scope of this method. A function $f : \mathcal{U}_m^{e,\bullet} \to \mathbb{R}$ invariant under this prolonged action is called a lifted differential invariant. Clearly, these invariants are functions on the orbit space $\mathcal{U}_m^{e,\bullet}/\sim$. However, $\mathcal{U}_m^{e,\bullet}/\sim$ is nothing but $\mathcal{U}_m^{e,e}/\theta_m(G_0)$! This is easily seen by mapping a representative of an orbit to some point in $\mathcal{U}_m^{e,e}$ by some m-arrow of G. Now the moving frame algorithm pulls back a complete set of lifted invariants to a set of invariants that separate the points of $\mathcal{U}_m^{e,e}/\theta_m(G_0)$, as demanded by Step 2. Therefore, other than the abstract Rosenlicht Theorem, the moving frame method gives an effective algorithm to find these separating set of rational functions. However, we remark here that this algorithm is a local method and gives the functions (19.4) only locally, at least for the present (see [KL] for a recent global result for algebraic groups).

We now want to linearize the nonlinear PDE $\mathcal{P}_r = (M, \mathcal{P}_r(M))$ defined by (19.3) (or (19.4)). There is a general theory of Lie groupoids and algebroids (see [MK1], [MK2], [Cr], [CF], and [F]), and $\mathcal{P}_r$ is a very special Lie groupoid. The process of constructing the Lie algebroid of $\mathcal{P}_r$ corresponds, of course, to this linearization and produces the required linear PDE. However, it is desirable for our purpose here to carry out this linearization directly in coordinates using (19.3). The idea is simple: We replace the r-arrow in (17.41) with the infinitesimal r-arrow

$$\left(x^i,\ x^i + t\xi^i,\ \delta^i_{j_1} + t\xi^i_{j_1},\ f^i_{j_2j_1} + t\xi^i_{j_2j_1},\ \ldots,\ f^i_{j_r\ldots j_1} + t\xi^i_{j_r\ldots j_1}\right) \tag{19.38}$$

substitute (19.38) into the left-hand side of (19.3), differentiate with respect to t at $t = 0$, and set the resulting expression to zero. This gives

$$\frac{\partial\Phi^j_r\left(x^i,\delta^i_j,0,\ldots,0\right)}{\partial y^a}\xi^a + \frac{\partial\Phi^j_r\left(x^i,\delta^i_j,0,\ldots,0\right)}{\partial f^a_b}\xi^a_b + \cdots$$
$$- \frac{\partial\Phi^j_r\left(x^i,\delta^i_j,0,\ldots,0\right)}{\partial f^a_{b_r\ldots b_1}}\xi^a_{b_r\ldots b_1} = 0 \tag{19.39}$$

or, in brief,

$$\frac{\partial\Phi^j_r(j_r(\mathrm{Id})^{x,x})}{\partial y^a_\mu}\xi^a_\mu = 0 \tag{19.40}$$

where $0 \leq |\mu| \leq r$, $1 \leq j \leq \dim \mathcal{U}_r - \dim \mathcal{P}_r$. Now, (19.40) gives independent linear equations for each x in the variables $\left(\xi^i, \xi^i_j, \xi^i_{j_2j_1}, \ldots, \xi^i_{j_r\ldots j_1}\right) = \left(\xi^i_\mu\right)$, $0 \leq |\mu| \leq r$. Recalling that the variables $\left(\xi^i_\mu\right)$ parametrize the fibers of $\pi_{r,0} : J_rT \to M$ in coordinates, the linear equations (19.40) define a subspace $\mathfrak{P}^x_r$ of the fiber $(J_rT)^x$ for each $x \in M$. Clearly, $\dim \mathfrak{P}^x_r = \dim \mathcal{P}^{e,\bullet}_r$ for each x. Since $\mathcal{P}_r \subset \mathcal{U}_r$ satisfies **A1**, the bundle of vectors $\cup_{x\in M}\mathfrak{P}^x_r \to M$ is a locally trivial vector subbundle of $J_rT \to M$, which we denote by $\mathfrak{B}_r \to M$ and also by $(M, \mathfrak{P}_r)$. By **A2**, we have $\pi_{m+1,k}(\mathfrak{P}_{m+1}) = \mathfrak{P}_k \subset J_kT$ and $\mathfrak{P}_{m+1} \approx \mathfrak{P}_m$. Clearly, $\mathfrak{P}_0 \to M$ is $T \to M$.

For instance, linearization of (19.22) and (19.23) gives

$$\xi^i_a\mathbf{g}^a_{jk} - \mathbf{g}^i_{bk}\xi^b_j - \mathbf{g}^i_{bj}\xi^b_k + \xi^i_{jk} - \frac{\partial\mathbf{g}^i_{jk}}{\partial x^a}\xi^a = 0 \tag{19.41}$$

$$\xi^a_i\mathbf{g}_{aj} + \xi^a_j\mathbf{g}_{ia} + \frac{\partial\mathbf{g}_{ij}}{\partial x^a}\xi^a = 0 \tag{19.42}$$

To define the concept of a solution of $\mathfrak{P}_{m+1} \to M$, let $\left(\overline{\xi}^i_\mu\right)$, $0 \leq |\mu| \leq r$, be a vector in the fiber $\mathfrak{P}_r^p$. A vector field $\xi = (\xi^i(x))$ is a local solution of $\mathfrak{P}_r \to M$ around p satisfying the initial condition $\left(\overline{\xi}^i_\mu\right)$ if $\partial\xi^i(x)/\partial x^\mu$ is a local section of $\mathfrak{P}_r \to M$; that is, $\partial\xi^i(x)/\partial x^\mu$ identically satisfies (19.40) for all x near p and $\partial\xi^i(p)/\partial x^\mu = \overline{\xi}^i_\mu$. We call $\mathfrak{P}_r \to M$ locally solvable if all initial conditions defined by its fibers integrate to local solutions. We want to define the linear curvature $\mathfrak{R}$ in such a way that $\mathfrak{R} = 0$ if and only if $\mathfrak{P}_{m+1} \to M$ is uniquely locally solvable. If $\mathcal{P}_{m+1}$ arises from the first column of (18.28), then $\mathfrak{P}_r = J_r(\overline{\mathfrak{g}})$ and $\mathfrak{P}_{m+1} \to M$ coincides with the second column of (18.32). In this case, the elements of $\overline{\mathfrak{g}}$ provide local (in fact global) solutions. Note that this problem of integration is vacuous for $J_rT \to M$ and $\mathcal{U}_r$ for all r, and there is no "curvature" in the universal setting.

As an aside, suppose that we have constructed a geometric object Θ on M of order $m + 1$ and expressed $\mathcal{P}_{m+1} \subset \mathcal{U}_{m+1}$ as the subgroupoid preserving this structure object as in the above examples. The equations (19.40) now have an obvious geometric interpretation: For any vector field X on M, we define the Lie derivative $\mathcal{L}_X\Theta$ as in [Y] and replace all derivatives of X by jet variables in the formula $\mathcal{L}_X\Theta = 0$. What we get is (19.40). The set of vector fields satisfying $\mathcal{L}_X\Theta = 0$ is known to be closed under bracket (see [Y]) and is therefore a Lie algebra, as we remarked at the end of Chapter 7. However, these vector fields may not satisfy all the initial conditions imposed by the fibers of $\mathfrak{P}_{m+1} \to M$. In particular, this Lie algebra need not be locally transitive on M.

In Chapter 17, we developed a "calculus" in the universal setting. For instance, (M, J_kT) is associated with $(M, \mathcal{U}_{k+1}(M))$, and more generally $(\mathcal{U}_r^{e,\bullet}, J_{r+k}T)$ is associated with $(\mathcal{U}_r^{e,\bullet}, \mathcal{U}_{r+k+1})$, sections of (M, J_kT) are endowed with the Spencer bracket, there is a 1–1 correspondence between $\mathcal{U}_r^{e,e}$-invariant sections of $(\mathcal{U}_r^{e,\bullet}, J_{r+k}T)$ and sections of $(M, J_{r+k}T)$, etc. We now want to restrict this calculus to the PDEs $(M, \mathcal{P}_{m+1})$ and $(M, \mathfrak{P}_m)$. It is a remarkable fact that the Spencer bracket on (M, J_rT) restricts to $(M, \mathfrak{P}_r)$, $0 \leq r \leq m+1$, and, with this restricted bracket, $(M, \mathfrak{P}_r)$ becomes the algebroid of $(M, \mathcal{P}_r)$. Using (19.3), (19.40), (17.48), and (17.49), we invite the interested reader to show this fact directly. For the other restrictions, however, we need a crucial assumption that will be the subject of the next chapter.

20

Curvature and Generalized PHGs

A PDE defined as a submanifold of some jet space has prolongations. This concept is the foundation stone of the formal theory of PDEs developed by Spencer and his co-workers. In order to prolong the nonlinear PDE (19.3), we formally substitute $\partial^{|\mu|}y^i/\partial x^\mu = f^i_\mu$, $0 \le |\mu| \le r$ into the left-hand side of (19.3), differentiate with respect to x^s, substitute back the jet variables, and finally set the resulting expression to zero. We call this process "formal differentiation" as in Chapter 14. This gives

$$\frac{\partial \Phi^j_r(x^i, f^i_\mu)}{\partial x^s} + \frac{\partial \Phi^j_r(x^i, f^i_\mu)}{\partial y^a} f^a_s + \frac{\partial \Phi^j_r(x^i, f^i_\mu)}{\partial f^a_b} f^a_{sb} + \cdots + \frac{\partial \Phi^j_r(x^i, f^i_\mu)}{\partial f^a_{b_m \ldots b_1}} f^a_{sb_r \ldots b_1} = 0 \quad (20.1)$$

We write (20.1) in brief as

$$\frac{\partial \Phi^j_r(x^i, f^i_\sigma)}{\partial x^s} + \frac{\partial \Phi^j_r(x^i, f^i_\sigma)}{\partial f^a_\mu} f^a_{\mu+1_s} = 0, \qquad 0 \le |\mu| \le r \quad (20.2)$$

with the obvious meaning of $\mu + 1_s$. The equations (20.2) *together with* (19.3) have a coordinate-free meaning; that is, in the same way as (19.3) is the local description of the globally defined subset $\mathcal{P}_r \subset \mathcal{U}_r$, (19.3) *and* (20.2) give the local description of a globally defined subset of $\mathcal{U}_{r+1}$. We denote this subset by $\mathcal{P}_{r,1} \subset \mathcal{U}_{r+1}$ and call it the first prolongation of $\mathcal{P}_r$. The idea is clear: If a local diffeomorphism solves (19.3), then it surely also solves (19.3) and (20.2). We warn the reader that (20.2) alone has no invariant meaning. Also a warning with the notation: Since $\mathcal{P}_r = \mathcal{P}_r(M) = (M, \mathcal{P}_r(M))$, we should denote $\mathcal{P}_{r,1}$ more precisely by $(M, \mathcal{P}_r(M))_1$ or $\rho_1(M, \mathcal{P}_r(M))$. This point will be important when we change the base manifold.

Similarly, we differentiate the left-hand side of (19.3) formally k times and join these k equations to (19.3). The set of all these $k+1$ equations defines the subset $\mathcal{P}_{r,k} \subset \mathcal{U}_{r+k}$, called the kth prolongation of $\mathcal{P}_r$.

Two important remarks can be made concerning these subsets:

1. Even though $\mathcal{P}_r \subset \mathcal{U}_r$ is a submanifold by **A1** of Chapter 19, $\mathcal{P}_{r,k} \subset \mathcal{U}_{r+k}$ need not be a submanifold without any further assumption. Therefore, it is not

An Alternative Approach to Lie Groups and Geometric Structures, Ercüment H. Ortaçgil (2018).

possible to define $\mathcal{P}_{r,k,l}$ in general. However, if $\mathcal{P}_{r,k} \subset \mathcal{U}_{r+k}$ is a submanifold, then $\mathcal{P}_{r,k,l} \subset \mathcal{U}_{r+k+l}$ is defined in the obvious way, and, of course, we have $\mathcal{P}_{r,k,l} = \mathcal{P}_{r,k+l}$.

2. The projection $\pi_{r+k,r} : \mathcal{U}_{r+k} \to \mathcal{U}_r$ restricts to $\mathcal{P}_{r,k}$, and we surely have $\pi_{r+k,r}(\mathcal{P}_{r,k}) \subset \mathcal{P}_r$ because (19.3) is among the defining equations of $\mathcal{P}_{r,k}$. However, the other k equations obtained by formal differentiations of (19.3) may impose restrictions on the solutions of (19.3), and the inclusion $\pi_{r+k,r}(\mathcal{P}_{r,k}) \subset \mathcal{P}_r$ may be proper. This fact indicates that prolongation is a very subtle process.

Since we use an arbitrary local diffeomorphism $y = y(x)$ in the definition of $\mathcal{P}_{r,k}$ without assuming local solvability of $\mathcal{P}_r$, $\mathcal{P}_{r,k}$ is defined for any PHG $\mathcal{P}_{m+1}$ for $0 \leq r \leq m$. Now, if $\mathcal{P}_{m+1}$ is the second column of (18.28) arising from the first column, then we have

$$\mathcal{P}_{m+1} \subset \mathcal{P}_{r,m-r+1}, \qquad 0 \leq r \leq m \tag{20.3}$$

because any transformation $y = y(x)$ coming from the first column of (18.28) satisfies (19.3) identically with arbitrary initial conditions, and therefore (20.3) holds by the definition of $\mathcal{P}_{r,m-r+1}$.

Definition 20.1 *Let $\mathcal{P}_{m+1}$ be a PHG of order m. Then $\mathcal{P}_{m+1}$ is r-torsion-free if (20.3) holds for some r, $0 \leq r \leq m$.*

Since $\mathcal{P}_0 = M \times M$, $\mathcal{P}_{0,k} = \mathcal{U}_k$, and therefore $\mathcal{P}_{m+1}$ is always 0-torsion-free. It is a nontrivial fact that (20.3) need not hold for an arbitrary PHG. For instance, if $\mathcal{P}_{m+1}$ is m-torsion-free, then $\mathcal{P}_{m,1}$ must surject onto $\mathcal{P}_m$, because $\mathcal{P}_{m+1} \subset \mathcal{P}_{m,1}$ and $\mathcal{P}_{m+1}$ surjects onto $\mathcal{P}_m$ by the definition of a PHG. However, the existence of the surjection $\mathcal{P}_{m,1} \to \mathcal{P}_m$ may be a very strong condition on $\mathcal{P}_m$, as we will see below.

Now our purpose is to understand the geometric meaning of (20.3). We recall the groupoid $\left(\mathcal{U}_r^{e,\bullet}, \mathcal{U}_{m+1}\right)$ of $(m+1-r)$-arrows on $\mathcal{U}_r^{e,\bullet}$ induced by $\mathcal{U}_{m+1}$, as explained in Chapter 17. Since $\mathcal{P}_{m+1} \subset \mathcal{U}_{m+1}$, the $(m+1)$-arrows of $\mathcal{P}_{m+1}$ on M also define $(m+1-r)$-arrows on $\mathcal{U}_r^{e,\bullet}$. However, these $(m+1-r)$-arrows need not restrict to the submanifold $\mathcal{P}_r^{e,\bullet} \subset \mathcal{U}_r$ and define $(m+1-r)$-arrows *on* $\mathcal{P}_r^{e,\bullet}$. We have the following proposition:

Proposition 20.2 *The following are equivalent:*

(i) *$(M, \mathcal{P}_{m+1})$ is r-torsion-free;*

(ii) *$(m+1-r)$-arrows of $\mathcal{P}_{m+1}$ on $\mathcal{U}_r^{e,\bullet}$ restrict to $\mathcal{P}_r^{e,\bullet} \subset \mathcal{U}_r$.*

Therefore, if $\mathcal{P}_{m+1}$ is r-torsion-free, then we get a groupoid of $(m+1-r)$-arrows on $\mathcal{P}_r^{e,\bullet}$. We denote this groupoid by $\left(\mathcal{P}_r^{e,\bullet}, \mathcal{P}_{m+1}\right) \subset \left(\mathcal{U}_r^{e,\bullet}, \mathcal{U}_{m+1}\right)$, where $\subset$ is an inclusion of pairs and $\left(\mathcal{P}_r^{e,\bullet}, \mathcal{P}_{m+1}\right)$ is easily seen to be a PHG of order $(m-r)$. Thus, we get the following corollary:

Corollary 20.3 *If $(M, \mathcal{P}_{m+1})$ is m-torsion-free, then it defines the PHG $(\mathcal{P}_r^{e,\bullet}, \mathcal{P}_{m+1})$ of order $m-r$. If $\{(G_0, G)\} = \{(M, \mathcal{P}_{m+1})\}$, then $\{(G_r, G)\} = \left\{(\mathcal{P}_r^{e,\bullet}, \mathcal{P}_{m+1})\right\}$, where G_r is defined by the filtration (16.21).*

Corollary 20.3 is the "curved analog" of Proposition 16.6. For $r = m$, something remarkable happens: For any $a, b \in \mathcal{P}_m^{e,\bullet}$, we get the unique 1-arrow $\varepsilon(a \circ b^{-1}) \in (\mathcal{P}_m^{e,\bullet}, \mathcal{P}_{m+1})$ from a to b. Thus, we get a splitting ε, or, equivalently, an absolute parallelism on $\mathcal{P}_m^{e,\bullet}$ as in Part I. Therefore, if $\mathcal{P}_{m+1}$ is m-torsion-free, *everything* we did in Part I applies to $\mathcal{P}_m^{e,\bullet}$! It is worth singling out this fact as the following corollary:

Corollary 20.4 *If $(M, \mathcal{P}_{m+1})$ is m-torsion-free, then we have the parallelizable manifold $\left(\mathcal{P}_m^{e,\bullet}, \varepsilon\right)$, and the splitting ε is defined by $\mathcal{P}_{m+1}$.*

Note that $(M, \mathcal{P}_m)_1$ that we denoted by $\mathcal{P}_{m,1}$ and $\left(\mathcal{P}_r^{e,\bullet}, \mathcal{P}_m\right)_1$ (assuming $\mathcal{P}_m$ is $(m-r)$-torsion-free, defined in the obvious way) are quite different objects for $r \neq 0$: In $(M, \mathcal{P}_m)_1$, the formal differentiation is carried out with respect to the parameters on M, whereas in $\left(\mathcal{P}_r^{e,\bullet}, \mathcal{P}_m\right)_1$, it is carried out with respect to the parameters on $\mathcal{P}_r^{e,\bullet}$.

We now want to understand the meaning of r-torsion-freeness in our examples from Chapter 19.

In Example 1, (20.3) is vacuous, since it always holds for $r = 0$.

In Example 2, $\mathcal{P}_1 = \mathcal{U}_1$, and therefore $\mathcal{P}_{1,1} = \mathcal{U}_2 \supset \mathcal{P}_2$. Therefore, (20.3) holds also for $r = 1$, and it is again vacuous. In particular, Corollary 20.4 applies to $\mathcal{P}_1^{e,\bullet}$, and we invite the reader to explicitly compute $\mathcal{R}$ and $\mathfrak{R}$ defined in Part I in local coordinates. He or she will immediately realize that it is quite difficult to do explicit local computations as we did in Part I, even in this simplest example.

Something interesting happens in Example 3. To find $\mathcal{P}_{1,1}$, we substitute $\left(f_j^i\right) = \left(\frac{\partial y^i}{\partial x^j}\right)$, $\left(f_j^i\right)^{-1} = \left(\overline{f}_j^i\right) = \left(\frac{\partial x^i}{\partial y^j}\right)$ into (19.23), differentiate with respect to x^r and replace the derivatives with jet variables. This gives

$$\overline{f}_{ci}^a f_r^c \overline{f}_j^b \, \mathbf{g}_{ab}(x) + \overline{f}_i^a \overline{f}_{cj}^b f_r^c \, \mathbf{g}_{ab}(x) + \overline{f}_i^a \overline{f}_j^b \frac{\partial \mathbf{g}_{ab}(x)}{\partial x^r} - \frac{\partial \mathbf{g}_{ij}(y)}{\partial y^a} f_r^a = 0 \tag{20.4}$$

Therefore, (19.23) and (20.4) define $\mathcal{P}_{1,1}$. Now the famous Gauss computation of shifting the indices shows that f_{jk}^i is determined by (20.4) if we assign f_j^i. Therefore, our assumption $\mathcal{P}_2 \subset \mathcal{P}_{1,1}$ forces $\mathcal{P}_2 = \mathcal{P}_{1,1}$, and from (19.21) and (20.4) we deduce after some computation the well known formula

$$\mathbf{g}_{jk}^i = \frac{1}{2}\mathbf{g}^{ia}\left(\frac{\partial \mathbf{g}_{ak}}{\partial x^j} + \frac{\partial \mathbf{g}_{aj}}{\partial x^k} - \frac{\partial \mathbf{g}_{jk}}{\partial x^a}\right) = \Gamma_{jk}^i \tag{20.5}$$

Therefore, $\mathcal{P}_2$ is 1-torsion-free if and only if $\mathbf{g}_{jk}^i = \Gamma_{jk}^i$ as defined by (20.5), in which case we have $\mathcal{P}_2 = \mathcal{P}_{1,1} \approx \mathcal{P}_1$.

An unexpected phenomenon occurs in Example 4. First, since $\mathcal{P}_1 = \mathcal{U}_1$, (20.3) holds trivially for $r = 1$ and $\mathcal{P}_3$ is 1-torsion-free. Even though Example 3 can be handled using only the Gauss trick, any reader who attempts this example with bare hands will soon realize that to study the surjectivity of $\mathcal{P}_{2,1} \to \mathcal{P}_2$ requires new ideas and methods. We refer to pages 130–2 of [P1] for a detailed study of this example using the highly developed machinery of formal integrability. We will single out the result of this computation as the following proposition:

Proposition 20.5 *$\mathcal{P}_{2,1}$ surjects onto $\mathcal{P}_2$ if and only if $W = 0$, where W is the Weyl projective curvature tensor. In this case, $\mathcal{P}_{2,1} \simeq \mathcal{P}_3$.*

Therefore, if $\mathcal{P}_3$ is 2-torsion-free, then $W = 0$, and $\mathcal{P}_3 = \mathcal{P}_{2,1} \simeq \mathcal{P}_2$ is uniquely locally solvable. Therefore, nothing is left of any curvature!

Proposition 20.5 warns us that we should stay away from the assumption of m-torsion-freeness if we attempt to define the curvature of $\mathcal{P}_{m+1}$. It is possible to define geometric structures as subgroupoids of the universal jet groupoids and study their local solvability by the methods of formal integrability without insisting on a single entity called curvature as we do in this book. This is the approach adapted in [Ku1], [P1], and [P2].

Proposition 20.5 raises a serious question: For which $\mathcal{P}_{m+1}$ does the surjectivity of $\mathcal{P}_{m,1} \to \mathcal{P}_m$ imply $\mathcal{P}_{m+1} = \mathcal{P}_{m,1} \simeq \mathcal{P}_m$ and the local solvability of $\mathcal{P}_{m+1}$? We do not know the answer, but surely this is so also in the conformal case.

Continuing with the general theory, we now we take a look at the prolongation of the linearization $(M, \mathfrak{P}_r)$ of $(M, \mathcal{P}_r)$. In the same way, we define the kth prolongation $\mathfrak{P}_{r,k} \to M$ as a bundle of vectors by formally differentiating (19.40) k times and adjoining the resulting equations to (19.40). Since linearization commutes with prolongation, (20.3) implies

$$\mathfrak{B}_{m+1} \subset \mathfrak{B}_{r,m-r+1} \tag{20.6}$$

Therefore, if $(M, \mathcal{P}_{m+1})$ is r-torsion-free, then (20.6) holds.

We recall from Chapter 17 that $j_{r+s}(\xi)^p \in (J_{r+s}T)^p$ defines the s-jet $j_s(\rho_r\xi)^a$ of the vector field $\rho_r\xi$ at $a \in \mathcal{U}_r^{e,\bullet}$ that projects to p. Thus, $J_{r+s}T \to M$ defines a vector bundle over $\mathcal{U}_r^{e,\bullet}$, which we denoted by $(\mathcal{U}_r^{e,\bullet}, J_{r+s}T)$ in Chapter 17. Now, for $a \in \mathcal{P}_r^{e,\bullet}$, the so-defined s-jet $j_s(\rho_r\xi)^a$ of the vector field $\rho_r\xi$ at $a \in \mathcal{P}_r^{e,\bullet}$ need not restrict to $\mathcal{P}_r^{e,\bullet}$. This is the case if and only if (20.6) holds.

We now come to the curvature. Rather than dealing with the nonlinear curvature of $(M, \mathcal{P}_{m+1})$, we will deal with the linear curvature of $(M, \mathfrak{P}_{m+1})$.

Since $\mathfrak{P}_m$ is given by (19.37) for $r = m$, $\mathfrak{P}_{m,1}$ is given by

$$\begin{aligned}\mathfrak{P}_{m,1} :\ & \frac{\partial \Phi_m(x, j_m(\mathrm{Id}))}{\partial y^a_\mu}\xi^a_\mu = 0 \\ :\ & \frac{\partial^2 \Phi_m(x, j_m(\mathrm{Id}))}{\partial x^r \partial y^a_\mu}\xi^a_\mu + \frac{\partial \Phi_m(x, j_m(\mathrm{Id}))}{\partial y^a_\mu}\xi^a_{\mu+1_r} = 0\end{aligned} \tag{20.7}$$

where $0 \leq |\mu| \leq m$, and $\mathfrak{P}_{m+1}$ is given by

$$\mathfrak{P}_{m+1} : \frac{\partial \Phi_{m+1}(x, j_m(\mathrm{Id}))}{\partial y^a_\sigma} \xi^a_\sigma = 0 \tag{20.8}$$

where $0 \leq |\sigma| \leq m+1$. We have the Spencer operator

$$\begin{aligned} D : J_{m+1}T &\longrightarrow T^* \otimes J_m T \\ : (\xi^i_\sigma) &\rightarrow \left(\frac{\partial \xi^i_\sigma}{\partial x^r} - \xi^i_{\sigma+1_r} \right) \end{aligned} \tag{20.9}$$

We now assume that $\mathcal{P}_{m+1}$ is m-torsion-free, so that (20.6) holds for $r = m$.

Proposition 20.6 *We define $\overline{D} : \mathfrak{P}_m \rightarrow T^* \otimes J_m T$ by $\overline{D}(\xi_{(m)}) \stackrel{\text{def}}{=} D(\varepsilon\xi_{(m)})$, where $\xi_{(m)}$ is a section of $(M, \mathfrak{P}_m)$ and $\varepsilon\xi_{(m)} = \xi_{(m+1)}$ is the unique lift of $\xi_{(m)}$ to a section of $(M, \mathfrak{P}_{m+1})$. Then the image of $\overline{D}$ is contained in $T^* \otimes \mathfrak{P}_m$ if and only if (20.6) holds for $r = m$, that is, $\mathfrak{P}_{m+1} \subset \mathfrak{P}_{m,1}$ (which holds if $\mathcal{P}_{m+1}$ is m-torsion-free).*

Proof This is easy: Let $\xi_{(m)} = (\xi^i_\mu)$, so that $\xi_{(m)}(x) = (\xi^a_\mu(x))$ identically satisfies the first equation in (20.7). Now $\varepsilon\xi_{(m)}$ satisfies (20.8). Since we assume $\mathfrak{P}_{m+1} \subset \mathfrak{P}_{m,1}$, we conclude that $\varepsilon\xi_{(m)}$ satisfies (20.9) identically. Now substituting $\xi_{(m)}(x) = (\xi^a_\mu(x))$ into the first equation in (20.7) and differentiating with respect to x^r gives

$$\frac{\partial^2 \Phi_m(x, j_m(\mathrm{Id}))}{\partial x^r \partial y^a_\mu} \xi^a_\mu + \frac{\partial \Phi_m(x, j_m(\mathrm{Id}))}{\partial y^a_\mu} \frac{\partial \xi^a_\mu}{\partial x^r} = 0 \tag{20.10}$$

Substituting $\varepsilon\xi_{(m)}$ into (20.7) and subtracting (20.7) from (20.10), we obtain

$$\frac{\partial \Phi_m(x, j_m(\mathrm{Id}))}{\partial y^a_\mu} \left(\frac{\partial \xi^a_\mu}{\partial x^r} - \xi^a_{\mu+1_r} \right) = 0 \tag{20.11}$$

Now (20.11) and the first equation in (20.7) show that $D(\varepsilon\xi_{(m)})^i_{r,\mu}$ is a section of $T^* \otimes \mathfrak{P}_m \rightarrow M$. Therefore, we obtain the first-order differential operator

$$\overline{D} : \mathfrak{P}_m \longrightarrow T^* \otimes \mathfrak{P}_m \tag{20.12}$$

as claimed. The converse also follows easily by reversing the steps. ■

We will denote $\overline{D}$ also by D and call it the restriction of the Spencer operator to $\mathfrak{P}_m$. Using the definitions, it is easy to see that D is a connection on the vector bundle $(M, \mathfrak{P}_m)$. As an intriguing fact, we will show in the Appendix that $(M, \mathcal{P}_{m+1})$ *does not define a connection on the principal bundle $\mathcal{P}^{e,\bullet}_m \rightarrow M$ in the same way in general!*

We now extend (20.12) one step to the right as

$$\mathfrak{P}_m \longrightarrow T^* \otimes \mathfrak{P}_m \longrightarrow \wedge^2(T^*) \otimes \mathfrak{B}_m \tag{20.13}$$

by acting with D on $\mathfrak{P}_m$ and by the exterior derivative on T^* as in Chapter 11. The composition $\mathfrak{R}$ of the operators in (20.13) defines a section of $\wedge^2(T^*) \otimes \mathrm{Hom}(\mathfrak{P}_m, \mathfrak{P}_m)$: For two tangent vectors θ, θ' at $p \in M$, $\mathfrak{R}(p)(\theta, \theta')$ is a linear homomorphism of the fiber $\mathfrak{P}_m^p$.

Definition 20.7 *Let $(M, \mathcal{P}_{m+1})$ be a PHG of order m that is m-torsion-free. Then the linear curvature $\mathfrak{R} \in \wedge^2(T^*) \otimes \mathrm{Hom}(\mathfrak{P}_m, \mathfrak{P}_m)$ of $(M, \mathcal{P}_{m+1})$ is defined by the composition of the operators in* (20.13).

Clearly, $\mathfrak{R}$ is the curvature of the linear connection (20.12). Since the second operator in (20.13) gives all the integrability conditions of the first operator, (20.13) is locally exact at the middle term if and only if $\mathfrak{R} = 0$. In this case, (20.13) extends to the right to a locally exact sequence called the linear Spencer sequence of $\mathcal{P}_{m+1}$, as in (11.22). Clearly, $\mathfrak{R} = 0$ if and only if $(M, \mathfrak{P}_{m+1})$ is uniquely locally solvable, in which case the kernel of the first operator in (20.13) gives the sheaf of local solutions as local infinitesimal generators.

If $\mathcal{P}_{m+1}$ is m-torsion-free, we can also construct a nonlinear differential operator

$$\mathcal{P}_m \longrightarrow T^* \otimes \mathfrak{P}_m \tag{20.14}$$

again using the Spencer operator. We will comment more on (20.14) in the Appendix. Now (20.14) extends to

$$\mathcal{P}_m \longrightarrow T^* \otimes \mathfrak{P}_m \longrightarrow \wedge^2(T^*) \otimes \mathfrak{P}_m \tag{20.15}$$

by taking the integrability conditions of the first operator. The definitions of the second operators in (20.13) and (20.14) use the restricted Spencer bracket on the algebroid $\mathfrak{P}_m \to M$. The nonlinear curvature $\mathcal{R}$ can be defined using the composition in (20.15) as in Chapter 8. Again (20.15) is locally exact at the middle term if and only if $\mathcal{R} = 0$. In this case, $\mathcal{P}_{m+1}$ is uniquely locally solvable, and the kernel of the first operator in (20.15) gives the pseudogroup of the local solutions of $\mathcal{P}_{m+1}$. This sequence is the nonlinear Spencer sequence of $\mathcal{P}_{m+1}$. For $m = 0$, (20.15) restricts to (8.14) as the reader can easily check.

We now state the following proposition:

Proposition 20.8 (Lie's Third Theorem) *Let $(M, \mathcal{P}_{m+1})$ be m-torsion-free. Then the following are equivalent:*

(i) $\mathcal{R} = 0$;

(ii) (20.15) *is locally exact at the middle term;*

(iii) $\mathcal{P}_{m+1}$ *is uniquely locally solvable;*

(iv) $\mathfrak{R} = 0$;

(v) (20.13) *is locally exact at the middle term*

(vi) $\mathfrak{P}_{m+1} \to M$ *is uniquely locally solvable.*

Let us interpret $\mathfrak{R}$ for our examples from Chapter 19.

The first two parts of this book have already been devoted to the study of Example 1.

In Example 2, we recall that $\mathcal{P}_2$ is 1-torsion-free because $\mathcal{P}_1 = \mathcal{U}_1$. We have $\mathfrak{R} \in \wedge^2 (T^*) \otimes \text{Hom}(\mathfrak{P}_1, \mathfrak{P}_1)$ and $\mathfrak{P}_1 = J_1 T$. Now, Γ defines a splitting $J_1 T = T \oplus V$, and $\mathfrak{R}$ preserves this direct sum and acts trivially on V The restriction of $\mathfrak{R}$ to T defines a section of $\wedge^2(T^*) \otimes \text{Hom}(T, T)$, which is the well-known curvature $\overline{\mathfrak{R}}$ of the connection Γ defined by (5.11) or (6.9). Therefore, $\mathfrak{R}$ and $\overline{\mathfrak{R}}$ are essentially the same objects. This implies that there is only one possible model in this example, namely, $\mu = 1$.

Something truly remarkable happens in Example 3. We recall that $\mathcal{P}_1$ is 1-torsion-free if and only if (20.5) holds, which we assume henceforth. Again we have $\mathfrak{R} \in \wedge^2 (T^*) \otimes \text{Hom}(\mathfrak{P}_1, \mathfrak{P}_1)$, and $\mathfrak{P}_1$ is defined by (19.22) and (19.23). Therefore, if $\xi = (\xi^i, \xi^i_j)$ solves (19.22) and (19.23) then so does $\mathfrak{R}(\cdot, \cdot)(\xi)$. By Proposition 20.8, $\mathfrak{R} = 0$ if and only if $\mathcal{P}_2$ is uniquely locally solvable, which is equivalent to the metric $\mathbf{g} = (\mathbf{g}_{ij})$ having constant curvature (see [P2] for the details). Consequently, it seems that we found a new curvature in the 160-year-old Riemannian geometry! We remark here that $\mathfrak{R}$ was noted by Blaom in coordinate-free form on page 6 of [B1] and is also implicit in some other works. Using (19.22) and (19.23), it is easy to compute the components of $\mathfrak{R}$ explicitly. If $\overline{\mathfrak{R}}$ is the Riemann curvature tensor, then clearly $\overline{\mathfrak{R}} = 0 \Rightarrow \mathfrak{R} = 0$, but the converse is false because $\mu = 3$ in this example. Note that $\overline{\mathfrak{R}}$ is a tensor, whereas $\mathfrak{R}$ is a second-order linear geometric object.

Now we have the following trivial but striking consequence:

Proposition 20.9 *If $\mathfrak{P}_{m+1} \to M$ is locally solvable, then the Pontryagin classes (P-classes) of $\mathfrak{P}_m \to M$ vanish.*

First note that $\mathfrak{P}_{m+1} \to M$ and $\mathfrak{P}_m \to M$ have the same P-classes, which can be defined using any connection on $\mathfrak{P}_m \to M$. However, D in (20.12) is such a connection, and its curvature $\mathfrak{R}$ vanishes if and only if $\mathfrak{P}_{m+1} \to M$ is locally solvable, by Proposition 20.9. It is standard to interpret the P-classes classes of vector bundles as topological obstructions to the triviality of the vector bundle. It was a great surprise when Bott showed in 1970 in [Bo1] that they are also obstructions to integrability of certain plane fields on manifolds. Now, at least for certain vector bundles, Proposition 20.9 asserts that P-classes are topological obstructions to integrability by their very definition!

Proposition 20.9 states something nontrivial already in Example 3: For a Riemannian PHG $(M, \mathcal{P}_2)$, it asserts that the P-classes of $\mathfrak{P}_1 \to M$ vanish if M is of constant curvature. On the other hand, we have $\mathfrak{P}_1 \to T \to M$, and Chern proved in [Ch] that if M has constant curvature, then the P-classes of $T \to M$ vanish. We do not know whether one assertion implies the other, the difficulty being that the P-classes of the unknown factor K in $\mathfrak{P}_1 = T \oplus K$.

The meaning of Proposition 20.9 is further clarified by the following corollary:

Corollary 20.10 *Let (G_0, G) be an effective Klein geometry and $\mathfrak{P} \to G/G_0$ be the associated vector bundle of the homogeneous principal bundle $G \to G/G_0$ formed by the adjoint representation of G_0 on the Lie algebra $\mathfrak{g}$ of G. Suppose **A** from Chapter 18 holds. Then the P-classes classes of $\mathfrak{P} \to G/G_0$ vanish.*

Note that there is no mention of jets in Corollary 20.10 except for the assumption **A** from Chapter 18. For the proof, if $o(G_0, G) = m$, then, using **A** we embed $G \to G/G_0$ as the second column of (18.28), linearize, and observe that the second column in (18.32) becomes the vector bundle described in Corollary 20.10 by the identification of the first and second columns.

It is now time to face Example 4 and ask the following question:

> **Q:** Is it possible to define the curvature of a PHG $\mathcal{P}_{m+1}$ without the assumption of m-torsion-freeness?

Since the question whether the arrows of $\mathcal{P}_{m+1}$ integrate has an invariant meaning, we believe that the answer to **Q** is affirmative. If this is so, there seems to be only one possible way of defining curvature: Let $\overline{\varepsilon} : \mathcal{U}_m \to \mathcal{U}_{m+1}$ be a map that restricts to the splitting $\varepsilon : \mathcal{P}_m \to \mathcal{P}_{m+1}$ and $\pi \circ \overline{\varepsilon} = \text{Id}$. Note that $\overline{\varepsilon}$ always exists, but cannot be chosen to be another splitting (homomorphism of groupoids) for $m, n \geq 2$, by our remark about the jet groups in Chapter 17. Thus, we get the linear lift $\overline{\varepsilon} : J_m T \to J_{m+1} T$ which restricts to $\varepsilon : \mathfrak{P}_m \to \mathfrak{P}_{m+1}$. Now let ξ be a section of $\mathfrak{P}_m \to M$. We extend ξ to a section $\overline{\xi}$ of $J_m T \to M$ (which is always possible), and map $\overline{\varepsilon}(\overline{\xi})$ by D in (20.9) to a section of $T^* \otimes J_m T$ and then to a section of $\wedge^2(T^*) \otimes J_m T$ using $\overline{\varepsilon}$ and D. If this final section is a section of $\wedge^2(T^*) \otimes \mathfrak{P}_m$ and is independent of the extension $\overline{\xi}$ and $\overline{\varepsilon}$, we get again $\mathfrak{R} \in \wedge^2(T^*) \otimes \text{Hom}(\mathfrak{P}_1, \mathfrak{P}_1)$, but now $\mathfrak{R}$ need not be the curvature of any connection on $\mathfrak{P}_m \to M$ unless $\mathcal{P}_{m+1}$ is m-torsion-free. As a remarkable fact, this prescription works out in Example 3, but we are unable to justify it in the general case.

Our purpose now is to circumvent **Q** by some construction inspired by an idea communicated to us by Blaom. Our starting point is Corollary 20.4. Since $\mathcal{P}_m^{e,\bullet}$ is parallelizable if $(M, \mathcal{P}_{m+1})$ is m-torsion-free, the idea is to start with some parallelizable (N, ε) as in Part I, keeping in mind that N is a substitute for $\mathcal{P}_m^{e,\bullet}$, and work "inside N" by going "from the top to the bottom" rather than "from the the bottom to the top" using jets. From this new perspective, the following question is fundamental:

> *How do we define a subgroup H of a parallelizable manifold (N, ε) in such a way that when $\mathfrak{R} = 0$ and the local Lie group $(N, \varepsilon, \mathcal{G})$ globalizes to the Lie group N according to Proposition 4.4, then H becomes a Lie subgroup of N?*

For trivial H, we will recover (N, ε), so this definition will be a direct generalization of Part I. Furthermore, we want this definition to generalize both an m-torsion-free PHG and a classical Cartan geometry whose model is fixed by the Cartan connection. Note that the Lie group N in the above question is not assigned a priori; that is, we do not fix a model (H, N) beforehand. However, it is natural to expect that the existence of H will impose certain restrictions on the possible Lie groups N.

Let (N, ε) be a parallelizable manifold as in Part I, so that *everything* done in Parts I and II applies to (N, ε). We assume the existence of a locally trivial fiber bundle $N \to M$ with connected fibers. We denote the foliation on N defined by the fibers $\pi^{-1}(p)$ by $\mathcal{F}$. If $a \in N$ with $\pi(a) = p$, we denote the fiber (or leaf) $\pi^{-1}(p)$ passing through a also by $\mathcal{F}(a)$. The leaves of $\mathcal{F}$ are now embedded submanifolds of N and are of dimension $0 \lneqq k \lneqq n = \dim N$. For $a, b \in N$, the 1-arrow $\varepsilon(a,b)$ defines the isomorphism $\varepsilon(a,b) : T_aN \to T_bN$ according to Lemma 1.3.

We start by assuming that

(i) $\varepsilon(a,b) : T_aN \to T_bN$ restricts to an isomorphism $\varepsilon(a,b) : T_a\mathcal{F}(a) \to T_b\mathcal{F}(b)$ for all $a, b \in N$.

In short, we require ε to restricted to the leaves of $\mathcal{F}$. Note that if $x, y \in \mathcal{F}(a)$, then $\varepsilon(x,y)$ has a double meaning, although this should not be confusing. By Lemma 1.3, all the leaves of $\mathcal{F}$ are parallelized by ε. Thus, we have the parallelizable manifolds $(\mathcal{F}(a), \varepsilon)$ for all $a \in N$, where $\varepsilon = \varepsilon_{|\mathcal{F}(a)}$ and everything done in Parts I and II applies to each leaf. Using our notation from Part I, let $\widetilde{\mathfrak{X}}_\varepsilon(N)$ and $\widetilde{\mathfrak{X}}_\varepsilon(\mathcal{F}(a))$ denote the spaces of ε-invariant vector fields on N and $\mathcal{F}(a)$, respectively. Clearly, $\dim \widetilde{\mathfrak{X}}_\varepsilon(\mathcal{F}(a)) = k$ for all $a \in N$. If some element of $\widetilde{\mathfrak{X}}_\varepsilon(N)$ is tangent to some leaf at some point, then it is tangent to all leaves at all points, by (i). We denote this subspace by $\widetilde{\mathfrak{X}}_\varepsilon(\mathcal{F}) \subset \widetilde{\mathfrak{X}}_\varepsilon(N)$. In particular, we have the *canonical* isomorphism $\widetilde{\mathfrak{X}}_\varepsilon(\mathcal{F}(a)) \simeq \widetilde{\mathfrak{X}}_\varepsilon(\mathcal{F}) \simeq \widetilde{\mathfrak{X}}_\varepsilon(\mathcal{F}(b))$. We recall that $\widetilde{\mathfrak{X}}_\varepsilon(N)$ is a Lie algebra if and only if $\mathfrak{R} = 0 \Leftrightarrow \mathcal{R} = 0$.

We observe the following "integrated form" of the infinitesimal condition(i): Suppose that $\mathfrak{R} = 0$ and the LLG $(N, \varepsilon, \mathcal{G})$ globalizes. Then any global transformation $f \in \mathcal{G}$ that maps a to b maps the leaf $\mathcal{F}(a)$ diffeomorphically onto the leaf $\mathcal{F}(b)$. This is shown first locally for the elements of the pseudogroup $\mathcal{G}$ by integrating the infinitesimal condition (i) and then globally using the connectedness of the leaves. In particular, $\mathcal{G}$ descends to some transitive transformation group of M.

We now strengthen (i) to

(ii) The parallelizable manifolds $(\mathcal{F}(a), \varepsilon)$ are *globalizable* local Lie groups for all $a \in N$.

According to Part I, all 1-arrows of $\varepsilon_{|\mathcal{F}(a)}$ and $\widetilde{\varepsilon}_a$ on $\mathcal{F}(a)$ integrate to *global* diffeomorphisms of those leaves, giving the commuting transformation groups $\mathcal{G}_a$ and $C(\mathcal{G}_a)$ of $\mathcal{F}(a)$. Observe our notation $\widetilde{\varepsilon}_a$ rather than $\widetilde{\varepsilon}_{|\mathcal{F}(a)}$, which gives the impression that $\widetilde{\varepsilon}(x,y)$ is defined for all $x, y \in N$ although this not true unless $\mathfrak{R} = 0$ and $(N, \varepsilon, \mathcal{G})$ is globalizable. Note that $\widetilde{\varepsilon}(x,y)$ is not defined for the moment as a 1-arrow of N even if x, y are in the same leaf $\mathcal{F}(a)$, in which case $\widetilde{\varepsilon}_a(x,y)$ is a 1-arrow of $\mathcal{F}(a)$ but not N. Obviously, $\mathcal{G}_a = \mathcal{G}_b$ if a and b are in the same leaf. According to Proposition 4.4, $\mathcal{F}(a)$ becomes a Lie group once we fix a basepoint and make a choice of left/right.

We now have the Lie algebras $\widetilde{X}_\varepsilon(\mathcal{F}(a))$ and $\mathfrak{X}_{\widetilde{\varepsilon}}(\mathcal{F}(a))$ defined in Chapter 6 such that $\widetilde{X}_\varepsilon(\mathcal{F}(a))$ integrates to $C(\mathcal{G}_a)$, and $\mathfrak{X}_{\widetilde{\varepsilon}}(\mathcal{F}_0(a))$ integrates to $\mathcal{G}_a$. The above canonical isomorphism of vector spaces now becomes the canonical isomorphism of Lie algebras $\widetilde{X}_\varepsilon(\mathcal{F}(a)) \simeq \widetilde{X}_\varepsilon(\mathcal{F}) \simeq \widetilde{X}_\varepsilon(\mathcal{F}(b))$. However, note that $\mathfrak{X}_{\widetilde{\varepsilon}}(\mathcal{F})$ is not defined like $\widetilde{X}_\varepsilon(\mathcal{F}_0)$ unless $\mathfrak{R} = 0$ and $(\mathcal{F}, \varepsilon, \mathcal{G})$ globalizes.

Now, a choice of some $f \in C(\mathcal{G}_a)$ does not determine some $g \in C(\mathcal{G}_b)$ if $\mathcal{F}_0(a) \neq \mathcal{F}_0(b)$, because the transformation groups $\{C(\mathcal{G}_a) \mid a \in N\}$ are not related to each other in the directions "transversal to the leaves." We want to "totally order" or "trivialize" the family $\{C(\mathcal{G}_a) \mid a \in \mathcal{F}\}$ in the transversal directions. In particular, we want all leaves of $\mathcal{F}$ to be isomorphic. It is crucial that we want to trivialize $\{C(\mathcal{G}_a) \mid a \in N\}$ but not $\{\mathcal{G}_a \mid a \in N\}$, for a reason to be made clear below. Therefore, we strengthen (ii) to

(iii) There exists a (smooth) transformation group $C(\mathcal{G})_v \subset \mathrm{Diff}(N)$ (with C standing for centralizer and v for vertical) preserving the leaves and satisfying the following condition: if $\phi \in C(\mathcal{G})_v$, then $\phi_{|\mathcal{F}(a)} \in C(\mathcal{G}_a)$ and the restriction map $C(\mathcal{G})_v \to C(\mathcal{G}_a)$ is an isomorphism, so that $C(\mathcal{G})_v \simeq C(\mathcal{G}_a)$ for all $a \in N$.

Therefore, $C(\mathcal{G})_v$ acts freely on N and simply transitively on each leaf. Note that $C(\mathcal{G})_v$ is by no means the centralizer of $\mathcal{G}$, which does not exist for the moment since we are not assuming $\mathfrak{R} = 0$. However, our purpose is to arrange the construction such that when $\mathfrak{R} = 0$ and $\mathcal{G}$ globalizes, we will have $C(\mathcal{G})_v \subset C(\mathcal{G})$.

Some inspection shows that the nonconstructive (iii) has another problem: iii) does not impose any restriction on $C(\mathcal{G})_v$ in the directions transversal to the leaves (or horizontal directions). We now fix $x, y \in \mathcal{F}(a)$ and choose $z \in \mathcal{F}(b)$. There exists a unique $\phi \in C(\mathcal{G})_v$ with $\phi(x) = y$. Choosing z in some neighborhood of x, ϕ defines a local diffeomorphism that maps x to y and therefore the 1-arrow $j_1(\phi)^{x,y}$. By choosing z in $\mathcal{F}(a)$ and using (iii), we see that $j_1(\phi)^{x,y}$ restricts to $\widetilde{\varepsilon}_a(x,y)$ in $\mathcal{F}(a)$. Therefore, we can now use the notation $\widetilde{\varepsilon}_a(x,y) = \widetilde{\varepsilon}_{|\mathcal{F}(a)}$ as long as x, y are in the same leaf. However, we can also fix x and z and choose $y \in \mathcal{F}(a)$ in some neighborhood of x. This gives a map $T_x(\mathcal{F}(a)) \to T_z(\mathcal{F}(b))$, but such a map is already given by (i)! This is the horizontal condition that we seek. We now strengthen (iii) to

(iv) For any $a, b \in N$, the map $T_a(\mathcal{F}(a)) \to T_b(\mathcal{F}(b))$ defined by $C(\mathcal{G})_v$ coincides with (i).

Strictly speaking, the bundle $\pi : N \to M$ with the "structure group" $C(\mathcal{G})_v$ is not a principal bundle. To define a Lie group structure on $C(\mathcal{G})_v$, we fix $a \in N$, choose left/right in the Lie group $\mathcal{F}(a)$ and use the 1–1 correspondence in (iii). With these choices, the other fibers becomes the "candidates" for the left cosets of $\mathcal{F}(a)$. If $\mathfrak{R} = 0$ and $\mathcal{G}$ globalizes, then $C(\mathcal{G})_v \subset C(\mathcal{G})$, as we desire. For the moment, all we can say about $\mathcal{G}$ is that $C(\mathcal{G})_v \subset C(\mathcal{G})$ and $\dim \mathcal{G} = \dim M + \dim \mathcal{F}(a)$. However, we will see below that there may be other "hidden" conditions on $\mathcal{G}$ coming from "higher-order jets."

Definition 20.11 *Let (N, ε) be a parallelizable manifold with the fibration $\pi : N \to M$ satisfying (iv). Then $(\mathcal{F}, N)$ is a Cartan pair.*

Our purpose now is to define an important vector bundle over M and a special connection on this vector bundle. We call some vector field $\xi \in \mathfrak{X}(N)$ $C(\mathcal{G})_v$-invariant if it is invariant with respect to the action of $C(\mathcal{G})_v$ on N. These vector fields are global sections of a certain vector bundle on M. Indeed, for $p \in M$, let us choose some tangent vector $\xi_x \in T_x(\mathcal{F})$ for all $x \in \pi^{-1}(p)$ in such a way that the vectors $(\xi_x)_{x \in \pi^{-1}(p)}$ are invariant under the action of $C(\mathcal{G})$. We call this data a $C(\mathcal{G})_v$-invariant vector over p. The $C(\mathcal{G})_v$-invariant vectors over p have an obvious vector space structure, which we denote by $\mathfrak{X}_{C(\mathcal{G})_v}(N)^p$. Clearly, $\dim \mathfrak{X}_{C(\mathcal{G})_v}(N)^p = \dim N$. Now the bundle of vectors $\mathfrak{X}_{C(\mathcal{G})_v}(N) \stackrel{\text{def}}{=} \cup_{p \in M} \mathfrak{X}_{C(\mathcal{G})_p}(N)^p \to M$ is a locally trivial vector bundle. The global sections of $\mathfrak{X}_{C(\mathcal{G})_v}(N) \rightarrow M$ (which we also denote by $\mathfrak{X}_{C(\mathcal{G})_v}(N)$ as in Part I) are in 1–1 correspondence with the $C(\mathcal{G})_v$-invariant vector fields on N and are closed under the usual bracket of vector fields. Thus, the vector bundle $\mathfrak{X}_{C(\mathcal{G})_v}(N) \rightarrow M$ is an algebroid and is the analog of the algebroid $\mathfrak{P}_m \rightarrow M$ endowed with the Spencer bracket. The algebroid $\mathfrak{X}_{C(\mathcal{G})_v}(N) \rightarrow M$ is also called the Atiyah algebroid of the "principal bundle" $N \rightarrow M$. Now $\xi, \eta \in \mathfrak{X}_{C(\mathcal{G})_v}(N)$ project to the same vector field on M if and only if $\xi - \eta$ is a vector field that restricts to the leaves and is $\widetilde{\varepsilon}$-invariant; that is, $(\xi - \eta)_{|\mathcal{F}(a)} \in \mathfrak{X}_{\widetilde{\varepsilon}}(\mathcal{F}(a))$ for all $a \in \mathcal{F}$.

We now recall the operator ∇ (not $\widetilde{\nabla}$ with $\widetilde{\mathfrak{R}} = 0$!) defined in Chapter 6, which acts on the vector fields on N. We will now show that ∇ descends to M and defines a linear connection ∇^B on the vector bundle $\mathfrak{X}_{C(\mathcal{G})_v}(N) \rightarrow M$. Equivalently, we will define the covariant derivative

$$\nabla^B_X : \mathfrak{X}_{C(\mathcal{G})_v}(N) \longrightarrow \mathfrak{X}_{C(\mathcal{G})_v}(N), \qquad X \in \mathfrak{X}(M) \tag{20.16}$$

First, we recall ∇. We assume $\mathfrak{R} = 0$, so that $\widetilde{\varepsilon}(x, y)$ is defined for x, y sufficiently close even if they are not on the same leaf. Let $a \in N$, $\eta \in \mathfrak{X}(N)$, $\eta(a) = (\eta^i) \in T_aN$, and $\xi(a) = (\xi^i) \in T_aN$. We choose a curve $c(t)$ that passes through a at $t = 0$ with tangent vector $\xi(a)$. Translating $\eta(c(t))$ to a by $\widetilde{\varepsilon}$ gives the tangent vector $\widetilde{\varepsilon}^i_s(c(t), a)\eta^s(c(t))$ at a. Now,

$$\begin{aligned}
\frac{d}{dt}\left[\widetilde{\varepsilon}^i_s(c(t), a)\eta^s(c(t))\right]_{t=0} &= \left[\frac{\partial \widetilde{\varepsilon}^i_s(c(t), a)}{\partial x^r}\frac{dc^r(t)}{dt}\eta(c(t))^s\right]_{t=0} \\
&\quad + \left[\widetilde{\varepsilon}^i_s(c(t), a)\frac{\partial \eta^s}{\partial x^r}\frac{dc^r(t)}{dt}\right]_{t=0} \\
&= -\Gamma^i_{sr}(a)\xi^r(a)\eta^s(a) + \delta^i_s\frac{\partial \eta^s(a)}{\partial x^r}\xi^r(a) \\
&= \left(\frac{\partial \eta^i(a)}{\partial x^r} - \Gamma^i_{sr}(a)\eta^s(a)\right)\xi^r(a) \\
&= (\nabla_\xi \eta)(a)
\end{aligned} \tag{20.17}$$

Thus, $(\nabla_\xi \eta)(a)$ measures the change in η at a in the direction of $\xi \in \mathfrak{X}_{\widetilde{\varepsilon}}(N)$, where ξ is the $\widetilde{\varepsilon}$-extension of $\xi(a)$ to the points near a. However, (20.17) shows the remarkable fact that $(\nabla_\xi \eta)(a)$ is defined without the assumption $\mathfrak{R} = 0$, as we already observed in Chapter 6. If $\mathfrak{R} = 0$, we recall that (20.17) can be interpreted as a Lie derivative as follows: Let ξ be the unique ε-extension of $\xi(a)$, so that it is the unique solution of $\widetilde{\nabla}\xi = 0$ with the initial condition $\xi(a)$ by Proposition 5.5. Therefore, we have (omitting the point a from our notation)

$$\widetilde{\nabla}_r \xi^i = \frac{\partial \xi^i}{\partial x^r} - \Gamma^i_{ra}\xi^a = 0 \tag{20.18}$$

Now,

$$\begin{aligned}
(\mathcal{L}_\xi \eta)^i = [\xi, \eta]^i &= \xi^a \frac{\partial \eta^i}{\partial x^a} - \eta^a \frac{\partial \xi^i}{\partial x^a} \\
&= \xi^b \frac{\partial \eta^i}{\partial x^b} - \eta^a \Gamma^i_{ab} \xi^b \\
&= \left(\frac{\partial \eta^i}{\partial x^b} - \eta^a \Gamma^i_{ab} \right) \xi^b \\
&= (\nabla_\xi \eta)^i
\end{aligned} \tag{20.19}$$

as stated.

Now let X be a vector field on M and η a section of $\mathfrak{X}_{C(\mathcal{G})_v}(N) \to M$, or, in brief, $\eta \in \mathfrak{X}_{C(\mathcal{G})_v}(N)$. For $a \in N$ with $\pi(a) = p$, we choose some $\xi \in T_pN$ that projects to $X(p)$ and define

$$(\nabla^B_X \eta)(a) \stackrel{\text{def}}{=} (\nabla_\xi \eta)(a) \tag{20.20}$$

If ξ' is another such choice, then $\nabla_\xi \eta - \nabla_{\xi'} \eta = \nabla_{\xi - \xi'} \eta = 0$, because η is $C(\mathcal{G})_v$-invariant, that is, vertically $\widetilde{\varepsilon}$-invariant or, equivalently, ∇-parallel by Chapter 6. Further, $\nabla_X \eta \in \mathfrak{X}_{C(\mathcal{G})_v}(N)$. Indeed, computing $\nabla_\xi \eta$ at some $a \in \mathcal{F}_0(a)$ and translating the result by $\widetilde{\varepsilon}$ to some $b \in \mathcal{F}(a)$ is the same as translating ξ to b and computing $\nabla_\xi \eta$ at b, as follows easily from (20.18).

Definition 20.12 *The connection ∇^B on $\mathfrak{X}_{C(\mathcal{G})_v}(N) \to M$ defined by (20.20) is the Blaom connection determined by ∇.*

The B-connection ∇^B is "compatible" with the algebroid bracket on the sections of $\mathfrak{X}_{C(\mathcal{G})_v}(N) \to M$ and is called a Cartan connection in [B1] and [B2]. Such connections are also called tractor connections (see [CG], [CS], and [Csa]). Now let $\mathfrak{R}^B$ be the curvature of ∇^B. What is the relation between $\mathfrak{R}^B$ and $\mathfrak{R}$? This question is answered by the following proposition:

Proposition 20.13 *$\mathfrak{R}$ is $C(\mathcal{G})_v$-invariant and horizontal over M. Therefore, it descends to a 2-form on M, which is $\mathfrak{R}^B$. In particular, $\mathfrak{R} = 0 \Leftrightarrow \mathfrak{R}^B = 0$.*

Corollary 20.14 *$\mathfrak{R}^{(k)} = \mathfrak{R} \wedge \mathfrak{R} \wedge \cdots \wedge \mathfrak{R}$ (k copies) $= 0$ as a $2k$-form on N for $2k \not\geq \dim M =$ codimension of $\mathcal{F}$.*

There is an interesting parallel between the above situation and the proof of Bott's vanishing theorem (see [Bo1]). The foliation $\mathcal{F}$ defines an integrable subbundle $T(\mathcal{F}) \to N$ of $T(N) \to N$, and we know that ∇ is also a connection ∇_B on $T(N) \to N$. Now, Bott defines in [Bo1] some partially flat connection in the direction of the leaves and extends it to some basic connection on the normal bundle $T(N)/T(\mathcal{F}) \to N$ of $T(\mathcal{F}) \to N$ (later called a Bott connection). The connection ∇_B has both these properties, "B" now standing for Bott. In fact, the projection $\pi : N \to M$ pulls back $\mathfrak{X}_{C(\mathcal{G})_v}(N) \to M$ and ∇^B to $T(N)/T(\mathcal{F}) \to N$ and ∇_B.

By (12.36), Corollary 20.14 gives

$$d(c_{2k-1}) = \operatorname{Tr} \mathfrak{R}^{(k)} = 0, \qquad 2k \ngeq \dim M \tag{20.21}$$

Let $C^*_{C(\mathcal{G})_v}(N) \subset C^*(N)$ denote the subcomplex of $C(\mathcal{G})_v$-invariant forms on N. Now c_{2k-1} belongs to $C^*_{C(\mathcal{G})_v}(N)$. The question whether c_{2k-1} is exact in $C^*(N)$ (in the range given by (20.21)) depends only on 1-jets and is a topological problem. However, whether it is exact in $C^*_{C(\mathcal{G})_v}(N)$ may strictly depend on "higher-order jets," to which we now turn.

We first change our notation to $N = \mathcal{F} = \mathcal{F}_0$, $M = \mathcal{F}/\mathcal{F}_0$, $C(\mathcal{G})_v = C(\mathcal{G}_0)_v$, and $k = k_0 = \dim \mathcal{F}_0$, so that we have the fibration $\pi : \mathcal{F} \to \mathcal{F}/\mathcal{F}_0$ defining the foliation $\mathcal{F}_0$ on $\mathcal{F}$ whose leaves are of dimension k_0. We define inductively

$$\widetilde{X}_\varepsilon(\mathcal{F}_{k+1}) \stackrel{\text{def}}{=} \{X \in \widetilde{X}_\varepsilon(\mathcal{F}_k) \mid [X, Y] \in \widetilde{X}_\varepsilon(\mathcal{F}_k),\ Y \in \widetilde{X}_\varepsilon(\mathcal{F})\}, \qquad k \geq 0 \tag{20.22}$$

The Jacobi identity shows that $[\mathfrak{X}_\varepsilon(\mathcal{F}_k), \mathfrak{X}_\varepsilon(\mathcal{F}_k)] \subset \mathfrak{X}_\varepsilon(\mathcal{F}_k)$ for all $k \geq 0$, even though $\widetilde{X}_\varepsilon(\mathcal{F})$ need not be a Lie algebra unless $\mathfrak{R} = 0$. Thus, $\widetilde{X}_\varepsilon(\mathcal{F}_k) \subset \widetilde{X}_\varepsilon(\mathcal{F})$ is a Lie algebra of vector fields. Let us take a closer look at $\widetilde{X}_\varepsilon(\mathcal{F}_1)$. At each point $a \in \mathcal{F}$, we define the subspace

$$\widetilde{X}_\varepsilon(\mathcal{F}_1)(a) \stackrel{\text{def}}{=} \operatorname{Span}\{X(a) \mid X \in \widetilde{X}_\varepsilon(\mathcal{F}_1)\} \subset T_a\mathcal{F}_0 \tag{20.23}$$

We define $k_1(a) \stackrel{\text{def}}{=} \dim \widetilde{X}_\varepsilon(\mathcal{F}_1)(a) \leq k_0$ and can easily show that $k_1(a)$ is independent of $a \in \mathcal{F}$. Let $\mathcal{F}_1(a)$ denote the maximal integral submanifold of $\mathcal{F}_0(a)$ that integrates the involutive distribution $\widetilde{X}_\varepsilon(\mathcal{F}_1) \subset \widetilde{X}_\varepsilon(\mathcal{F}_0)$ of constant rank $k_1 \leq k_0$. We assume that the leaves $\{\mathcal{F}_1(a),\ a \in \mathcal{F}\}$ are the fibers of a fibration $\mathcal{F} \to \mathcal{F}/\mathcal{F}_1$. Now, ε and $\widetilde{\varepsilon}$ restrict to $\mathcal{F}_1(a)$, and $\mathcal{F}_1(a)$ is a LLG for all $a \in \mathcal{F}$. Thus, we get the transformation group $C(\mathcal{G}_1)_v \subset C(\mathcal{G}_0)_v$, and $(\mathcal{F}_1, \mathcal{F})$ is another Cartan pair like $(\mathcal{F}_0, \mathcal{F})$. Iterating this process and assuming that the leaves $\{\mathcal{F}_k(a),\ a \in \mathcal{F}\}$ are the fibers of a fibration $\mathcal{F} \to \mathcal{F}/\mathcal{F}_1$, a Cartan pair $(\mathcal{F}_0, \mathcal{F})$ determines the filtration of foliations

$$\cdots \subset \mathcal{F}_2 \subset \mathcal{F}_1 \subset \mathcal{F}_0 \subset \mathcal{F} \tag{20.24}$$

and the filtration of normal subgroups

$$\{1\} \subset C(\mathcal{G}_{m-1})_v \subset \ldots \subset C(\mathcal{G}_1)_v \subset C(\mathcal{G}_0)_v \tag{20.25}$$

such that $C(\mathcal{G}_k)_v$ acts freely on $\mathcal{F}$ and simply transitively on the leaves of $\mathcal{F}_k$. Note the meaning of $\mathcal{F}_{k+1} \subset \mathcal{F}_k$: The leaves $\mathcal{F}_{k+1}(x)$ for $x \in \mathcal{F}_k(a)$ foliate $\mathcal{F}_k(a)$; that is, $\mathcal{F}_k(a) = \cup_{x \in \mathcal{F}_k(a)} \mathcal{F}_{k+1}(x)$. This situation is a special case of a more general theory proposed by Tanaka (see [T] and [Mo]). We call such a Cartan pair $(\mathcal{F}_0, \mathcal{F})$ regular. If $\mathcal{F}_{m+1}(a) = \{a\}$ for some (hence all) $a \in \mathcal{F}$ and m is the smallest such integer, then $(\mathcal{F}_0, \mathcal{F})$ is effective of order m.

Definition 20.15 *A generalized PHG (g-PHG) is a regular and effective Cartan pair* $(\mathcal{F}_0, \mathcal{F})$.

An effective Klein pair (G_0, G) defines a g-PHG with $\mathfrak{R} = 0$ that is globalizable. An m-torsion-free PHG $\mathcal{P}_{m+1}$ and a classical Cartan geometry give nonflat examples. One can produce other nonflat examples by perturbing the horizontal arrows of (G_0, G) by a gauge transformation that acts as the identity in the vertical directions. However, it seems futile at this stage to elaborate on this point before clarifying **FP1** and **FP2**.

Now, a g-PHG of order m gives the tower

$$\mathcal{F} \longrightarrow \mathcal{F}/\mathcal{F}_{m-1} \longrightarrow \ldots \longrightarrow \mathcal{F}/\mathcal{F}_1 \longrightarrow \mathcal{F}/\mathcal{F}_0 \tag{20.26}$$

as the curved analog of (16.14) and the vector bundles

$$\mathfrak{X}_{C(\mathcal{G}_k)_v}(\mathcal{F}) \longrightarrow \mathcal{F}/\mathcal{F}_k \tag{20.27}$$

whose sections are $C(\mathcal{G}_k)_v$-invariant vector fields on $\mathcal{F}$. We have the analogy

$$\left(\mathfrak{X}_{C(\mathcal{G}_k)_v}(\mathcal{F}) \rightarrow \mathcal{F}/\mathcal{F}_k\right) \leftrightsquigarrow \left(\mathfrak{P}_m \rightarrow \mathcal{P}_k^{e,\bullet}\right) \tag{20.28}$$

Let $\nabla^B(k)$ denote the B-connection on $\mathfrak{X}_{C(\mathcal{G}_k)_v}(\mathcal{F}) \rightarrow \mathcal{F}/\mathcal{F}_k$ determined by $\nabla^B = \nabla^B(0)$. We observe that we need two hypotheses to construct the analog of $\nabla(k)$ on the right-hand side of (20.28) using $\mathfrak{P}_{m+1}$: First, $(M, \mathcal{P}_{m+1})$ must be k-torsion-free so that $(\mathcal{P}_k^{e,\bullet}, \mathcal{P}_{m+1})$ is a PHG of order $m - k$ by Corollary 20.3. Second, we must have $(\mathcal{P}_k^{e,\bullet}, \mathcal{P}_{m+1}) \subset (\mathcal{P}_k^{e,\bullet}, \mathcal{P}_m)_1$ so that $(\mathcal{P}_k^{e,\bullet}, \mathfrak{P}_{m+1})$ defines a connection on the vector bundle $(\mathcal{P}_k^{e,\bullet}, \mathfrak{P}_m)$. This shows the flexibility and generality of the definition of a g-PHG together with its obvious connection $\nabla^B(0)$.

We would now like to express a personal conviction that is open to criticism and correction. We hope to have made clear in Part III (see also **A**) that the concepts of principal bundles and connections have their origins in jets. We believe that this is a fact. We remark here that this fact is stated by many mathematicians long before us, above all by Ehresmann, who was the creator of the theory of jets, principal bundles, and groupoids. However, it is also a fact that the theory of connections on principal bundles has produced the beautiful theory of characteristic classes, which has occupied (and still does) geometers and topologists since 1940. The recent spectacular achievements of gauge theory have contributed greatly to the glory of this theory. On the other hand, jet theory still remains an essentially local theory and has not produced any new global invariants. We believe that this is the main reason why, in differential geometry today, jet theory has no place at the front of the stage. Since jets are essentially Taylor polynomials, it is clear that the route to such invariants is first to establish global higher-order structures that are relevant in geometry, or equivalently, to clarify **FP1**. Of course, it is possible to define higher-order PDEs as submanifolds of the jet extensions of some fibered manifolds and study them using the Spencer machinery, but clearly this approach, although adopted by some authors, is not very realistic unless these PDEs represent relevant geometric structures. When we started to learn jet theory, we always wondered whether such structures do not exist for some reason or whether geometers simply had no time to think about them since they were busy with more concrete and fundamental examples. Therefore, assuming that we have satisfactorily answered **FP1** and **FP2** and expanded our list of g-PHGs, we pose the following problem:

FP3: Find cohomological invariants of g-PHGs that depend strictly on higher-order jets.

In an attempt to make a start with **FP3**, we observe that (20.26) gives the filtration

$$C^*_{C(\mathcal{G}_0)_v}(\mathcal{F}) \subset C^*_{C(\mathcal{G}_1)_v}(\mathcal{F}) \subset \ldots \subset C^*_{C(\mathcal{G}_{m-1})_v}(\mathcal{F}) \subset C^*(\mathcal{F})$$

where $C^*_{C(\mathcal{G}_k)_v}(\mathcal{F})$ is the complex of $C(\mathcal{G}_k)_v$-invariant forms on $\mathcal{F}$. If the closed forms c_{2k-1} in the range given by (20.21) are not exact in the subcomplex $C^*_{C(\mathcal{G}_r)_v}(\mathcal{F})$, then they define cohomology classes that depend strictly on $(m-r+1)$-jets by construction. If $\mathfrak{R} = 0$, then we know from Chapter 12 that c_{2k-1} is defined for all $k \geq 1$ and that these cohomology classes localize. The purely algebraic problem now is to clarify the relation between $[c_{2k-1}] \in H^{2k-1}(\mathfrak{g}, \mathbb{R})$ and the filtration (16.2). We hope that illustrating the nontriviality of these characteristic classes with particular examples may be helpful in elevating the present status of jet theory.

APPENDIX
TORSION-FREE CONNECTIONS

The purpose of this appendix is to shed some light on the concept of a torsion-free connection using jets.

In Chapter 17, we defined the jet group $G_k(n)$ in terms of the chain rule formulas (17.5). Let $A(i)$ denote the ith formula in (17.5), so that $A(1), \ldots, A(k)$ define the group operation of $G_k(n)$. Let $\widehat{J}_1 G_k(n)$ be the set whose elements are of the form

$$(f^i_{j_1}, f^i_{j_2 j_1}, \ldots, f^i_{j_k \ldots j_1} \mid f^i_{r,j_1}, \ldots, f^i_{r,j_k \ldots j_1}) \tag{A.1}$$

which we write in brief as $(f^i_\mu \mid f^i_{r,\mu})$, $1 \leq |\mu| \leq k$. Note that $f^i_{r,\mu}$ need not be symmetric; that is, the index r need not commute with the indices in μ. We define the operation $(h^i_\mu \mid h^i_{r,\mu}) = (g^i_\mu \mid g^i_{r,\mu}) \circ (f^i_\mu \mid f^i_{r,\mu})$ as follows: We recall that $A(i+1)$ is obtained from $A(i)$ by differentiation. Let $B(i+1)$ be the formula obtained from $A(i)$ by putting a comma whenever we differentiate a term in $A(i)$. As an important point to be clarified below, we will write f^a_r rather than $f^a_{r,}$. In this way, we get the formulas

$$
\begin{aligned}
B(2) \quad &: \quad h^i_{r,j} = g^i_{a,b} f^a_r f^b_j + g^i_a f^a_{r,j} \\
B(3) \quad &: \quad h^i_{r,mj} = g^i_{a,bc} f^a_r f^b_m f^c_j + g^i_{ab} f^a_{r,m} f^b_j + g^i_{ab} f^a_m f^b_{r,j} + g^i_a f^a_{r,mj} \\
B(4) \quad &: \quad h^i_{r,lmj} = \cdots \\
&\vdots \\
B(k+1) \quad &: \quad h^i_{r,j_k \ldots j_1} = g^i_{a_{k+1},a_k \ldots a_1} f^{a_1}_{j_1} \cdots f^{a_k}_{j_k} f^{a_{k+1}}_r + \cdots + g^i_a f^a_{r,j_k,\ldots j_1}
\end{aligned} \tag{A.2}
$$

Now the formulas $A(1), \ldots, A(k)$ *together* with $B(2), \ldots, B(k+1)$ define an operation on the set $\widehat{J}_1 G_k(n)$. We will explain below why $\widehat{J}_1 G_k(n)$ is a group.

The definition of the group $\widehat{J}_1 G_k(n)$ has three immediate consequences. First, we have the canonical projection homomorphism

$$
\begin{aligned}
\pi : \widehat{J}_1 G_k(n) &\longrightarrow G_k(n) \\
: (f_\mu \mid f_{r,\mu}) &\longrightarrow (f_\mu)
\end{aligned} \tag{A.3}
$$

Second, the kernel $\Gamma_k = (\mathrm{Id}^i_\mu \mid \Gamma^i_{r,\mu})$ of π is a vector group, which is easily seen from (A.2) by setting $(f^i_{j_1}, f^i_{j_2 j_1}, \ldots, f^i_{j_k \ldots j_1}) = (\delta^i_j, 0, \ldots, 0) = j_k(\mathrm{Id})^o$. Third, we have the canonical splitting

$$
\begin{aligned}
\gamma_0 : G_k(n) &\longrightarrow \widehat{J}_1 G_k(n) \\
: (f_\mu) &\longrightarrow (f_\mu \mid 0)
\end{aligned} \tag{A.4}
$$

with the obvious meaning of $(f_\mu \mid 0)$.

Definition A.1 *A connection γ on $G_k(n)$ is a splitting $\gamma : G_k(n) \to \widehat{J}_1 G_k(n)$, and γ_0 is the canonical connection.*

Omitting the indices, a connection γ is of the form

$$(f) \longrightarrow (f \mid \overline{\gamma}(f)) \tag{A.5}$$

Using the notation of (A.5), we define $\widehat{\gamma}(f)$ by

$$(\mathrm{Id} \mid \widehat{\gamma}(f)) \stackrel{\text{def}}{=} (f \mid \overline{\gamma}(f)) \circ (f \mid 0)^{-1} = \gamma(f)\gamma_0(f)^{-1} \tag{A.6}$$

identifying $\widehat{\gamma}(f)$ with $(\mathrm{Id} \mid \widehat{\gamma}(f)) \in \Gamma_k$.

Definition A.2 $\widehat{\gamma}(f) = (\gamma^i_{r,\mu})$ *in* (A.6) *are the components of the connection γ at f.*

Therefore, the components of the canonical connection γ_0 vanish at all f. To understand the relation between the components of γ at $f \in G_k(n)$ and at $fg = f \circ g \in G_k(n)$,

$$\begin{aligned}
(\mathrm{Id} \mid \widehat{\gamma}(fg)) &= (fg \mid \overline{\gamma}(fg)) \circ (fg \mid 0)^{-1} \\
&= (f \mid \overline{\gamma}(f)) \circ (g \mid \overline{\gamma}(g)) \circ (g \mid 0)^{-1} \circ (f \mid 0)^{-1} \\
&= (f \mid \overline{\gamma}(f)) \circ (\mathrm{Id} \mid \widehat{\gamma}(g)) \circ (f \mid 0)^{-1}
\end{aligned} \tag{A.7}$$

Therefore, the components of γ at some f determine its components on $G_k(n)$ by (A.7). Now

$$(f^i_\mu \mid f^i_{r,\mu}) = (f^i_\mu \mid f^i_{r,\mu}) \circ (f^i_\mu \mid 0)^{-1} \circ (f^i_\mu \mid 0) \stackrel{\text{def}}{=} (\mathrm{Id}^i_\mu \mid \Gamma^i_{r,\mu}) \circ (f_\mu \mid 0) \tag{A.8}$$

From (A.8), we arrive at the semidirect product

$$\widehat{J}_1 G_k(n) = \gamma_0(G_k(n)) \ltimes \Gamma_k \tag{A.9}$$

Therefore, there is a 1–1 correspondence between the elements of Γ_k and the elements of the left coset space $\widehat{J}_1 G_k(n)/\gamma_0(G_k(n))$. In particular, $\widehat{J}_1 G_k(n)/\gamma_0(G_k(n))$ is an affine space and is parametrized by Γ_k as in our examples in Chapter 19. The left action of $\widehat{J}_1 G_k(n)$ on $\widehat{J}_1 G_k(n)/\gamma_0(G_k(n)) \simeq \Gamma_k$ is computed as in (19.11) and is given by

$$(\mathrm{Id}^i_\mu \mid \overline{\Gamma}^i_{r,\mu}) = (f^i_\mu \mid f^i_{r,\mu}) \circ (\mathrm{Id}^i_\mu \mid \Gamma^i_{r,\mu}) \circ (f^i_\mu \mid 0)^{-1} \tag{A.10}$$

Observe the difference between (A.7) and (A.10): There is no connection in (A.10), but all possible connection components. We cannot use (A.10), for the moment, to define a geometric object Γ with the components $(\Gamma^i_{r,\mu})$, because $(f^i_\mu \mid f^i_{r,\mu})$ does not correspond to a coordinate change, which should be (first of all) symmetric.

For $k = 1$, if we choose $(f^i_j \mid f^i_{r,j})$ in (A.10) to be symmetric, then (A.10) gives (19.12), but $\Gamma^i_{r,\mu}$ and $\overline{\Gamma}^i_{r,j}$ are now not necessarily symmetric. This observation raises the following important question:

Q: What is a torsion-free connection?

One is tempted to define $\Gamma = (\Gamma^i_{r,\mu})$ as torsion-free if the components $(\Gamma^i_{r,\mu})$ are symmetric for all $1 \leq |\mu| \leq k$. However, an inspection of $B(3)$ shows that if $(g^i_j, g^i_{jk} \mid g^i_{r,j}, g^i_{r,jk})$ and $(f^i_j, f^i_{jk} \mid f^i_{r,j}, f^i_{r,jk})$ are symmetric, then $h^i_{r,jk}$ in $(h^i_j, h^i_{jk} \mid h^i_{r,j}, h^i_{r,jk}) = (g^i_j, g^i_{jk} \mid g^i_{r,j}, g^i_{r,jk}) \circ (f^i_j, f^i_{jk} \mid f^i_{r,j}, f^i_{r,jk})$ need not be symmetric even though $h^i_{r,j}$ is symmetric. Therefore, the symmetric elements of $\widehat{J}_1 G_k(n)$ do not form a subgroup for $k \geq 2$. In particular, even if we choose $(\Gamma^i_{r,\mu})$ and $(f^i_\mu \mid f^i_{r,\mu})$ to be symmetric in (A.10), $(\overline{\Gamma}^i_{r,\mu})$ need not be symmetric for $k \geq 2$. Consequently, it is not possible to define a torsion-free connection in this way except for $k = 1$, which has been the starting point of the concept of a torsion-free connection! However, if we choose $f^i_{r,\mu}$ in $(f^i_\mu \mid f^i_{r,\mu})$ to be symmetric *and equal to* $f^i_{\mu+1_r}$ for $k \geq 2$, then something interesting happens. To see this, we observe that the definition of $\widehat{J}_1 G_k(n)$ has another consequence not as immediate as the above ones: We have the canonical splitting

$$\sigma : G_{k+1} \longrightarrow \widehat{J}_1 G_k(n) \tag{A.11}$$

defined by

$$\left(f^i_{j_1}, f^i_{j_2 j_1}, \ldots, f^i_{j_{k+1} \ldots j_1}\right) \longrightarrow \left(f^i_{j_1}, f^i_{j_2 j_1}, \ldots, f^i_{j_k \ldots j_1} \mid f^i_{j_2 j_1}, \ldots, f^i_{j_{k+1} \ldots j_1}\right) \tag{A.12}$$

because if we omit the commas in $B(i+1)$, then $B(i+1)$ becomes identical with $A(i+1)$.

Assume for the moment that there exists some splitting

$$\varepsilon : G_k(n) \longrightarrow G_{k+1}(n) \tag{A.13}$$

which is the case for $k = 1$. Composing ε with σ, we get the connection

$$\sigma \circ \varepsilon : G_k(n) \longrightarrow \widehat{J}_1 G_k(n) \tag{A.14}$$

Unfortunately, this argument is vacuous for $n \geq 2$ unless $k = 1$, because, as we remarked in Chapter 19, (A.13) does not exist for $k, n \geq 2$. Therefore, for $n \geq 2$, "torsion-free connections" in the universal setting exist only in the exceptional case $k = 1$. However, recall that there are many splittings inside (A.13) defined by Klein geometries according to Chapter 16.

To justify the name "connection" in Definition A.1, we want to express $\widehat{J}_1 G_k(n)$ as the vertex groups of a transitive groupoid (or, equivalently, a principal bundle) on a smooth manifold M, $\dim M = n$. Now observing that a local section f of the principal bundle $\mathcal{U}^{e,\bullet}_k \to M$ with the structure group $\mathcal{U}^{e,e}_k$ is of the form $f(y) = \left(e^i, y^i, f^i_j(y), f^i_{j_2 j_1}(y), \ldots, f^i_{j_k \ldots j_1}(y)\right) = (e^i, y^i, f^i_\mu(y))$, $1 \leq |\mu| \leq k$, we can easily check that the bundle $\widehat{J}_1(\mathcal{U}^{e,\bullet}_k) \to M$ of 1-jets of local sections of $\mathcal{U}^{e,\bullet}_k \to M$ is another principal bundle with structure group $\widehat{J}_1(\mathcal{U}^{e,\bullet}_k)^e \simeq \widehat{J}_1 G_k(n)$, and the connections on $\mathcal{U}^{e,\bullet}_k \to M$ are in 1–1 correspondence with the $\mathcal{U}^{e,e}_k \simeq G_k(n)$-invariant sections of the Γ_k-principal bundle $\widehat{J}_1(\mathcal{U}^{e,\bullet}_k) \to \mathcal{U}^{e,\bullet}_k$. Surprisingly, it is not also immediate from this well-known construction why $\widehat{J}_1 G_k(n)$ is a group, because there is no obvious composition defined on these local sections.

Now let $\mathcal{P}_{m+1}$ be a PHG on M and consider the principal bundle $\mathcal{P}^{e,\bullet}_m \to M$. A local section f of $\mathcal{P}^{e,\bullet}_m \to M$ over (y^i) is of the form $f(y) = (e^i, y^i, f^i_\mu(y))$, $1 \leq |\mu| \leq m$, and is subject to the constraint

$$\Phi_m(e^i, y^i, f^i_\mu(y)) = 0 \tag{A.15}$$

according to (19.3). Therefore, $j_1(f)^y = (e^i, y^i, f^i_\mu(y) \mid f^i_{r,\mu}(y))$ is subject to

$$\frac{\partial \Phi_m(e^i, y^i, f^i_\mu)}{\partial f^a_b} f^a_{r,b} + \cdots + \frac{\partial \Phi_m(e^i, y^i, f^i_\mu)}{\partial f^a_{b_m \ldots b_1}} f^a_{r,b_m \ldots b_1} = 0 \tag{A.16}$$

where $f^a_{r,\mu}(y) = \partial f^a_{r,\mu}(y)/\partial y^r$, $1 \leq |\mu| \leq m$. Now, there are three main differences between (A.16) and (20.1) for $r = m$:

1. The first two terms in (20.1) are absent in (A.16), since we fixed the source.
2. The symmetric $f^a_{sb_j \ldots b_1}$ in (20.1) is replaced by the nonsymmetric $f^a_{s,b_j \ldots b_1}$ in (A.16).
3. The solutions of (A.15) and (A.16) *surject onto* the solutions of (A.15) at any point $y = q$, by the definition of (A.16), whereas the solutions of (19.3) and (20.1) need not surject onto the solutions of (19.3) unless $\mathcal{P}_{m+1}$ is m-torsion-free.

Let $\widehat{J}_1(\mathcal{P}_m)$ denote also the groupoid obtained from the principal bundle $\widehat{J}_1(\mathcal{P}_m) \to \mathcal{P}_m$.

Definition A.3 *$\mathcal{P}_{m+1}$ is a torsion-free connection on $\mathcal{P}_m$ if*

$$\mathcal{P}_{m+1} \subset \widehat{J}_1(\mathcal{P}_m) \tag{A.17}$$

We now have two concepts of torsion-freeness for a given PHG $(M, \mathcal{P}_{m+1})$:

(i) $(M, \mathcal{P}_{m+1})$ is torsion-free according to Definition A.3.

(ii) $(M, \mathcal{P}_{m+1})$ is m-torsion-free as defined in Chapter 20: $\mathcal{P}_{m+1} \subset \mathcal{P}_{m,1}$.

Note that both conditions are vacuous for $m = 0$, whereas we defined T in Chapter 5 and called it torsion! For $m \geq 1$, these two concepts are *not* equivalent, owing to the two missing terms in (A.16). However, something remarkable happens in Riemannian geometry where $m = 1$. In this case, (A.15) is given by

$$\mathbf{g}_{ab}(y) f^a_i(y) f^a_j(y) = \mathbf{g}_{ij}(e) \tag{A.18}$$

and therefore (A.16) is given by

$$\frac{\partial \mathbf{g}_{ab}(q)}{\partial y^r} f^a_i(q) f^a_j(q) + \mathbf{g}_{ab}(q) f^a_{r,i}(q) f^a_j(q) + \mathbf{g}_{ab}(q) f^a_i(q) f^a_{r,j}(q) = 0 \tag{A.19}$$

where $f^a_{r,i}(q) = \left[\partial f^a_i(y)/\partial y^r\right]_{y=q}$. Observe that the zero on the right-hand side of (A.19) is due to the fact that we fixed the source and has nothing to do with $\mathbf{g}$, since we would have the same zero if we worked with any other geometric object. We multiply both sides of (A.19) with $g^i_l(q) g^j_k(q)$, where $g = f^{-1}$, sum over i, j, and set $\Gamma^i_{r,l}(q) \overset{\text{def}}{=} f^a_{r,i}(q) g^i_l(q)$. Omitting q, we get

$$\frac{\partial \mathbf{g}_{lk}}{\partial y^r} + \Gamma^a_{r,l} \mathbf{g}_{ak} + \Gamma^a_{r,k} \mathbf{g}_{al} = 0 \tag{A.20}$$

On the other hand $\mathcal{P}_{1,1}$ is given by (19.3) and (20.2) and is obtained by differentiating $\mathbf{g}_{ab}(y)(\partial y^a/\partial x^i)(\partial y^b/\partial x^j) = \mathbf{g}_{ij}(x)$ and substituting the jet variables as in (19.23) and (20.4). We rewrite $\mathcal{P}_{1,1}$ here:

$$
\begin{aligned}
\mathbf{g}_{ab}(y)f^a_i f^b_j &= \mathbf{g}_{ij}(x)\\
\frac{\partial \mathbf{g}_{ab}(y)}{\partial y^c} f^c_s f^a_i f^b_j + \mathbf{g}_{ab}(y) f^a_{si} f^b_j + \mathbf{g}_{ab}(y) f^a_i f^b_{sj} &= \frac{\partial \mathbf{g}_{ij}(x)}{\partial x^s}
\end{aligned} \tag{A.21}
$$

As we know, $\mathcal{P}_{1,1}$ surjects onto $\mathcal{P}_1$ if and only (20.5) holds. Now, the metric $\mathbf{g}$ is very special: Around any p, there exists a coordinate system (x) such that $\mathbf{g}_{ij}(p) = \delta_{ij}$ and $\left[\partial \mathbf{g}_{ij}(x)/\partial x^r\right]_{x=p} = 0$. Choosing (x) in (A.21) in this way and evaluating (A.21) at $x = p, y = q$, (A.21) becomes

$$
\begin{aligned}
\mathbf{g}_{ab} f^a_i f^b_j &= \delta_{ij}\\
\frac{\partial \mathbf{g}_{ab}}{\partial y^a} f^c_s f^a_i f^b_j + \mathbf{g}_{ab} f^a_{si} f^b_j + \mathbf{g}_{ab} f^a_i f^b_{sj} &= 0
\end{aligned} \tag{A.22}
$$

Multiplying both sides of the second equation of (A.22) by $g^s_r g^i_l g^j_k$, summing over s, i, j, and setting $\Gamma^a_{rl} = f^a_{si} g^s_r g^i_l$, we again obtain (A.20), but this time with symmetric Γ!! At this point, we are tempted to define a torsion-free connection and start searching for it, not only in Riemannian geometry but also in more sophisticated geometries (see, e.g., [CS]).

We now "correct" the group $\widehat{J}_1 G_k(n)$ as follows: When we differentiate $A(i)$, we keep the comma in $f^a_{r,}$, which need not be equal to f^a_r. We denote the set of equations obtained in this way by $C(i)$. Since $A(0)$ is $z(x) = z(y(x))$, $C(1)$ is $h^i_{,r} = g^i_{a,} f^a_{r,}$. Now $A(0), \ldots, A(k)$ together with $C(1), \ldots, C(k+1)$ define the set $J_1(G_k(n))$, whose elements are of the form

$$
\left(f^i_{j_1}, f^i_{j_2 j_1}, \ldots, f^i_{j_k \ldots j_1} \mid f^i_{r,}, f^i_{r, j_1}, \ldots, f^i_{r, j_k \ldots j_1}\right) \tag{A.23}
$$

The subset of $J_1(G_k(n))$ obtained by setting $f^i_j = f^i_{j,}$ can be identified with $\widehat{J}_1 G_k(n)$. With this identification, we have

$$
\widehat{J}_1 G_k(n) \subset J_1 G_k(n) \tag{A.24}
$$

We now want to define the groupoid $J_1(\mathcal{U}_k)$ on M whose vertex groups can be identified with $J_1 G_k(n)$ so that $J_1 G_k(n)$ will have a natural group structure. A local bisection of $\mathcal{U}_k \to M \times M$ is the data consisting of a local diffeomorphism $f : U \to f(U) = V$ and a smooth choice of k-arrows of $\mathcal{U}_k$ from x to $y = f(x)$ for all $x \in U$. If $U = M$, then f is a global bisection. We do not require these k-arrows to be the k-arrows of f. In coordinates, a bisection of $\mathcal{U}_k$ is of the form

$$
\left(x^i, y^i, f^i_{j_1}(x), f^i_{j_2 j_1}(x), \ldots, f^i_{j_k \ldots j_1}(x)\right) = \left(x^i, f^i_\mu(x)\right), \qquad 0 \le |\mu| \le k \tag{A.25}
$$

In coordinate-free language, we have the source projection $p_1 : \mathcal{U}_k \to M$, and a bisection is a special section of this bundle. Therefore, a bisection can be defined for more general groupoids. The bisections can be composed and inverted in the obvious way because their values are the usual k-arrows defining the groupoid $\mathcal{U}_k$. The object that we called a gauge transformation or a vertical diffeomorphism in Chapters 1 and 8 is a special *global* bisection that projects to $\mathrm{Id} \in \mathrm{Diff}(M)$. The horizontal diffeomorphisms of $\mathcal{U}_k$ are special bisections of the form $(x^i, y^i, \partial y^i/\partial x^j, \ldots, \partial^k y^i/\partial x^{j_k} \cdots \partial x^{j_1})$ obtained by prolonging (local) diffeomorphisms of M.

The bisection (A.25) now defines the object

$$\left(p^i, y(p)^i, f^i_{j_1}(p), f^i_{j_2j_1}(p), \ldots, f^i_{j_k\cdots j_1}(p) \,\middle|\, \frac{\partial y^i(p)}{\partial x^r}, \frac{\partial f^i_{j_1}(p)}{\partial x^r}, \ldots, \frac{\partial f^i_{j_k\cdots j_1}(p)}{\partial x^r}\right)$$

$$= (p^i, f^i_\mu(p) \mid f^i_{r,\mu}(p)) \tag{A.26}$$

Note that $f^i_{r,}(p) = \partial y^i(p)/\partial x^r \neq f^i_r(p)$. We call (A.26) a $(k, 1)$-arrow from p to $q = y(p)$. Thus, we obtain the groupoid $J_1(\mathcal{U}_k) \supset \widehat{J}_1(\mathcal{U}_k)$ and the diagram

$$\begin{array}{ccccc} J_1(\mathcal{U}_k) & \longrightarrow & \mathcal{U}_k & \longrightarrow & M \times M \\ \cup & & \| & & \| \\ \widehat{J}_1(\mathcal{U}_k) & \longrightarrow & \mathcal{U}_k & \longrightarrow & M \times M \end{array} \tag{A.27}$$

Note that $\mathcal{U}_{k+1} \subset J_1(\mathcal{U}_k)$ and the inclusion of the arrows is given by

$$\left(x^i, y^i, f^i_{j_1}, f^i_{j_2j_1}, \ldots, f^i_{j_{k+1}\cdots j_1}\right) \longrightarrow \left(x^i, y^i, f^i_{j_1}, f^i_{j_2j_1}, \ldots, f^i_{j_k\cdots j_1} \mid f^i_{j_1}, f^i_{j_2j_1}, \ldots, f^i_{j_{k+1}\cdots j_1}\right) \tag{A.28}$$

which should be compared with (A.12). We have used the bottom row of (A.27) and its restriction (A.17) to interpret $\mathcal{P}_{m+1}$ as a torsion-free connection. Now we will use the top row of (A.27) to give a quite different interpretation of $\mathcal{P}_{m+1}$ by asking a very natural question: When is a bisection a horizontal diffeomorphism? We will first handle the case of the universal groupoid $(M, \mathcal{U}_{m+1})$. This question amounts to solving the PDE

$$\left(x^i, y^i, \frac{\partial y^i}{\partial x^j}, \ldots, \frac{\partial^k y^i}{\partial x^{j_{m+1}} \cdots \partial x^{j_1}}\right) = (x^i, y^i, f^i_j(x), f^i_{j_2j_1}(x), \ldots, f^i_{j_{m+1}\cdots j_1}(x)) \tag{A.29}$$

of order $m + 1$. The idea is to replace (A.29) with the equivalent first-order PDE

$$\left(x^i, y^i, f^i_j, \ldots, f^i_{j_m\cdots j_1} \,\middle|\, \frac{\partial y^i}{\partial x^r}, \frac{\partial f^i_j}{\partial x^r}, \ldots, \frac{\partial f^i_{j_m\cdots j_1}}{\partial x^r}\right)$$

$$= (x^i, y^i, f^i_j, \ldots, f^i_{j_m\cdots j_1} \mid f^i_r, f^i_{rj}, \ldots, f^i_{rj_m\cdots j_1}) \tag{A.30}$$

Since $\mathcal{U}_{m+1} \subset J_1(\mathcal{U}_m)$ in view of (A.28), we observe that both sides of (A.30) define arrows of the groupoid $J_1(\mathcal{U}_m) \rightarrow M \times M$ that project to the same arrows of $\mathcal{U}_m$. We rewrite (A.30) as $a = b$ and require $b^{-1} \circ a = \mathrm{Id}$ to hold in $J_1(\mathcal{U}_m)$. The interesting fact is that $b^{-1} \circ a - \mathrm{Id}$ turns out to be a section of $T^* \otimes J_m T \rightarrow M$ over (x), and this is the definition of the first operator $\mathcal{D}_1$ in the universal nonlinear Spencer sequence

$$\mathcal{U}_{m+1} \xrightarrow{\mathcal{D}_1} T^* \otimes J_m T \xrightarrow{\mathcal{D}_2} \wedge^2(T^*) \otimes J_{m-1} T, \qquad m \geq 1 \tag{A.31}$$

Therefore, the bisection f of $\mathcal{U}_{m+1}$ on the right-hand side of (A.29) satisfies (A.30) and becomes the horizontal diffeomorphism in (A.29) if and only if $\mathcal{D}_1(f) = 0$. We refer to [P2] for an equivalent definition of $\mathcal{D}_1$ using the Spencer operator. As we observed above, $\partial y^i/\partial x^r \neq f^i_r$ in a, and we can easily check that $b^{-1} \circ a - \mathrm{Id}$ is in the form $(x, x, 0, \ldots, 0 \mid (\partial y^i/\partial x^a)g^a_r - \delta^i_r, \ldots), f^i_a g^a_j = \delta^i_j$. We invite the reader to compute the next two terms using the group operation of $J_1 G_m(n)$ and refer to [P2] for further details. If $(M, \mathcal{P}_{m+1})$ is m-torsion-free, then $\mathcal{D}_1$ restricts to (20.14) and $\mathcal{P}_{m+1}$ stabilizes the order of jets in (A.31), now for $m \geq 0$. In Chapter 8, we constructed the restriction of (A.31) to $\mathcal{P}_1 = \varepsilon(\mathcal{P}_0) = \varepsilon(M \times M)$ directly (with the choice $a \circ b^{-1} - \mathrm{Id}$ over (y)). The operator $\mathcal{D}_2$ in (A.31) gives the integrability conditions for $\mathcal{D}_1$ and is defined using the Spencer bracket (17.49), $D = D_1$ in (20.13) is the linearization of $\mathcal{D}_1$, and all the operators in the linear Spencer sequence are consequences of D_1, since they resolve D_1 with the assumption $\mathfrak{R} = 0$ as in (11.21). Therefore, all the operators in the nonlinear and linear Spencer sequences are consequences of $\mathcal{D}_1$ defined above.

REFERENCES

[AO1] E. Abadoğlu, E. Ortaçgil, F. Öztürk: Klein geometries, parabolic geometries and differential equations of finite type. *J. Lie Theory* **18** (2008) 67–82.

[AO2] E. Abadoğlu, E. Ortaçgil: Intrinsic characteristic classes of a local Lie group. *Port. Math.* **67** (2010) 453–83.

[B1] A. D. Blaom: Geometric structures as deformed infinitesimal symmetries, *Trans. Amer. Math. Soc.* **358** (2006) 3651–71.

[B2] A. D. Blaom: Lie algebroids and Cartan's method of equivalence. *Trans. Amer. Math. Soc.* **364** (2012) 3071–135.

[Bo1] R. Bott: On topological obstructions to integrability. *Actes du Congrès International des Mathématiciens (Nice, 1970)*, Tome 1, pp. 27–36. Gauthier-Villars, Paris, 1971.

[Bo2] R. Bott: Homogeneous vector bundles. *Ann. Math.* **66** (1957) 203–48.

[BT] R. Bott, L. Tu: *Differential Forms in Algebraic Topology*. Graduate Texts in Mathematics 82. Springer, New York, 1982.

[C] S. S. Chern: On the curvature and characteristic classes of a Riemannian manifold. *Abh. Sem. Univ. Hamburg* **20** (1956) 117–126.

[CE] C. Chevalley, S. Eilenberg: Cohomology of Lie groups and Lie algebras. *Trans. Amer. Math. Soc.* **63** (1948) 85–124.

[CF] M. Crainic, R. L. Fernandes: Secondary characteristic classes of Lie algebroids. In *Quantum Field Theory and Noncommutative Geometry*. Lecture Notes in Physics 662, pp. 157–76. Springer, Berlin, 2005.

[CG] A. Čap, A. R. Gover: Tractor calculus for parabolic geometries. *Trans. Amer. Math. Soc.* **354** (2002) 1511–48.

[Cr] M. Crainic: Differentiable and algebroid cohomology, Van Est isomorphisms, and characteristic classes. *Comment. Math. Helv.* **78** (2003) 681–721.

[CS] A. Čap, J. Slovák: *Parabolic Geometries I: Background and General Theory*. Mathematical Surveys and Monographs 154. American Mathematical Society, Providence, RI, 2009.

[CSa] M. Crampin, D. Saunders: *Cartan Geometries and their Symmetries: A Lie Algebroid Approach*. Atlantis Studies in Variational Geometry 4. Atlantis Press, Paris, 2016.

[CSi] S. S. Chern, J. Simons: Characteristic forms and geometric invariants. *Ann. Math.* **99** (1974) 48–69.

[D] D. DeTurck: Deforming metrics in the direction of their Ricci tensors. *J. Differential Geom.* **18** (1983) 157–62.

[Dr] J. Draisma: Lie Algebras of Vector Fields. PhD thesis, TU/e, Eindhoven, Netherlands, 2002.

[EM] S. Eilenberg, S. McLane: Cohomology theory in abstract groups I. *Ann. Math.* **48** (1947) 51–78.

[Ei] L. P. Eisenhart: *Non-Riemannian Geometry*. American Mathematical Society, Providence, RI, 1927 (reprinted by Dover Publications, New York, 2005).

[Es] W. T. Van Est: Group cohomology and Lie algebra cohomology in Lie groups. *Proc. Kon. Ned. Akad. Ser. A* **56** (1953) 484–504.

[F] R. L. Fernandes: Lie algebroids, holonomy and characteristic classes. *Adv. Math.* **170** (2002) 119–79.

[FO1] M. Fels, P. Olver: Moving coframes. I. A practical algorithm. *Acta Appl. Math.* **51** (1998) 161–213.

[FO2] M. Fels, P. Olver: Moving coframes. II. Regularization and theoretical foundations. *Acta Appl. Math.* **55** (1999) 127–208.

[G] H. Goldschmidt: Integrability criteria for systems of nonlinear partial differential equations. *J. Differential Geom.* **1** (1969) 269–307.

[GHV] W. Greub, S. Halperin, R. Vanstone: *Connections, Curvature, and Cohomology*, Vol. III. Academic Press, New York, 1976.

[GS1] V. Guillemin, S. Sternberg: *Deformation Theory of Pseudogroup Structures*. Memoirs of the American Mathematical Society 64, Providence, RI, 1966.

[GS2] V. Guillemin, S. Sternberg: An algebraic model of transitive differential geometry. *Bull. Amer. Math. Soc.* **70** (1964) 16–47.

[Ho] G. Hochschild: *The Structure of Lie Groups*. Holden-Day, San Francisco, 1965.

[Kb] S. Kobayashi: *Transformation Groups in Differential Geometry*. Ergebnisse der Mathematik und ihrer Grenzgebiete 70. Springer, Berlin, 1972.

[KN1] S. Kobayashi, K. Nomizu: *Foundations of Differential Geometry*, Vol. 1. Wiley-Interscience, New York, 1963.

[KN2] S. Kobayashi, T. Nagano: On filtered Lie algebras and geometric structures, 1. *J. Math. Mech.* **13** (1964) 875–907.

[Kd] K. Kodaira: *Complex Manifolds and Deformations of Complex Structures*. Grundlehren der mathematischen Wissenschaften 283. Springer, Berlin, 1986.

[KMS] I. Kolar, P. W. Michor, J. Slovak: *Natural Operations in Differential Geometry*. Springer, Berlin, 1993.

[Ks] B. Kostant: Lie algebra cohomology and the generalized Borel–Weil Theorem. *Ann. Math.* **74** (1960) 329–87.

[KL] B. Kruglikov, V. Lychagin: Global Lie–Tresse theorem. *Sel. Math. New Ser.* **22** (2016) 1357–411.

[Ku1] A. Kumpera: On the equivalence problem for geometric structures, I and II. *J. Math. Sci. Adv. Appl.* 2015, **35** (2015) 1–71 and **36** (2015) 47–93.

[Ku2] A. Kumpera: An introduction to Lie groupoids. arXiv:1512.01454 [math.DG].

[KS] A. Kumpera, D. C. Spencer: *Lie Equations*, Vol. I: *General Theory*. Annals of Mathematics Studies 73. Princeton University Press, Princeton, NJ, 1972.

[LSU] O. A. Ladyzenskaja, V. A. Solonnikov, N. N. Uralceva: *Linear and Quasilinear Equations of Parabolic Type*. Translations of Mathematical Monographs 23. American Mathematical Society, Providence, RI, 1967.

[L] D. Laugwitz: *Differential and Riemannian Geometry*. Academic Press, New York, 1965.

[Ly] V. Lychagin: Homogeneous structures on manifolds. *Math. Zametki* **52**(4) (1992) 54–68.

[MK1] K. MacKenzie: *Lie Groupoids and Lie Algebroids in Differential Geometry*. London Mathematical Society Lecture Note Series 124. Cambridge University Press, Cambridge, 1987.

[MK2] K. MacKenzie: *General Theory of Lie Groupoids and Lie Algebroids*. London Mathematical Society Lecture Note Series 213. Cambridge University Press, Cambridge, 2005.

[Mo] V. V. Morozov: On primitive groups. *Mat. Sb.* **5**(47) (1939) 355–90.

[Mr] T. Morimoto: Geometric structures on filtered manifolds. *Hokkaido Math. J.* **22** (1983) 263–347.

[O1] P. J. Olver: Nonassociative local Lie groups. *J. Lie Theory* **6** (1996) 23–51.

[O2] P. J. Olver: *Applications of Lie Groups to Differential Equations*. Graduate Texts in Mathematics 107. Springer, New York, 1986.

[O3] P. J. Olver: Lectures on moving frames. In *Symmetries and Integrals of Difference Equations*. London Mathematical Society Lecture Note Series 381, pp. 207–46. Cambridge University Press, Cambridge, 2011.

[O4] P. J. Olver: Private communication.

[On] A. L. Onishchik: On Lie groups transitive on compact manifolds. II. *Mat. Sb. (N.S.)* **74(116)**(3) (1967) 398–416 [English translation: *Math. USSR-Sb.*, **3**(3) (1967) 373–88].

[OP] P. Olver, J. Pohjanpelto: Maurer–Cartan forms and the structure of Lie pseudogroups. *Sel. Math. (N.S.)* **11** (2005) 99–126.

[P1] J. F. Pommaret: *Systems of Partial Differential Equations and Lie Pseudogroups*. Gordon & Breach, New York, 1978.

[P2] J. F. Pommaret: *Lie Pseudogroups and Mechanics*. Gordon & Breach, New York, 1988.

[Pn] L. Pontryagin: Homologies in compact Lie groups. *Mat. Sb. (N.S.)* **6** (1939) 389–42.

[Ps] G. Post: On the structure of graded transitive Lie algebras. Memorandum 1557, Department of Applied Mathematics, University of Twente, Enschede, 2000.

[Sa] H. Samelson: Topology of Lie groups. *Bull. Amer. Math. Soc.* **58** (1952) 2–37.

[Sh] R. W. Sharpe: *Differential Geometry: Cartan's Generalization of Klein's Erlangen Program*. Graduate Texts in Mathematics 166. Springer, New York, 1997.

[Sp] D. C. Spencer: Deformations of structures on manifolds defined by transitive continuous pseudogroups. *Ann. Math.* **76** (1962) 306–44.

[SS] I. M. Singer, S. Sternberg: The infinite Lie groups of Lie and Cartan. Part I (The transitive groups). *J. Anal. Math.* **15** (1965) 1–114.

[T] N. Tanaka: On the equivalence problems associated with simple graded Lie algebras. *Hokkaido Math. J.* **8** (1979) 23–84.

[V] A. Verbovetsky: Notes on the horizontal cohomology. In *Proceedings of Conference on Secondary Calculus and Cohomological Physics, August 24–31, 1997, Moscow*, pp. 211–23. Contemporary Mathematics 219. American Mathematical Society, Providence, RI, 1998.

[Vn] A. M. Vinogradov: A spectral sequence associated with a nonlinear differential equation and algebro-geometric foundations of Lagrangian field theory with constraints. *Sov. Math. Dokl.* **19** (1978) 144–8.

[W] H. C. Wang: Complex parallelizable manifolds. *Proc. Amer. Math. Soc.* **5** (1954) 771–6.

[Wg] G. Weingart: Holonomic and semiholonomic geometries. *Sémin. Congrès* **4** (2000) 307–28.

[Wn1] A. Weinstein: The modular automorphism group of a Poisson manifold. *J. Geom. Phys.* **23** (1997) 379–94.

[Wn2] A. Weinstein: Groupoids: unifying internal and external symmetry—A tour through some examples. *Notices Amer. Math. Soc.* **43** (1996) 744–52.

[Wi] J. Winkelmann: *Complex Analytic Geometry of Complex Parallelizable Manifolds*. Mémoirs de la Société Mathématique de France 72/73, Paris, 1998.

[Y] K. Yano: *The Theory of Lie Derivatives and its Applications*. Bibliotheca Mathematica III, North-Holland, Amsterdam, 1955.

[YB] K. Yano, S.Bochner: Curvature and Betti Numbers, Annals of Mathematics Studies, No. 32, Princeton, 1953.

INDEX

The manufacturer's authorised representative in the EU for product safety is Oxford University Press España S.A. of el Parque Empresarial San Fernando de Henares, Avenida de Castilla, 2 – 28830 Madrid (www.oup.es/en or product.safety@oup.com). OUP España S.A. also acts as importer into Spain of products made by the manufacturer.

www.ingramcontent.com/pod-product-compliance
Lightning Source LLC
Chambersburg PA
CBHW070459060426
42429CB00045B/122
* 9 7 8 0 1 9 8 8 2 1 6 5 6 *